Natural Computing Series

Series Editors: G. Rozenberg
Th. Bäck A.E. Eiben J.N. Kok H.P. Spaink

Leiden Center for Natural Computing

For further volumes:
www.springer.com/series/4190

Bogdan Aman · Gabriel Ciobanu

Mobility in Process Calculi and Natural Computing

Springer

Dr. Bogdan Aman
Institute of Computer Science
Romanian Academy
Iaşi, Romania
and
Alexandru Ioan Cuza University
Iaşi, Romania
baman@iit.tuiasi.ro

Prof. Dr. Gabriel Ciobanu
Institute of Computer Science
Romanian Academy
Iaşi, Romania
and
Alexandru Ioan Cuza University
Iaşi, Romania
gabriel@info.uaic.ro

ISSN 1619-7127 Natural Computing Series
ISBN 978-3-642-43711-3 ISBN 978-3-642-24867-2 (eBook)
DOI 10.1007/978-3-642-24867-2
Springer Heidelberg Dordrecht London New York

ACM Computing Classification (1998): F.1, F.4, I.2, I.6, J.3

Printed on acid-free paper

Springer is part of Springer Science+Business Media (www.springer.com)

Preface

The design of formal calculi in which the fundamental concepts underlying interactive systems can be described and studied has been a central theme of theoretical computer science over the last two decades. In this book we refer to the formal description of *mobility* in computer science by using π-calculus, ambient calculus, bioambients, brane calculi and systems of mobile membranes.

In process algebra the moving entities are the links (π-calculus), the ambients (ambient calculus and bio-ambients) and the branes (brane calculi). In membrane systems the movement is provided by rules inspired by endocytosis and exocytosis. Cell movement is a dynamic phenomenon that is essential to a variety of biological processes (e.g., immune response).

In **Chapter 1: Mobility in Process Calculi** we refer to the formal description of *mobility* in process calculi [52]. When expressing mobility, we should mention what entities move and in what space they move. The *π-calculus* [121] is a formalism where links are the moving entities, and they move in a virtual space of linked processes (the network of web pages is a good example for this approach). This option is powerful enough to express moving processes both in a physical space of computing locations and in a virtual space of linked processes [121]. The π-calculus has a simple semantics and a tractable algebraic theory [121]; it is a widely accepted model of interacting systems with dynamically evolving communication topology and (channel) mobility. Its mobility increases the expressive power enabling the description of many high-level concurrent features.

Timed distributed π-calculus ($tD\pi$) [67] is a rigorous framework for describing distributed systems with time constraints. The timers on channels define timeouts for communications, and timers on the channel types restrict the channels' availability. Whenever the timer of either a channel or a channel type expires, the corresponding channel is discarded, and respectively the channel type is lost. $tD\pi$ combines temporal constraints with types and locations in order to give the possibility of modelling *located* and *timed* interactions between distributed processes with *time-restricted resource access*.

Another formalism able to express mobility is the *ambient calculus* [42]; it describes computation carried out on mobile devices (i.e. networks having a dynamic topology), and mobile computation (i.e. executable code able to move around the network). The primitive concept of the calculus is the ambient defined as a bounded place in which computation can occur. Ambients can be nested inside other ambients. Each ambient has a name used to control access to it. Computation is represented as the movement of ambients: they can be moved as a whole, changing their location by consuming certain capabilities: in, out, open.

Mobile ambients with timers (tMA) [9, 10, 15] represent a conservative extension of the ambient calculus. Inspired by [41], we introduce types for ambients in tMA. The type system associates to each ambient a set of types in order to control its communication by allowing only well-typed messages. For instance, if a process inside an ambient sends a message of a type which is not included in the type system of the ambient, then the process fails. In tMA the process may continue its execution after the timer of the corresponding output communication expires.

The biological inspiration is predominant in the case of *brane calculi* [40]. The operations of the two basic brane calculi, namely *pino, exo, phago* (for the PEP fragment) and *mate, bud, drip* (for the MBD fragment) are directly inspired by the biologic processes of *endocytosis, exocytosis* and *mitosis*. Since some proteins are embedded in cell membranes, and can act on both sides of the membrane simultaneously, brane calculi use both sides of the membrane, emphasizing that computation happens also on the membrane surface.

On the other hand, in **Chapter 2: Mobility in Membrane Computing**, we study mobility in the framework of natural computing. Natural computing refers to both computational models inspired by nature and biological processes. When complex natural phenomena are analyzed in terms of computational processes, our understanding of both nature and computation is enhanced. Natural computing is looking for concepts, principles and mechanisms underlying natural systems.

Membrane computing is part of natural computing, being a rule-based formalism inspired by biological cells [128]. Mobile membranes represent a formalism describing the movement of membranes inside a spatial structure by applying specific rules from a given set. We define several systems of mobile membranes: simple, enhanced and mutual mobile membranes, as well as mutual mobile membranes with objects on surface. When membrane systems are considered as computing devices, two main research directions are considered: the computational power in comparison with the classical notion of Turing computability, and the efficiency in algorithmically solving hard problems (e.g., NP-problems) in polynomial time. In this chapter we present mobile systems which are both powerful (mostly equivalent to Turing machines) and efficient (membrane system algorithms provide efficient solutions to NP-complete problems through the generation of an exponential space in polynomial time).

Reachability is the problem of deciding whether a system may reach a given configuration during its execution. This is one of the most critical properties in the verification of systems; most of the safety properties of computing systems can be reduced to the problem of checking whether a system may reach an "unintended state". We investigate the problem of reaching a certain configuration in systems of mobile membranes with replication rules, starting from a given configuration. We prove that reachability in systems of mobile membranes can be decided by reducing it to the reachability problem of a version of pure and public ambient calculus from which the open capability has been removed.

In **Chapter 3: Encodings** we establish several links between process calculi and membrane computing in order to be able to use techniques from one area in the other one. The difference between these two research areas is the fact that process algebras provide a tool for the high-level description of interactions, communications, and synchronizations between a collection of independent agents or processes, providing also algebraic laws that allow process descriptions to be manipulated and analyzed, and permit formal reasoning about equivalences between processes (e.g., using bisimulation), while membrane computing uses techniques from languages, automata, complexity, and dynamical systems. We consider our encodings as the first efforts towards bridging the gap between process calculi and mobile membranes.

In order to study the expressive power of $tD\pi$ we use a method of *embeddings among languages* introduced in [148]. The method is based on a tuple composed of a set of process expressions $\mathscr{P}$, a partial operation over $\mathscr{P}$ (in process calculi we choose the parallel composition operator) and an observational equivalence. To compare two formalisms by looking at their sets of syntactic expressions (languages) L_1 and L_2, we are required to identify the corresponding *algebraic languages* $(\mathscr{P}; |; \simeq)$ respectively $(\mathscr{P}'; |'; \simeq')$. We adapt this method and use it to show that $tD\pi$ is more expressive than the underlining $D\pi$.

Although both the π-calculus and the calculus of mobile ambients are Turing-complete [42, 121] and they have almost the same field of application (mobile computations), it is widely believed (see [77]) that the π-calculus does not directly model phenomena such as the distribution of processes within different localities, their migrations, or their failures. We present a translation of mobile ambients into the asynchronous π-calculus: in order to imitate the spatial structure of mobile ambients we impose some very rigid restrictions on the structural congruence rules of the π-calculus. A key idea of the encoding is based on the separation of the spatial structure of mobile ambients from their operational semantics.

Membrane systems [127, 128] and mobile ambients [42] have similar structures and common concepts. Both have a hierarchical structure representing locations, and are used to model various aspects of biological systems. Mobile ambients are suitable to represent the movement of ambients through ambients and the communication which takes place inside the boundaries of ambients. Membrane systems are suitable to represent the movement of objects and membranes through membranes. We consider these new computing models used in describing various biological phenomena [40, 65], and encode the ambients into membrane systems [12, 19]. We present such an encoding, and provide an operational correspondence between the safe ambients and their encodings, as well as various related properties of the membrane systems [14].

Some work has been done trying to relate membrane systems and brane calculi [35, 37, 47, 105, 106]. Inspired by brane calculi, a model of membrane systems having objects attached to membranes has been introduced in [45]. In [31], a class of membrane systems containing both free floating objects and objects attached to membranes has been proposed. We are continuing this research line, and simulate a fragment of brane calculi by using systems of mutual membranes with objects on surface. By defining an encoding of the PEP fragment of brane calculi into systems of mutual membranes with objects on surface, we show that the difference between the two models is not significant.

A relation can be established between mobile membranes and coloured Petri nets by providing an encoding of the first formalism into the second one. By considering the endocytic pathway for low-density lipoprotein degradation, we show how mobile membranes can be used to model such a biological phenomenon, while coloured Petri nets can be used to analyze and verify automatically some behavioural properties of the pathway. Some connections between membrane systems and Petri nets are presented for the first time in [78] and [137]. In [101, 102], a direct structural relationship between these two formalisms is established by defining a new class of Petri nets called Petri nets with localities. This new class of Petri nets has been used to show how maximal evolutions from membrane systems are faithfully reflected in the maximally concurrent step sequence semantics of their corresponding Petri nets with localities.

The book is devoted to researchers. However, since it contains examples and exercises, it can be used as a course support. The dependencies of its chapters and sections are represented by the following graph

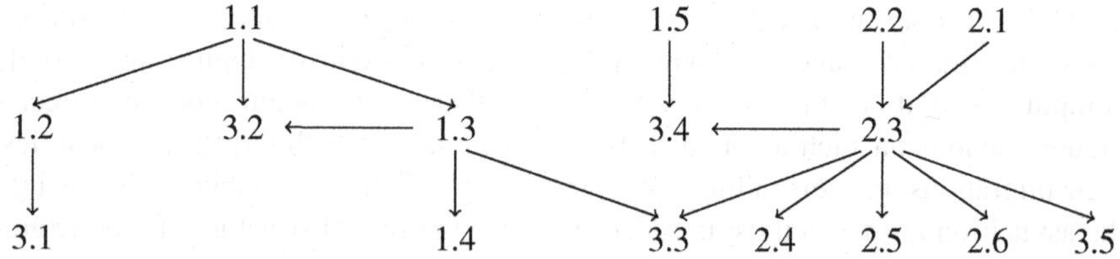

The book is designed primarily for computer scientists working in concurrency (process calculi, Petri nets), in biologically inspired formalisms (brane calculi, membrane systems), and also for the mathematically inclined scientists interested in formalizing moving agents and biological phenomena. As far as we know, the book is the first monograph that treats mobility as its central topic.

Iaşi, 2011

Bogdan Aman
Gabriel Ciobanu

Acknowledgements

Mobility was described and studied formally in computer science by Robin Milner. Robin Milner introduced and studied mobility in his π-calculus, a theoretical framework for concurrent systems derived from his calculus of communicating systems. We express our gratitude to Robin Milner not only as a pioneer of studying mobility in computer science and his remarkable contributions (Turing Award 1991), but also as a supervisor and a mentor of one of the authors (Gabriel Ciobanu) during his academic year spent at University of Edinburgh (1991–1992).

Mobility as a primitive operation was used by Luca Cardelli in his ambient calculus and various brane calculi. Mobile membranes describe the movement of membranes in membrane computing, a branch of natural computing aiming to bridge formal languages and biology; membrane computing was initiated by Gheorghe Păun and it has developed quickly during the last decade.

We are also grateful to:

- our colleagues: Oana Agrigoroaiei, Oana Andrei, Cosmin Bonchiş, Laura Cornăcel, Daniel Dumitriu, Mihai Gontineac, Dorin Huzum, Cornel Izbaşa, Călin Juravle, Dorel Lucanu, Gabriel Moruz, Florentin Olariu, Dorin Paraschiv, Dana Petcu, Cristian Prisacariu, Andreas Resios, Mihai Rotaru, Dănuţ Rusu, Gheorghe Ştefănescu, Bogdan Tanasă, Daniela Zaharie,
- our (international) collaborators: Hugh Anderson, Daniela Besozzi, Rahul Desai, Maciej Koutny, Shankara Narayanan Krishna, Akash Kumar, Gheorghe Păun, Mario J. Pérez-Jiménez, Vladimir Zakharov, Guo Wenyuan,
- all who helped us to advance in our research approach, including anonymous reviewer and editors. A part of the research reported in this book was carried out over the last five years with their valuable help.

Many thanks to Ronan Nugent and his team at Springer for their nice and kind collaboration in writing this book, improving its layout, and polishing its English.

We would like to thank the Romanian National Authority for Scientific Research, the National Research Council (CNCS) and other public authorities for funding our research projects; many results presented in this book have been obtained during these projects.

Last but not least, the warmest thanks go to our families for their important support.

Contents

Chapter 1
Mobility in Process Calculi

Abstract Mathematical models are useful in different fields to provide a deeper and more insightful understanding of various systems and notions. We refer here to the formal description of *mobility* in computer science [52]. The first formalism in computer science able to describe mobility is the π-calculus [121]. It was followed by ambient calculus [42]. A biologically-inspired version of ambient calculus is given by bioambients [138] and several brane calculi [40].

When expressing mobility, we should mention what entities move and in what space they move. There are several possibilities: processes moving in a physical space of computing locations, processes moving in a virtual space of linked processes, links moving in a virtual space of linked processes, etc.

1.1 π-calculus

The π-calculus is a formalism where links are the moving entities, and they move in a virtual space of linked processes (the network of web pages is a good example of this approach). This option is powerful enough to express moving processes both in a physical space of computing locations and in a virtual space of linked processes [121].

The π-calculus was developed as a calculus of communicating systems that allows the representation of concurrent computations whose configuration may change during the computation. The computational world of the π-calculus contains just processes (also called agents) and channels (also called ports). In contrast to the λ-calculus which represents computations through functions, the π-calculus uses the process as an abstraction of an independent thread of control. A channel is an abstraction of the communication link between processes, and processes interact by sending information through these channels. Since variables may be channel names, computation can change the channel topology and process mobility is supported. Milner emphasized the importance of identifying the "elements of interaction" [120], and his π-calculus extends the Church-Turing model by adding the

B. Aman, G. Ciobanu, *Mobility in Process Calculi and Natural Computing*,
Natural Computing Series, DOI 10.1007/978-3-642-24867-2_1,

interaction between a sender and a receiver to the algebraic elegance of λ-calculus. The π-calculus has a simple semantics and a tractable algebraic theory [121]. Actually the π-calculus is a widely accepted model of interacting systems with dynamically evolving communication topology and (channel) mobility. Its mobility increases the expressive power enabling the description of many high-level concurrent features. The π-calculus can model networks in which messages are sent from one site to another site and may contain links to active processes or to other sites; it is a general model of computation which takes interaction as primitive. This formalism is also used in modelling biological systems [53, 72].

1.1.1 Syntax

We briefly present the monadic version of the π-calculus ("monadic" means that the messages sent between processes consist of exactly one name). Let $\mathscr{X}$ be an infinite countable set of *names*. The elements of $\mathscr{X}$ are denoted by $x, y, z, \ldots$ The terms (expressions) of this formalism are called processes, and they are denoted by $P, Q, R, \ldots$.

The **processes** are defined over the set $\mathscr{X}$ of names as follows:

Table 1.1 Syntax of π-calculus

P ::=		processes
	0	empty process
	$\overline{x}\langle z\rangle.P$	output
	$x(y).P$	input
	$P \mid Q$	parallel composition
	$P+Q$	choice
	$!P$	replication
	$\nu x\, P$	restriction

The π-calculus expressions are defined by guarded processes $\overline{x}\langle z\rangle.P$ and $x(y).P$, parallel composition $P \mid Q$, nondeterministic choice $P+Q$, replication $!P$ and a restriction $\nu x\, P$ creating a local fresh channel x for process P. The π-calculus replication $!P$ can also be expressed by recursive equations of parametric processes. 0 is the empty process.

Input guards and output guards represent sending and receiving a channel name along a link. The output guarded process $\overline{x}\langle z\rangle.P$ sends z along x and then, after the output has completed, continues as P. An input guarded process $x(y).Q$ waits until a name is received along x, substitutes it for the bound variable y and continues as Q[1].

[1] There is an important distinction between input and output guards. The output guard is a simple sending of a name z along a channel x, but the input guard has a more complex action: the name received along the channel x replaces y in the process following the input guard. The input guard

The parallel composition $\overline{x}\langle z\rangle.P \mid x(y).Q$ may thus synchronize on x, and so the processes can interact by using channels they share. A name received in one interaction can be used in another; by receiving a channel name, a process can interact with processes which are unknown to it, but now they share the same channel name. This aspect is important in defining mobility in the π-calculus, together with the scope of names defined by $\nu x\, P$ and extrusion of names from their scopes.

Over the set of processes is defined a structural congruence relation $\equiv$ providing a static semantics. The structural congruence is defined as the smallest congruence over the set of processes which satisfies the following equalities involving the set $fn(P)$ of the free occurrences in a process P and standard α-conversion denoted by $=_\alpha$:

Table 1.2 Structural Congruence of π-calculus

$P \equiv Q$ if $P =_\alpha Q$		
$P+0 \equiv P,$	$P+Q \equiv Q+P,$	$(P+Q)+R \equiv P+(Q+R),$
$P \mid 0 \equiv P,$	$P \mid Q \equiv Q \mid P,$	$(P \mid Q) \mid R \equiv P \mid (Q \mid R),$
$!P \equiv P \mid !P$		
$\nu x 0 \equiv 0,$	$\nu x \nu y P \equiv \nu y \nu x P,$	$\nu x(P \mid Q) \equiv P \mid \nu x Q$ if $x \notin fn(P).$

The rule $\nu x(P \mid Q) \equiv P \mid \nu x Q$ whenever $x \notin fn(P)$ describes the extrusion of names from their scope, and it plays an important role in defining mobility in π-calculus.

1.1.2 Operational Semantics

The evolution of a process is described in the π-calculus by a reduction relation over processes called *reaction*. This relation contains those transitions which can be inferred from a set of rules. The reduction relation over processes is defined as the smallest relation $\rightarrow$ satisfying the following rules:

Table 1.3 Operational Semantics of π-calculus

(com)	$(\overline{x}\langle z\rangle.P+R_1) \mid (\,x(y).Q+R_2) \rightarrow P \mid Q\{z/y\}$
(par)	$P \rightarrow Q$ implies $P \mid R \rightarrow Q \mid R$
(res)	$P \rightarrow Q$ implies $\nu x\, P \rightarrow \nu x\, Q$
(str)	$P \equiv P'$, $P' \rightarrow Q'$ and $Q' \equiv Q$ implies $P \rightarrow Q$

is a *binding* operator involving substitutions: in $x(y).P$, the name y binds free occurrences of y in P. In a second binding operator $\nu x\, P$, the name x binds free occurrences of x in P.

Example 1.1. We give an example describing a simple interaction between a mobile phone carried in a car and two base stations. The connections between the car (mobile phone) and a base station can change as the car moves around.

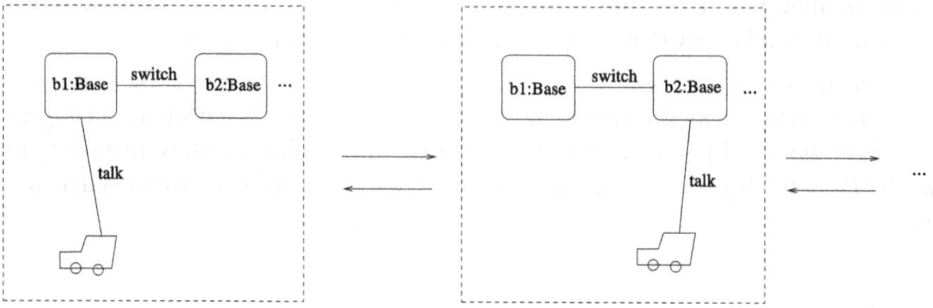

Fig. 1.1 A Car and Two Base Stations Interaction in π-calculus

We consider three processes B_1, B_2 and C corresponding to the two base stations and the car, respectively. We start with their parallel composition $B_1 \mid C \mid B_2$ described by the left square of the picture. The base B_1 and the car C are connected by a channel *talk*, and B_1 and B_2 by a channel *switch*. This means that *talk* is free in both B_1 and C, and *switch* is a free name in both B_1 and B_2. By the process expression $\nu\, talk\, (B_1 \mid C) \mid B_2$, the name *talk* is restricted to B_1 and C, and we interpret that B_1 and C have an exclusive communication along the channel *talk*. If $B_1 = \overline{switch}\langle talk\rangle.B_1'$, then base B_1 wishes to send the name of channel *talk* to base B_2 along the channel *switch*. Moreover, if *talk* is not free in B_1' ($talk \notin fn(B_1')$), then B_1' will lose its link to C. Base B_2 is waiting for a channel name sent by B_1, namely $B_2 = switch(y).B_2'$. Applying the reduction rule (com) and the extrusion of name *talk* from its previous scope given by B_1, we get the transition

$$\nu\, talk\, (B_1 \mid C) \mid B_2 \longrightarrow B_1' \mid \nu\, talk\, (C \mid B_2'')$$

where $B_2'' = B_2'\{talk/y\}$. The initial process $\nu\, talk\, (B_1 \mid C) \mid B_2$ changes its communication topology, and it becomes as it is described in the right square of the figure above. Now B_2'' and C have an exclusive communication along the channel *talk*. This is essentially the mobility mechanism offered by the π-calculus. More details are in [121].

Various forms of behavioural equivalence in process algebras are based on the notion of bisimulation. There are several definitions in the literature for bisimulation; their definitions are given by using the labelled transition system defined by the reduction rules. Systems can be checked automatically by studying the bisimilarity between two processes, namely the model and its specification. The properties of finite state transition systems can be specified in a very powerful logic called μ-calculus. Thus it is possible to use various verification techniques for proving properties about the mobile concurrent systems modelled in the π-calculus. Modelling with π-calculus and verifying with the μ-calculus and some of its proper subsets have been thoroughly investigated in the literature. Model checking π-calculus processes is discussed in several papers, and the Mobility Workbench [152] is a software tool supporting this model checking.

1.1.3 Extensions

There are several variants and extensions of the π-calculus: Spi [1], Dpi [91], tDpi [67], appliedPi [2], bigraphs [122]. An important change is introduced in the distributed version Dpi of the π-calculus presented in [91]: mobility is expressed in a simpler way, by using an explicit migration primitive *goto* $l.P$ enabling mobility between explicit location names. Regev and Shapiro use the π-calculus in describing biochemical systems by abstracting "cell-as-computation", and using processes as abstractions of molecules in biomolecular systems [139]; the authors use these abstractions for representation, simulation, and analysis of metabolic pathways.

1.1.4 Computational Power

The π-calculus is a universal model of computation as stated by Milner in [119], in which he presents two encodings of the λ-calculus in the π-calculus. The features of the π-calculus that make these encodings possible are name-passing and replication (or, equivalently, recursively defined agents). In the absence of replication/recursion, the π-calculus ceases to be Turing-powerful [145].

1.2 Timed Distributed π-calculus

We take up $D\pi$, extending it with decreasing timers attached to communication channels and to channel types. The new formalism is called *timed distributed π-calculus* ($tD\pi$) [67], and it is presented as a rigorous framework for describing distributed systems with time and resource constraints. The timers on channels define timeouts for communications, and timers on the channel types restrict the channels' availability. Whenever the timer of either a channel or a channel type expires, the corresponding channel is discarded, and respectively the channel type is lost. $tD\pi$ combines temporal constraints with types and locations in order to give the possibility of modelling *located* and *timed* interactions between distributed processes with *time-restricted resource access*. Following the method introduced in [153], we prove that the typing system of $tD\pi$ is sound with respect to the equivalence and reduction relations of the π-calculus. Moreover, time does not interfere with the typing system.

1.2.1 Syntax

By adding timers to communication channels, communication along a channel is no longer available for an indefinite time (like in $D\pi$). If no interaction happens in the predefined interval of time determined by the timer value, the process goes to another state. Each channel has two alternatives: one when the communication is achieved, and another when we have no communication. Channel timers are created once with the channel, but started only when the channel becomes active (available for communication).

1.2.1.1 $tD\pi$ Syntax

The syntax of a $D\pi$ channel a is extended by tagging it with a timer Δt; this means that the channel $a^{\Delta t}$ waits for communication only for the period of time determined by the timer value t (namely t units of time). We use a discrete time domain; this is related to the fact that we have synchronous communications in the standard π-calculus. If we want to model asynchronous systems, then a model based on dense time [8] would be more appropriate.

Table 1.4 Syntax of $tD\pi$

u ::=	x	Variable Name	P, Q ::=	$stop$	Termination
|	$a^{\Delta t}$	Timed Channel	|	$P \mid Q$	Composition
l ::=	x	Variable Name	|	$(\nu u : A)P$	Channel Restriction
|	k	Location Name	|	$go\, l.P$	Movement
v ::=	bv	Base Value	|	$u!\langle v\rangle.(P,Q)$	Output
|	$u \mid l$	Name	|	$u?(X:T).(P,Q)$	Input
|	$u@l$	Located Name	|	$^*P$	Replication
|	$(v_1,..,v_n)$	Tuple of Values	M, N ::=	$M \mid N$	Composition
X ::=	x	Variable	|	$(\nu u@l : T)N$	Located Restriction
|	$X@l$	Located Variable	|	$l[[P]]_{\Gamma}$	Located Process
|	$(X_1,..,X_n)$	Tuple of Variables			

The syntax of *Input* and *Output* communication uses a pair of processes (P,Q). For instance, the *Input* expression $a^{\Delta t}?(X:T).(P,Q)$ evolves to P whenever a communication is established on channel a during the interval of time given by Δt, otherwise it evolves to Q. In this expression, the variable X of type T is considered bound only in P. We consider timers for both input and output channels. In general, in synchronous systems an input process waits for a resource for a certain period of time, and an output process offers a resource for a certain period of time.

Table 1.4 defines in order the *channel names* and *location names*, *values*, *variables*, *processes* and *tagged located processes* of $tD\pi$. For a variable X of the *Input* expression $a^{\Delta t}?(X:T).(P,Q)$ we must also provide its type T, and for the channel name u in the *Channel Restriction* expression we have to provide its channel type A (types are presented in Subsection 1.2.2). Note that we may have a variable x in place of a channel name (in the *Input*, *Output* or *Located Restriction*) or in place of a location (in the *Located Processes*, *Located Names* or *Movement*). On an output channel, a process can send a value consisting of either a channel name (together with its timer), a location name, a name of a variable, a name of a channel (or variable) located at some location or a tuple of values. Note that with the located restriction $(\nu a@k : T)N$ we specify a new private channel a and its location k. For example, in the process

$$(\nu a@k : T)(l[[P]] \mid k[[Q]]) \mid k[[Q']]$$

the channel a is private for P and Q, and it is located at the current location k of Q. Moreover, Q' does not have any knowledge about channel a even though it also runs at location k. This means that process P must move to location k before communicating on the private channel a. Also note that channel restriction refers only to names.

The interaction between processes is given through the input and output process expressions which must have the same channel name; the channel timers play a secondary role in such an interaction.

Example 1.2. The following two processes running in parallel can interact along the common channel a.

$$a^{\Delta t}!\langle v\rangle.(P,Q) \mid a^{\Delta t'}?(X:T).(P',Q') \longrightarrow P \mid P'\{^{v}/_{X}\}$$

Intuitively, the process on the left of the reduction arrow evolves to the process on the right after such an interaction. The output process (the process on the left of the parallel composition operator) sends the value v on the channel named a and then behaves as P. When receiving the value v, in the input process (the process on the right of the parallel composition operator) all the occurrences of the bound variable X are replaced by v in P'.

Waiting indefinitely on a channel a is allowed by considering Δt as ∞. An output process expression $a^{\infty}!\langle v\rangle.(P,Q)$ awaits forever to send the value v, simulating the behaviour of an output process in untimed synchronous π-calculus.

1.2.2 Typing System

Each located process is tagged with a type environment Γ which is a set of *location types* denoted by K in Table 1.5. Formally the type environment is a mapping from free location names k to location types K. A location type K may contain location capabilities denoted by κ; these capabilities may express either capabilities of using *channel names* $\tilde{a}$ with their corresponding *channel types* $\tilde{A}$ ($\tilde{a}$:$\tilde{A}$), or *move* capabilities *go*, or *channel restriction* capabilities (i.e., permissions to create private channels) *newch*. A *channel type* A may contain the following channel capabilities generically denoted by α: *reading/writing/restricted reading* messages of type T respectively denoted by $r\langle T\rangle/w\langle T\rangle/ro\langle T\rangle$. A type T may contain tuples $(T_1,\ldots,T_n)$ of types, and channel types $A_1,\dots,A_n@K$ corresponding to channel names $a_1,\dots,a_n$ located at a location of type K. $\mathscr{B}$ represents the set of *base types*.

Table 1.5 Type System and Subtyping Relation of $tD\pi$-calculus

Types:			Subtyping:			
K	::=	$\mathrm{loc}\{\tilde{\kappa}\}$	κ	$<:$	κ	
A	::=	$\mathrm{res}\{\tilde{\alpha}\}\Delta t$	$a:A$	$<:$	$a:B$	if $A <: B$
E	::=	$A \mid K \mid \mathscr{B}$	K	$<:$	L	if $\forall\lambda \in L: \exists\kappa \in K: \kappa <: \lambda$
T	::=	$E \mid (T_1,\dots,T_n)$	A	$<:$	B	if $\forall\beta \in B: \exists\alpha \in A: \alpha <: \beta$
	$\mid$	$A_1,\dots,A_n@K$	$\tilde{A}@K$	$<:$	$\tilde{B}@L$	if $K <: L$ and $\tilde{A} <: \tilde{B}$
Capabilities:			$\tilde{S}$	$<:$	$\tilde{T}$	if $\forall i: S_i <: T_i$
κ	::=	$a:A$	$r\langle T\rangle$	$<:$	$r\langle T'\rangle$	if $T <: T'$
	$\mid$	$go \mid newch$	$w\langle S\rangle$	$<:$	$w\langle S'\rangle$	if $S' <: S$
α	::=	$r\langle T\rangle \mid w\langle T\rangle \mid ro\langle T\rangle$	$ro\langle T\rangle$	$<:$	$ro\langle T'\rangle$	if $T <: T'$

We may have only one instance of the capabilities *go* and *newch* in a location type K; they represent respectively the capability of a process to move to a location of type K, and the capability to create private channel names at a location of type K.

In order to exemplify, let us consider a process which has in its type environment Γ a channel name a with a channel type $res\{r\langle T\rangle, w\langle T'\rangle, ro\langle T''\rangle\}$. This means that along this channel a the process can receive messages of type T, and send messages of type T'. The *ro* capability is similar to an r capability, with the difference that the types of the received messages are not added to the type environment of the process. Types are accumulated when a name is received along an input channel with capability $r\langle\rangle$.

Having $ro\langle\rangle$ capabilities, we can describe processes which may use the data received in a message through an input channel with capability $ro\langle\rangle$ only if there exists a proper type for the new data within their type environments. More precisely, let us consider a process P at location k which receives a located channel name $b@k$ on the input channel a of type $res\{ro\langle T\rangle\}$. The located process $k[[P]]_\Gamma$ can use the new channel name b to communicate without generating errors only if its type environment Γ contains at location k the corresponding type of b, i.e., $\Gamma(k,b)$ should be defined. Runtime errors are presented at the end of Section 1.2.3, where Table 1.12 contains the rules of the error system.

In Dπ resources are accumulated, but can never be discarded. We extend the channel types of $D\pi$ with timers of form Δt. These timers define the existence of the channel types inside the type environment. We assume that we have a universal clock, and the timers decrease with each "tick" of the universal clock. Communication actions can be performed along a channel until the timer on its type has expired. After expiration, the channel capabilities are discarded and any communication would generate a runtime error. Timers are created once with the channel types, and they are activated when types are added to the type environment. For a clearer presentation, we write only the channel types $res\{\alpha\}$ instead of the channel types with the attached timers $res\{\alpha\}\Delta t$ whenever we are not interested in the timing aspects.

In our approach a process can move to a certain location, and wait for a period of time to establish a communication on a channel (a fixed local resource) with a complementary process. It is necessary to offer capabilities as $r\langle T\rangle$, $w\langle T\rangle$, and $ro\langle T\rangle$ for these fixed resources in order to restrict the actions performed by a process. The capabilities of the locations, and the capabilities of the channels from the type environment define what actions are allowed to be executed at each location. An example of a type environment is:

$$\Gamma = \{l : loc\{a : A, b : B\}, k : loc\{a : A'\}\}$$

where we denote by $\Gamma(k)$ the type $loc\{a : A'\}$ of location k, and by $\Gamma(l,b)$ the channel type B of the channel b located at l. The process of accumulating capabilities is made explicit by using *environment extensions*. We denote by $\Gamma\{k : K\}$ an environment Γ extended with a new location k of type K. Moreover, considering Γ as above, we can extend the type environment with a new type B' of a channel c located at k by

$$\Gamma\{c@k : B'\} = \{l : loc\{a : A, b : B\}, k : loc\{a : A', c : B'\}\}$$

When a process receives new channel names together with their associated types, capabilities of the new names become available (are added to the type environment of the process). As an example, let us suppose a process receiving a name of a located channel $c@k$ with channel type B' through an input channel with *reading* capability. The type of the new channel is added to the type environment at the corresponding location type of $k : loc\{\ldots\}$. It means that now the process knows about the new channel, and gains the capability to communicate through the accumulated channel c according to type B'.

A *subtyping relation* ($<:$) is introduced to compare type environments. If we consider $\Gamma = \{l : loc\{\tilde{a} : \tilde{A}, \tilde{b} : \tilde{B}\}\}$ and $\Gamma' = \{l : loc\{\tilde{a} : \tilde{A}\}\}$ as type environments, then we have $\Gamma <: \Gamma'$ according to the definition in the second column of Table 1.5. Comparing type environments Γ and Γ', we see that an environment with more capabilities (Γ) is a subtype of an environment with fewer capabilities (Γ'). The reason for such an interpretation of the subtyping relation is that Γ' is more restrictive than Γ. The subtyping relation represents the inverse of the subset relation from the set theory; if we consider type environments as sets of location types, the relation above becomes $\Gamma \supseteq \Gamma'$.

We extend both the *partial meet* $\sqcap$ and *partial join* $\sqcup$ operators of $D\pi$ with the new channel capability $ro\langle\rangle$. Intuitively $\sqcap$ behaves as the union operator of set theory, and $\sqcup$ behaves as the intersection operator. The partial meet operator for location types $K \sqcap K'$ is undefined if and only if there exists a channel name a such that $a : A \in K$, $a : A' \in K'$ and $A \sqcap A'$ is undefined (see Table 1.7 for the definition of $\sqcap$ for channel types).

We denote by γ any of the location capabilities *go* or *newch*. By $a : - \notin K$ we denote the fact that in the location type K there is no channel type A for channel a such that $a : A \in K$.

Table 1.6 Partial Meet Operator for Locations in $tD\pi$-calculus

$$
\begin{array}{rll}
K \sqcap K' = & \{\gamma & \mid \gamma \in K \text{ or } \gamma \in K'\} \\
\cup & \{a:A & \mid a:A \in K \text{ and } a:- \notin K'\} \\
\cup & \{a:A' & \mid a:- \notin K \text{ and } a:A' \in K'\} \\
\cup & \{a:A'' & \mid a:A \in K \text{ and } a:A' \in K' \text{ and } A'' = A \sqcap A'\}
\end{array}
$$

The method of removing capabilities is formalised by a binary *subtraction operator* $\backslash_\Delta$ defined by using a *join* operator $\sqcup$ (see Table 1.8), and a *symmetrical difference* operator denoted by $\backslash$ similar to the one defined in set theory (in our case it is applied to type environments). We write $\backslash_\Delta$ for the operation of removing from the first type environment all the types contained in the second type environment. We denote by $\mathscr{E}$ the set of type environments. The subtraction operator $\backslash_\Delta$ described above is defined as $\backslash_\Delta : \mathscr{E} \times \mathscr{E} \to \mathscr{E}$ where $\Gamma \backslash_\Delta \Gamma' = \Gamma \sqcup (\Gamma \backslash \Gamma')$.

Table 1.7 Partial Meet Operator for Channel Types in $tD\pi$-calculus

Partial meet operator for channel types ($A \sqcap A'$) is undefined iff:

$$
\begin{array}{lllll}
r\langle T\rangle \in A & \text{and} & r\langle T\rangle \in A' & \text{and} & T \sqcap T' \text{ undefined} \\
ro\langle T\rangle \in A & \text{and} & ro\langle T\rangle \in A' & \text{and} & T \sqcap T' \text{ undefined} \\
w\langle S\rangle \in A & \text{and} & w\langle S'\rangle \in A' & \text{and} & S \sqcup S' \text{ undefined} \\
r\langle T\rangle \in A & \text{and} & w\langle S'\rangle \in A' & \text{and} & S' \not<: T \\
w\langle S\rangle \in A & \text{and} & r\langle T'\rangle \in A' & \text{and} & S \not<: T' \\
ro\langle T\rangle \in A & \text{and} & w\langle S'\rangle \in A' & \text{and} & S' \not<: T \\
w\langle S\rangle \in A & \text{and} & ro\langle T'\rangle \in A' & \text{and} & S \not<: T' \\
ro\langle T\rangle \in A & \text{and} & r\langle T'\rangle \in A' & \text{and} & T' \backslash_\Delta T \text{ undefined}
\end{array}
$$

The definition

$$
\begin{array}{rll}
A \sqcap A' = & \{ro\langle T\rangle & \mid ro\langle T\rangle \in A \text{ and } ro\langle -\rangle \notin A'\} \\
\cup & \{ro\langle T''\rangle & \mid ro\langle T\rangle \in A \text{ and } ro\langle T'\rangle \in A' \text{ and } T'' = T \sqcap T'\} \\
\cup & \{w\langle S\rangle & \mid w\langle S\rangle \in A \text{ and } w\langle -\rangle \notin A'\} \\
\cup & \{w\langle S''\rangle & \mid w\langle S\rangle \in A \text{ and } w\langle S'\rangle \in A' \text{ and } S'' = S \sqcup S'\} \\
\cup & \{ro\langle T'\rangle & \mid r\langle T\rangle \in A \text{ and } ro\langle T'\rangle \in A' \} \\
\cup & \{r\langle T\rangle & \mid r\langle T\rangle \in A \text{ and } ro\langle -\rangle \notin A' \text{ and } r\langle -\rangle \notin A' \text{ or} \\
& & \quad r\langle T\rangle \in A \text{ and } ro\langle T'\rangle \in A', r\langle -\rangle \notin A' \text{ and } T \sqcup T' = \emptyset \text{ or} \\
& & \quad \text{undefined}\} \\
\cup & \{r\langle T''\rangle & \mid r\langle T\rangle \in A \text{ and } ro\langle -\rangle \notin A' \text{ and } r\langle T'\rangle \in A' \text{ and } T'' = T \sqcap T' \text{ or} \\
& & \quad r\langle T\rangle \in A \text{ and } ro\langle S\rangle \in A' \text{ and } r\langle T'\rangle \in A' \text{ and } T'' = T \sqcap T' \text{ and } T \sqcup S = \emptyset \text{ or} \\
& & \quad \text{undefined} \} \\
\cup & \{r\langle T''\rangle & \mid r\langle T\rangle \in A \text{ and } ro\langle T'\rangle \in A' \text{ and } r\langle -\rangle \notin A' \text{ and } T'' = T \backslash_\Delta T' \text{ or} \\
& & \quad r\langle T\rangle \in A \text{ and } ro\langle T'\rangle \in A' \text{ and } r\langle S\rangle \in A' \text{ and } T'' = T \backslash_\Delta T' \text{ and } T \sqcup S = \emptyset \text{ or} \\
& & \quad \text{undefined} \}
\end{array}
$$

plus all other natural cases resulted from swapping A with A'

If we consider two type environments

$$\Gamma = \{loc\{a:A, b:B\}\} \quad \text{and} \quad \Gamma' = \{loc\{b:B, c:C\}\},$$

each composed of one location type with two channel types, then by applying the subtraction operator $\backslash_\Delta$ we obtain

$$\Gamma \setminus_\Delta \Gamma' = loc\{a:A,b:B\} \sqcup (loc\{a:A,b:B\} \setminus loc\{b:B,c:C\}) = loc\{a:A\}$$

A process which has a channel type with a capability $ro\langle T\rangle$ can receive only messages of type T (or any subtype of T) without generating errors. When the type of the channel is extended with the capability $ro\langle T'\rangle$, then the process is able to receive messages of a less restrictive type $T'' = T \sqcap T'$. We solve the possible conflict between $r\langle\rangle$ and $ro\langle\rangle$ by providing a higher priority to $ro\langle\rangle$ capability (because it is more restrictive than $r\langle\rangle$). In consequence, $ro\langle\rangle$ keeps its types and $r\langle\rangle$ loses them in favour of $ro\langle\rangle$ whenever $r\langle T\rangle$ and $ro\langle T'\rangle$ overlap (i.e., $T \sqcup T' \neq \emptyset$). When extending the writing capability $w\langle S\rangle$ with a new capability $w\langle S'\rangle$, the channel becomes more restricted, having the capability $w\langle S''\rangle$ where $S''=S \sqcup S'$.

We denote by $r\langle -\rangle \notin A$ the fact that there is no type T such that $r\langle T\rangle \in A$. The notations $w\langle -\rangle \notin A$ and $ro\langle -\rangle \notin A$ are defined analogously.

Table 1.8 Partial Join Operator in $tD\pi$

$K \sqcup K'$	$=$	$\{\gamma$	$\mid \gamma \in K$ and $\gamma \in K'\}$
	$\cup$	$\{a:A''$	$\mid a:A \in K$ and $a:A' \in K'$ and $A'' = A \sqcup A'\}$
$A \sqcup A'$	$=$	$\{r\langle T''\rangle$	$\mid r\langle T\rangle \in A$ and $r\langle T'\rangle \in A'$ and $T'' = T \sqcup T'\}$
	$\cup$	$\{ro\langle T''\rangle$	$\mid ro\langle T\rangle \in A$ and $ro\langle T'\rangle \in A'$ and $T'' = T \sqcup T'\}$
	$\cup$	$\{w\langle S''\rangle$	$\mid w\langle S\rangle \in A$ and $w\langle S'\rangle \in A'$ and $S'' = S \sqcap S'\}$

Note that (i) $(\mathscr{E}, \sqcup)$ is a commutative monoid; (ii) $(\mathscr{E}, \setminus)$ is a commutative group; (iii) $\sqcup$ is distributive over $\setminus$, and $(\mathscr{E},\setminus,\sqcup)$ is a ring. The proofs of these remarks are based on the facts that $\sqcup$ and $\setminus$ are commutative, and the empty environment is the identity element. The distributivity of $\sqcup$ over $\setminus$ can be simply verified by translating the set operators into boolean operators, and using truth tables.

We define a cleanup function ψ which changes the type environments according to the passage of time. It decreases the timers of the channel types, and removes the types with an expired timer. It also removes location types with only *go* capability.

Definition 1.1 (Cleanup Function). $\psi : LP_\Gamma \to LP_\Gamma$ is defined over the set of tagged located processes LP_Γ by:

$$\psi(l[[P]]_\Gamma) = l[[P]]_{\Gamma'}$$

where l can be any location of a distributed system, Γ' is obtained from Γ such that every channel type $res\{\tilde{\alpha}\}\Delta t$ with $t > 1$ and $t \neq \infty$ is changed to $res\{\tilde{\alpha}\}\Delta(t-1)$, and every $res\{\tilde{\alpha}\}\Delta 1$ is removed. Moreover, location types $loc\{go\}$ are removed.

By removing channel types from Γ, we get Γ' where it is possible to have location types having only *go* capabilities. We consider these location types as *empty* because the only allowed action is a movement. Even if we have $k : loc\{go\}$ in Γ', and a sequence of movements for a process $go\,k.go\,l.P$, this process can be reduced to $go\,l.P$ because we can avoid the intermediary code migration to location k without losing any useful effect. Therefore ψ removes $k : loc\{go\}$ from Γ'. A process moving to a location l having the type $loc\{go\}$ has no other capability, thus when performing any action (communication or channel creation) it gives rise to *runtime errors*.

To simulate the passage of time we use a *time-stepping function* ϕ defined over the set $\mathscr{P}_l$ of processes running at an arbitrary location l. The possible communications are performed at each tick of the universal clock; active channels are those which could be involved in these communications. The time-stepping function affects the active channels which do not communicate at that tick; the timers of the affected channels are decreased by one unit of time. The channels involved in communication disappear together with their timers. In the definition of the time-stepping function ϕ, we omit the channel type and the transmitted message in the input and output processes in order to simplify the presentation.

Definition 1.2 (Time-Stepping Function $\phi : \mathscr{P}_l \rightarrow \mathscr{P}_l$).

$$\phi(P) = \begin{cases} a^{\Delta(t-1)}.(R,Q) & \text{if } P = a^{\Delta t}.(R,Q),\ t > 1 \text{ and } t \neq \infty \\ \text{Q} & \text{if } P = a^{\Delta t}.(R,Q),\ t \leq 1 \\ \phi(R) \mid \phi(Q) & \text{if } P = R \mid Q \\ (\nu a : A)\phi(R) & \text{if } P = (\nu a : A)R \\ \text{P} & \text{otherwise} \end{cases}$$

We also define a *tagged time-stepping function* ϕ_Δ taking care of the missing types. ϕ_Δ is a *global* function defined by using the *local* function ϕ.

The tagged time-stepping function ϕ_Δ is applied to tagged located processes ($l[[P]]_\Gamma$); it also changes the type environment of the located process by applying the cleanup function ψ.

Definition 1.3 (Tagged Time-Stepping Function $\phi_\Delta : LP_\Gamma \rightarrow LP_\Gamma$).

$$\phi_\Delta(l[[P]]_\Gamma) = \begin{cases} l[[\phi(P)]]_{\Gamma'} & \text{if } P = a^{\Delta t}.(R,Q),\ t > 1 \text{ and t} \neq \infty \\ l[[Q]]_{\Gamma'} & \text{if } P = a^{\Delta t}.(R,Q),\ t \leq 1 \\ & \quad \text{or if } P = a^{\Delta t}.(R,Q),\ t > 1 \text{ and } \Gamma \not<: \Gamma(l,a) \\ \phi_\Delta(l[[R]]_\Gamma) \mid \phi_\Delta(l[[Q]]_\Gamma) & \text{if } P = R \mid Q \\ (\nu a@l : A)\phi_\Delta(l[[R]]_{\Gamma\{a@l:A\}}) & \text{if } P = (\nu a : A)R \\ l[[\phi(P)]]_{\Gamma'} & \text{otherwise} \end{cases}$$

where Γ' is obtained by applying the cleanup function ψ.

The static semantics of $tD\pi$ is defined as a set of inference rules which describe the relationship between expressions and their corresponding types. Here we consider the type environment as a mapping from free names to types. A type environment is associated with each located process to restrict the range of resources it may access. The typing rules describe the behaviour of a process with respect to its types. A typing system is used to decide the *well-typedness* of the processes. Syntactically we write $\Gamma \vdash P$, and say that *a process P is well-typed with respect to a type environment* Γ. We also write $\Gamma \vdash_k P$ and say that *P is well-typed to run at location k*.

In Table 1.9 we give the rules for the typing system of $tD\pi$. Considering the rules (T-R$_{new}$) and (T-W$_{new}$), we observe that the intuitive notion of *well-typedness*

Table 1.9 Typing Rules in $tD\pi$

Processes

(T-R)
$$\frac{\begin{array}{l}\Gamma \vdash_l a : res\{r\langle T\rangle\}\Delta t\\ fv(X)\cap fv(\Gamma)=\emptyset\\ \Gamma\{X@l:T\}\vdash_l P\\ \Gamma\vdash_l Q\end{array}}{\Gamma \vdash_l a^{\Delta t}?(X:T).(P,Q)}$$

(T-RO)
$$\frac{\begin{array}{l}\Gamma \vdash_l a : res\{ro\langle T\rangle\}\Delta t\\ fv(X)\cap fv(\Gamma)=\emptyset\\ \Gamma\vdash_l P\\ \Gamma\vdash_l Q\end{array}}{\Gamma \vdash_l a^{\Delta t}?(X:T).(P,Q)}$$

(T-W)
$$\frac{\begin{array}{l}\Gamma \vdash_l a : res\{w\langle T\rangle\}\Delta t\\ \Gamma\vdash_l v:T\\ \Gamma\vdash_l P\\ \Gamma\vdash_l Q\end{array}}{\Gamma \vdash_l a^{\Delta t}!\langle v\rangle.(P,Q)}$$

(T-NEWCH)
$$\frac{\begin{array}{l}\Gamma(l) <: loc\{newch\}\\ a\notin fn(\Gamma)\\ \Gamma\{a@l:A\}\vdash_l P\end{array}}{\Gamma\vdash_l (\nu a:A)P}$$

(T-STR)
$$\frac{\begin{array}{l}\Gamma\vdash_l P\\ \Gamma\vdash_l Q\end{array}}{\Gamma\vdash_l stop, P\,|\,Q, {*P}}$$

(T-GO)
$$\frac{\begin{array}{l}\Gamma(k) <: loc\{go\}\\ \Gamma\vdash_k P\end{array}}{\Gamma\vdash_l go\,k.P}$$

(T-R$_{new}$)
$$\frac{a:-\notin\Gamma(l) \qquad \Gamma\vdash_l Q}{\Gamma \vdash_l a^{\Delta t}?(X:T).(P,Q)}$$

(T-W$_{new}$)
$$\frac{a:-\notin\Gamma(l) \qquad \Gamma\vdash_l Q}{\Gamma \vdash_l a^{\Delta t}!\langle v\rangle.(P,Q)}$$

Located Processes

(N-RUN)
$$\frac{\begin{array}{l}\Delta\vdash_l P\\ \Gamma <: \Delta\end{array}}{\Gamma \Vdash l[[P]]_\Delta}$$

(N-SRT)
$$\frac{\begin{array}{l}\Gamma\Vdash M\\ \Gamma\Vdash N\end{array}}{\Gamma\Vdash 0,\ M\,|\,N}$$

(N-NEWCH)
$$\frac{\begin{array}{l}\Gamma(l) <: loc\{newch\}\\ a\notin fn(\Gamma)\\ \Gamma\{a@l:A\}\Vdash N\end{array}}{\Gamma\Vdash(\nu a@l:A)N}$$

from $D\pi$ is no longer valid in $tD\pi$. In our calculus we accept tagged located processes with missing channel types (the types are removed with the passage of time), and these processes do not generate errors.

In order to say that $a^{\Delta t}!\langle v\rangle.(R,Q)$ is well-typed to run at location k with respect to type environment Γ, the following statements should hold:

- $\Gamma\vdash_k v:T$ which means that v is a value of type T at location k;
- $\Gamma\vdash_k a : res\{w\langle T\rangle\}\Delta t'$ which means that channel a exists at location k, and may send values of type T for t' units of time;
- $\Gamma\vdash_k R; \Gamma\vdash_k Q$ which means that both R and Q are well-typed to run at location k.

For a tagged located process $k[[P]]_\Delta$, the well-typedness relation is denoted by $\Vdash$ and is defined by using the well-typedness relation $\vdash_k$ for a process P running at location k (see rule (N-RUN) in Table 1.9).

If a process communicates on a channel for which it has no capability, it can still be well-typed if the alternative process Q is well-typed. We call this second process the *safety process*. This behaviour is reflected in one of the cases in the definition of ϕ_Δ.

We can imagine the process action flow as a binary decision tree because of the decision-like syntax of the channels. At each time step one of the following alternatives must be chosen for an action: communication action, timer expiration or move action (see Section 1.2.3 for the extension of the *go* operator with a choice syntax). An alternate definition for well-typedness of processes is: A process is well-typed if in the action flow tree there exists a path from the root to a leaf which does not generate a runtime error.

Since the cleanup function ψ changes the type environment Δ by removing channel and location types, we are interested in whether the process is still well-typed under the new type environment Δ'.

Lemma 1.1 (Well-Typedness is Preserved by the Cleanup Function).
If $\Gamma \Vdash l[[P]]_\Delta$, then $\Gamma \Vdash \psi(l[[P]]_\Delta)$. In other words, if $\Gamma \Vdash l[[P]]_\Delta$, then $\Gamma \Vdash l[[P]]_{\Delta'}$ where Δ' is obtained by removing channel and location types from the type environment Δ.

Proof. The proof proceeds by induction on the structure of P, having a case for each process expression. We give here only the most interesting and significant cases. For a complete proof see [67].

Case 1.1 (*Composition: $R \mid Q$*). By the equivalence rule (S_Γ-SPLIT) we have that $\Gamma \Vdash l[[R]]_\Delta \mid l[[Q]]_\Delta$ which, by rule (N-STR) for located processes, is transformed into $\Gamma \Vdash l[[R]]_\Delta$ and $\Gamma \Vdash l[[Q]]_\Delta$. Applying the induction hypothesis, we obtain $\Gamma \Vdash \psi(l[[R]]_\Delta)$ and $\Gamma \Vdash \psi(l[[Q]]_\Delta)$ which, by applying ψ, become $\Gamma \Vdash l[[R]]_{\Delta'}$ and $\Gamma \Vdash l[[Q]]_{\Delta'}$. For both processes we have the same Δ' because the application of ψ to the tagged located processes takes into account only the type environment, and in our case the type environment is the same Δ. By applying the relation (S_Γ-SPLIT) we get the result $\Gamma \Vdash l[[R \mid Q]]_{\Delta'}$ which means $\Gamma \Vdash \psi(l[[R \mid Q]]_\Delta)$.

Case 1.2 (*Restriction: $(\nu\, a{:}A)Q$*). From (N-RUN) we have that $\Delta \vdash_l (\nu a : A)Q$ and $\Gamma <: \Delta$. By (T-NEWCH) we get $\Delta\{a@l : A\} \vdash_l Q$ and $\Gamma\{a@l : A\} <: \Delta\{a@l : A\}$. By applying the weakening property, we infer that $\Gamma\{a@l : A\} \Vdash l[[Q]]_{\Delta\{a@l:A\}}$. Applying the induction hypothesis, we get $\Gamma\{a@l : A\} \Vdash \psi(l[[Q]]_{\Delta\{a@l:A\}})$ which is equivalent to $\Gamma\{a@l : A\} \Vdash l[[Q]]_{\Delta'\{a@l:A\}}$ because the application of the function ψ does not affect the new name a. We apply again the (T-NEWCH) rule obtaining $\Gamma \Vdash (\nu a : A)l[[Q]]_{\Delta'\{a@l:A\}}$ which is structurally equivalent to $\Gamma \Vdash l[[(\nu a : A)Q]]_{\Delta'}$.

Case 1.3 (*Movement: $go\, k.Q$*). By the same line of reasoning as before we have that $\Delta \vdash_l go\ k.Q$ and $\Gamma <: \Delta$. By (T-GO) we get that $\Delta \vdash_k Q$ and $\Delta(k) <: loc\{go\}$. Using (N-RUN) we have $\Gamma \Vdash k[[Q]]_\Delta$ which by induction implies $\Gamma \Vdash k[[Q]]_{\Delta'}$. We now infer that $\Delta' \vdash_k Q$ and $\Gamma <: \Delta'$ are true. By application of the ψ function, the capability of the process to move to location k cannot be lost. This means that $\Delta'(k) <: loc\{go\}$ holds and, together with what we obtained above and by using the rule (T-GO), we have $\Delta' \vdash_l go\, k.Q$ and again $\Gamma \Vdash l[[go\, k.Q]]_{\Delta'}$. This is another syntactic form of what we were looking for, namely $\Gamma \Vdash \psi(l[[Q]]_\Delta)$.
The proof proceeds in the same manner if instead of (T-GO) we use the new rules (T-GO1) and (T-GO2) defined in Section 1.2.3.

Case 1.4 (*Input: $a^{\Delta t}?(X : T).(R, Q)$*). If we consider that channel a has the type $ro\langle\rangle$, then from $\Delta \vdash_l a^{\Delta t}?(X : T).(R, Q)$ and by using (T-RO) we have the following statements: $\Delta \vdash_l a : res\{ro\langle T\rangle\}$, $fv(X) \cap fv(\Delta) = \emptyset$, $\Delta \vdash_l R$ and $\Delta \vdash_l Q$. Applying the induction hypothesis, the last two statements are transformed into $\Gamma \Vdash l[[R]]_\Delta$

and $\Gamma \Vdash l[[Q]]_\Delta$ which provide the following two true statements: $\Gamma \Vdash \psi(l[[R]]_\Delta)$ and $\Gamma \Vdash \psi(l[[Q]]_\Delta)$. This means that three ($fv(X) \cap fv(\Delta) = \emptyset$, $\Delta' \vdash_l R$ and $\Delta' \vdash_l Q$) of the four statements needed by (T-RO) are true. If the cleanup function does not remove the type of the input channel from the capability set, then it is valid in the new environment Δ'. Thus by (T-RO), we obtain that $\Gamma \Vdash l[[a^{\Delta t}?(X:T).(R,Q)]]_{\Delta'}$. On the other hand, if the type of the active channel is missing, we can use the rule (T-R$_{new}$) and obtain the same result as before which is equivalent to the desired result, namely $\Gamma \vdash \psi(l[[a^{\Delta t}?(X:T).(R,Q)]]_\Delta)$.

The cases for *Output*, *Replication* and *Termination* are natural, and they follow the proof steps of the cases presented above. □

Exercise 1.1. Prove the *Output*, *Replication* and *Termination* cases of Lemma 1.1.

The following lemma shows that the passage of time does not interfere with the typing system. The lemma states that if a tagged located process is well-typed with respect to a type environment Γ, then the application of the tagged time-stepping function ϕ_Δ preserves its well-typedness property.

Lemma 1.2 (Tagged Time Passage).

$$\textit{If } \Gamma \Vdash l[[P]]_\Delta, \textit{ then } \Gamma \Vdash \phi_\Delta(l[[P]]_\Delta).$$

Proof. We use induction on the inference depth of $\Gamma \Vdash l[[P]]_\Delta$. From the hypothesis we derive that $\Delta \vdash_l P$ by (N-RUN), and $\Gamma <: \Delta$. We get $\Gamma \vdash_l P$ by using the weakening property. The proof continues by considering a case for each line in the definition of ϕ_Δ.

Case 1.5 ($P = R\,|\,Q$). Using (T-STR) we have $\Delta \vdash_l R$ and $\Delta \vdash_l Q$ which is equivalent to $\Delta \vdash l[[R]]$; by applying the weakening property we get $\Gamma \Vdash l[[R]]_\Delta$. The same result is obtained for process Q. By applying the induction hypothesis, we get that $\Gamma \Vdash \phi_\Delta(l[[R]]_\Delta)$ and $\Gamma \Vdash \phi_\Delta(l[[Q]]_\Delta)$. These lead to the desired result, by the application of ϕ_Δ to $R\,|\,Q$, *i.e.* $\Gamma \Vdash \phi_\Delta(l[[P]]_\Delta)$.

Case 1.6 ($P = a^{\Delta t}.(R,Q), t \leq 1$). We have two subcases, one when a is an input channel, and another when a is an output channel. The result of the application of ϕ_Δ to P is $l[[Q]]_{\Delta'}$ (with Δ' obtained by applying the cleanup function ψ) because $t \leq 1$. Let us suppose that a is an *output* channel, and thus $\Delta \vdash_l a^{\Delta t}!\langle v\rangle.(R,Q)$ and $\Gamma <: \Delta$. Using (T-W), we get $\Delta \vdash_l Q$, and by Lemma 1.1 we get $\Delta' \vdash_l Q$. Since $\Gamma <: \Delta <: \Delta'$, we infer $\Gamma \Vdash l[[Q]]_{\Delta'}$. A similar proof is obtained when we consider an *input* channel, by using the rules corresponding to the type of the input channel.

Case 1.7 ($P = a^{\Delta t}.(R,Q)$, $t > 1$ *and* $\Gamma \not<: \Gamma(l,a)$). This case is similar to the previous one, but instead of using the normal typing rules we use (T-R$_{new}$) and (T-W$_{new}$) because the capabilities of a are not included in the type environment.

Case 1.8 ($P = a^{\Delta t}.(R,Q)$, $t > 1$ *and* $t \neq \infty$). For this case we consider the *input* expression, namely $\Delta \vdash_l a^{\Delta t}?(X:T).(R,Q)$. In this case ϕ_Δ decreases the channel timer from $a^{\Delta t}$ to $a^{\Delta t-1}$. From the point of view of the typing system, the processes $a^{\Delta t}?(X:T).(R,Q)$ and $a^{\Delta t-1}?(X:T).(R,Q)$ are the same, and we can apply Lemma 1.1 and get $\Delta' \vdash_l a^{\Delta t}?(X:T).(R,Q)$. Since $\Gamma <: \Delta <: \Delta'$, we get the conclusion $\Gamma \Vdash l[[a^{\Delta t}?(X:T).(R,Q)]]_{\Delta'}$.

The case of channel restriction is similar, and uses the typing rule (T-NEWCH). □

Exercise 1.2. Prove the channel restriction case of Lemma 1.2.

Definition 1.4 (Structural equivalence $\equiv$ over timed channels). Is defined by

$$a_1^{\Delta t_1} \equiv a_2^{\Delta t_2} \text{ if and only if } a_1 = a_2 \text{ and } t_1 = t_2.$$

If the timers of the same channel name have different values, the corresponding processes have different behaviour. This aspect must be considered when defining timed bisimulations [25].

Table 1.10 Tagged Structural Equivalence in $tD\pi$

(S_Γ-GARBAGE)	$l[[stop]]_\Gamma$	$\equiv stop$
(S_Γ-SPLIT)	$l[[P \mid Q]]_\Gamma$	$\equiv l[[P]]_\Gamma \mid l[[Q]]_\Gamma$
(S_Γ-COPY)	$l[[*P]]_\Gamma$	$\equiv l[[P]]_\Gamma \mid l[[*P]]_\Gamma$
(S_Γ-NEW)	$l[[(\nu a:A)P]]_\Gamma$	$\equiv (\nu a@l:A)l[[P]]_\Delta$ if $a \notin fn(\Gamma)\cup\{l\}$ and $\Delta = \Gamma\{a@l:A\}$
(S_Γ-EXTR)	$M\vert(\nu a@k:A)N$	$\equiv (\nu a@k:A)(M\vert N)$ if $a \notin fn(M)$
(S_Γ-ASSOC)	$l[[P]]_\Gamma \vert l[[Q\vert R]]_\Gamma$	$\equiv l[[P\vert Q]]_\Gamma \vert l[[R]]_\Gamma$
(S_Γ-COMMU)	$l[[P]]_\Gamma \vert l[[Q]]_\Gamma$	$\equiv l[[Q]]_\Gamma \vert l[[P]]_\Gamma$
(S_Γ-NEUTR)	$l[[P]]_\Gamma \vert stop$	$\equiv l[[P]]_\Gamma$

The subject reduction property states that well-typedness is preserved by the reduction relation. This is a general approach in functional programming frameworks [153, 124]. We are also interested in proving that the well-typedness property is preserved by the structural equivalence relation. We present now such a result related to the structural equivalence relation. A more general subject reduction theorem is presented in Section 1.2.4.

If we have two tagged located processes which are structurally equivalent, and one of them is well-typed with respect to a type environment Γ, then the other process is also well-typed with respect to type environment Γ.

Theorem 1.1 (Subject Reduction for Tagged Equivalence Relation). *For all tagged located processes N, N' such that $N \equiv N'$,*

$$\Gamma \Vdash N \textit{ if and only if } \Gamma \Vdash N'.$$

Proof. We must consider all the equivalences given in Table 1.10.

Case 1.9 (S_Γ-NEW). From the hypothesis we have $\Gamma \Vdash l[[(\nu a:A)P]]_\Delta$, which means that $\Gamma <: \Delta$ and $\Delta \vdash_l (\nu a:A)P$. By using the rule (T-NEWCH) we get $\Delta\{a@l:A\} \vdash_l P$. By the rule (N-RUN) we get $\Delta\{a@l:A\} \vdash l[[P]]$, and with $\Gamma\{a@l:A\} <: \Delta\{a@l:A\}$ we have $\Gamma\{a@l:A\} \Vdash l[[P]]_{\Delta\{a@l:A\}}$. We again apply the rule (T-NEWCH) for tagged processes and get $\Gamma \Vdash (\nu a@l:A)l[[P]]_{\Delta\{a@l:A\}}$.

Case 1.10 (S_Γ-SPLIT). We start from $\Gamma <: \Delta$ and $\Delta \vdash_l P \,|\, Q$, and by (T-STR) we get that $\Delta \vdash_l P$ and $\Delta \vdash_l Q$. From $\Gamma <: \Delta$ and by (N-RUN) we obtain that $\Gamma \Vdash l[[P]]_\Delta$ and $\Gamma \Vdash l[[Q]]_\Delta$. We apply again (N-STR) and obtain the conclusion $\Gamma \Vdash l[[P]]_\Delta \,|\, l[[Q]]_\Delta$.

Case 1.11 (S_Γ-COPY). This case follows the steps of the previous one, and we leave it as an exercise for the reader.

Case 1.12 (S_Γ-EXTR). For this case we use the rules for located processes. Starting from $M \,|\, (\nu a@k : A)N$ and by (N-STR), we get $\Gamma \Vdash M$ and $\Gamma \Vdash (\nu a@k : A)N$. Both $\Gamma \Vdash (\nu a@k : A)N$ and (N-NEWCH) infer $\Gamma\{a@k : A\} \Vdash N$. By weakening, and because $a \notin fn(\Gamma)$, we get $\Gamma\{a@k : A\} \Vdash M$. We again apply (N-STR) and then (N-NEWCH), and we get the desired result $\Gamma \Vdash (\nu a@k : A)(M \,|\, N)$.

The cases inferred from (S_Γ-GARBAGE) and other rules are similar to the monoid laws of the π-calculus. □

Exercise 1.3. Prove the other cases of Theorem 1.1.

1.2.3 Operational Semantics

We consider the tagged located processes ranged over by N and M, namely N and M can be thought as process expressions of the form $l[[P]]_\Gamma$. We denote by $\not\rightarrow$ the fact that rules (R_Γ-COM1) and (R_Γ-COM2) cannot be applied. Using these notations, we give the following reduction rules in Table 1.11 providing an operational semantics for $tD\pi$.

Table 1.11 Reduction Relation of $tD\pi$

$$(\text{R}_\Gamma\text{-GO}) \; \frac{}{l[[go\,k.P]]_\Gamma \rightarrow \psi(k[[P]]_\Gamma)} \qquad (\text{R}_\Gamma\text{-IDLE}) \; \frac{l[[P]]_\Gamma \not\rightarrow}{l[[P]]_\Gamma \rightarrow \phi_\Delta(l[[P]]_\Gamma)}$$

$$(\text{R}_\Gamma\text{-COM1}) \; \frac{\Gamma(l,a) <: res\{r\langle T\rangle\}}{\begin{array}{c} l[[a^{\Delta t}!\langle v\rangle.(P,Q)]]_\Delta \;|\; l[[a^{\Delta t'}?(X:T).(P',Q')]]_\Gamma \rightarrow \\ \psi(l[[P]]_\Delta) \;|\; \psi(l[[P'\{^v/_X\}]]_{\Gamma\{v@l:T\}}) \end{array}}$$

$$(\text{R}_\Gamma\text{-COM2}) \; \frac{\Gamma(l,a) <: res\{ro\langle T\rangle\}}{\begin{array}{c} l[[a^{\Delta t}!\langle v\rangle.(P,Q)]]_\Delta \;|\; l[[a^{\Delta t'}?(X:T).(P',Q')]]_\Gamma \rightarrow \\ \psi(l[[P]]_\Delta) \;|\; \psi(l[[P'\{^v/_X\}]]_\Gamma) \end{array}}$$

$$(\text{R}_\Gamma\text{-PAR}) \; \frac{N \rightarrow N' \quad M \rightarrow M'}{N \;|\; M \rightarrow N' \;|\; M'} \qquad (\text{R}_\Gamma\text{-RES}) \; \frac{N \rightarrow N'}{(\nu a@l:A)N \rightarrow (\nu a@l:A)N'}$$

$$(\text{R}_\Gamma\text{-CONG}) \; \frac{N \equiv N' \quad N \rightarrow M \quad M \equiv M'}{N' \rightarrow M'}$$

We have two communication rules which depend on the type of the communication channel. In (R_Γ-COM2) we consider $ro\langle\rangle$ channels, and the process may use the received information without adding the new type to its type environment Γ, contrary to the behaviour of rule (R$_\Gamma$-COM1). The communication rules and the rule (R$_\Gamma$-GO) do not enter under the scope of ϕ_Δ. In this case the type environments are affected by the cleanup function ψ. In (R$_\Gamma$-IDLE) the function ϕ_Δ decreases the timers on channels, and for the expired timers it discards the channels and changes the state of the process. At each tick of the universal clock, the rule (R$_\Gamma$-IDLE) is applied to processes which do not enter any communication. When applying the rule (R$_\Gamma$-PAR), if process M does not have an internal communication reduction, then it is transformed into M' by rule (R$_\Gamma$-IDLE). The same argument is valid for N as well.

Removing location types from the type environment can lead to errors generated by *go* actions. We solve this problem by extending the syntax of *go* with a choice syntax similar to the one given for channels; therefore $go\,l.P$ becomes $go\,l.(P,Q)$. If $\Gamma(l)$ is not defined, then Q is executed. If the location type of l contains a capability *go*, then P is executed; otherwise, if the location type of l does not contain a capability *go*, an error is generated. We must change the corresponding typing rules where the operator *go* appears. Thus (T-GO) is translated into (T-GO1) and (T-GO2).

$$\text{(T-GO1)}\ \frac{k \notin dom(\Gamma) \quad \Gamma \vdash_l Q}{\Gamma \vdash_l go\,k.(P,Q)} \qquad \text{(T-GO2)}\ \frac{\Gamma(k) : loc\{go\} \quad \Gamma \vdash_k P}{\Gamma \vdash_l go\,k.(P,Q)}$$

A process P generating an error is denoted by $P \xrightarrow{err}$. The cases when a process generates a runtime error are defined by a set of rules in Table 1.12.

Table 1.12 Runtime Errors in $tD\pi$

$$\text{(E-GO)}\ \frac{\Gamma(k) \text{ is defined and } \Gamma(k) \not<: loc\{go\}}{l[[go\,k.(P,Q)]]_\Gamma \xrightarrow{err}}$$

$$\text{(E-SUBC)}\ \frac{\Gamma(l) \not<: loc\{newch\}}{l[[(\nu a:A)P]]_\Gamma \xrightarrow{err}}$$

$$\text{(E-SND)}\ \frac{\Gamma(l,a) \text{ is defined and } \Gamma_l(v) \not<: wobj(\Gamma(l,a))}{l[[a^{\Delta t}!\langle v\rangle.(P,Q)]]_\Gamma \xrightarrow{err}}$$

$$\text{(E-RCV)}\ \frac{\Gamma(l,a) \text{ is defined and } robj(\Gamma(l,a)) \not<: T \text{ or } roobj(\Gamma(l,a)) \not<: T}{l[[a^{\Delta t}?(X:T).(P,Q)]]_\Gamma \xrightarrow{err}}$$

$$\text{(E-COM)}\ \frac{\Gamma(l,a) \text{ and } \Delta(l,a) \text{ are defined and } \quad wobj(\Gamma(l,a)) \not<: robj(\Delta(l,a)) \text{ or } wobj(\Gamma(l,a)) \not<: roobj(\Delta(l,a))}{l[[a^{\Delta t}!\langle v\rangle.(P,Q)]]_\Gamma \mid l[[a^{\Delta t'}?(X:T).(P',Q')]]_\Delta \xrightarrow{err}}$$

$$\text{(E-NEW)}\ \frac{N \xrightarrow{err}}{(\nu a@k:T)N \xrightarrow{err}} \qquad \text{(E-PAR)}\ \frac{N \xrightarrow{err}}{N \mid M \xrightarrow{err}} \qquad \text{(E-STR)}\ \frac{M \equiv N \quad N \xrightarrow{err}}{M \xrightarrow{err}}$$

The partial functions *robj( ), roobj( ), wobj( )* are defined over the set of channel types, and return the type of the corresponding channel capabilities. For example, considering a channel type $a : res\{w\langle T\rangle\}$ in the type environment Γ at location l, the application of $wobj(\Gamma(l,a))$ returns T. In order to derive a runtime error, the channel type or location type must be in the type environment. A runtime error appears when a process tries to do something against the types accumulated in its type environment. When a type is not in the type environment of the process, the safety process is chosen by ϕ_Δ.

The reduction rule (R$_\Gamma$-GO) cannot check if the type of the location is in the type environment, and consequently we change the time-stepping function ϕ_Δ by adding two more lines to its definition:

$$\begin{cases} k[[R]]_{\Gamma'} & \text{if } P = go\,k.(R,Q) \text{ and } \Gamma(k) <: loc\{go\} \\ l[[Q]]_{\Gamma'} & \text{if } P = go\,k.(R,Q) \text{ and } k \notin dom(\Gamma) \end{cases}$$

The rule (R$_\Gamma$-GO) is changed into $l[[go\,k.(P,Q)]]_\Gamma \rightarrow \phi_\Delta(l[[go\,k.(P,Q)]]_\Gamma)$ which is captured by the (R$_\Gamma$-IDLE) rule. A process of the form $go\,k.(P,Q)$ is beyond the scope of any of the reduction rules R_Γ, excepting (R$_\Gamma$-IDLE), and so ϕ_Δ is applied. One of the above new lines is applied, and ϕ_Δ changes the process either by allowing the movement to the new location, or by choosing the safety process.

Regarding the behaviour of the $tD\pi$ system, we can say that a nondeterministic method is applied to select two interacting processes for each communication channel at each location of a distributed system. Afterwards the reduction rules are applied, and the communications are performed. ϕ_Δ is applied to the processes which do not enter in any communication. The type environments of the communicating processes are affected by the application of the ψ function.

A system described with $tD\pi$ satisfies the following properties [92]:

- **Time Determinism**: at each time only one reduction rule can be applied. A possible problem could appear only if we apply R$_\Gamma$-IDLE when we can apply a communication rule. However this is not possible because R$_\Gamma$-IDLE is applied only if the process does not enter in any communication ($\not\rightarrow$).
- **Maximal Progress**: a process cannot delay if it can enter in a communication. This property is sometimes referred to as *urgency*.

Regarding the global time aspect in distributed systems, we consider a global clock synchronising all the timers. Recent work [118] on Network Time Synchronisation Protocol (NTP) shows that it is possible to achieve time synchronisation in real applications. Having this technology we can suppose that the theoretical assumption about a universal clock is practical rather than speculation. Our global timing function ϕ_Δ has to apply the local time-stepping function ϕ for the locations of the distributed system. If we adopt the NTP synchronisation model, we can get a guaranteed frequency and local oscillator phase precision of no more than a few milliseconds, which in many cases is acceptable.

1.2.4 Soundness of tDπ

Regarding the soundness of $tD\pi$, we follow a method based on *subject reduction* and *type safety* [153] used also in proving the soundness of $D\pi$. This is a syntactic approach, in contrast to other approaches based on denotational semantics or structural operational semantics.

Theorem 1.2 (Subject Reduction). *For all tagged located processes N and N'*

(a) If $N \equiv N'$ then $\Gamma \Vdash N$ if and only if $\Gamma \Vdash N'$.
(b) If $N \rightarrow N'$ then $\Gamma \Vdash N$ if and only if $\Gamma \Vdash N'$.

Proof. Part *(a)* is in fact Theorem 1.1; its proof is in Section 1.2.2.
Part *(b)* is similar to the result presented in [124] which asserts consistency between the static and the dynamic semantics. We use the same technique, and proceed by induction on the depth of inference for $N \rightarrow N'$. We also use Lemma 1.1 which relates time and type environments, and Lemma 1.2 which relates time and communication channels. More details can be found in [67].

Case 1.13 (R_Γ-IDLE). This is covered by Lemma 1.2.

Case 1.14 (R_Γ-RES). From the hypothesis we know that $\Gamma \Vdash (\nu a@k : A)N$. This means that $\Gamma\{a@k : A\} \Vdash N$, and according to the induction hypothesis we have that $\Gamma \Vdash N'$. Since $\Gamma\{a@k : A\} <: \Gamma$, by applying the weakening property we get $\Gamma\{a@k : A\} \Vdash N'$. Simply applying again (N-NEWCH) we get $\Gamma \Vdash (\nu a@k : A)N'$.

Case 1.15 (R_Γ-COM1 *or* R_Γ-COM2). These rules can be treated in the same way. For R_Γ-COM1, starting from $\Gamma \Vdash l[[a^{\Delta t}!\langle v\rangle.(P,Q)]]_\Delta \mid l[[a^{\Delta t'}?(X:T).(P',Q')]]_{\Delta'}$ and after applying the rule (N-STR) we obtain that $\Gamma \Vdash l[[a^{\Delta t}!\langle v\rangle.(P,Q)]]_\Delta$ *(*)* and $\Gamma \Vdash l[[a^{\Delta t'}?(X:T).(P',Q')]]_{\Delta'}$ *(**)*. By using the rule (N-RUN) and *(*)* we have $\Delta \vdash_l a^{\Delta t}!\langle v\rangle.(P,Q)$ and with *(**)* we have $\Delta' \vdash_l a^{\Delta t'}?(X:T).(P',Q')$. By applying (T-W) we get $\Delta \vdash_l P$ which together with (N-RUN) give the statement $\Gamma \Vdash l[[P]]_\Delta$. We also have $\Delta \vdash_l v : T$ and the subtyping reactions $\Gamma <: \Delta$, $\Gamma <: \Delta'$ which means that $\Delta(l,u)$ and $\Delta'(l,u)$ must agree on the type they use. So by weakening we get $\Delta'\{v@l : T\} \vdash_l v : T$.

The difference between (R_Γ-COM1) and (R_Γ-COM2) is given by the typing rule used for the type of the input channel. By applying the rule (T-R) we get $\Delta'\{X@l : T\} \vdash_l P'$. We denote by Δ'' the type environment $\Delta'\{v@l : T\}$. Thus, by weakening we get $\Delta''\{X@l : T\} \vdash_l P'$, and we can use the substitution lemma of [93] to obtain $\Delta'' \vdash_l P'\{{}^v/_X\}$. However $\Gamma <: \Delta''$ and so, by applying (N-RUN), we get $\Gamma \Vdash l[[P'\{{}^v/_X\}]]_{\Delta'\{v@l:T\}}$. We apply Lemma 1.1 twice, and (N-STR) to get the result $\Gamma \Vdash \psi(l[[P]]_\Gamma) \mid \psi(l[[P'\{{}^v/_X\}]]_{\Delta'\{v@l:T\}})$.

It is easy to prove the second inference for (R_Γ-COM2), but we have to pay attention to the rules we use, because the type of the channel is now different.

Case 1.16 (R_Γ-PAR). We have $\Gamma \Vdash N \mid M$ which by applying (N-STR) gives us $\Gamma \Vdash N$ and $\Gamma \Vdash M$. We can also infer by induction that $\Gamma \vdash N'$. By Lemma 1.2 we have that $\Gamma \Vdash \phi(M)$, and we can apply again (N-STR) obtaining the result $\Gamma \Vdash N' \mid \phi(M)$. For the case when M reduces to M' by a rule other than (R_Γ-IDLE) (i.e., it is not affected by the passage of time), the proof steps are easy to find (and left to the reader). □

Subject reduction assures us that once it is well-typed, a process remains well-typed during its evolution. Note that well-typedness must be preserved by both equivalence rules and reduction rules. In the following we give a result of *type safety* which is necessary to have a complete proof of the soundness property of $tD\pi$. The *type safety* property states that if a system is well-typed, then it cannot generate runtime errors, and this is denoted by $P \overset{err}{\not\longmapsto}$.

Theorem 1.3 (Type Safety). *We have $N \overset{err}{\not\longmapsto}$ for all tagged located processes N, and all type environments Γ such that $\Gamma \Vdash N$.*

Proof. The outline of the proof follows a method which proves the contrapositive, namely *if N gives rise to a runtime error ($N \overset{err}{\longmapsto}$) then N cannot be well-typed under any type environment Γ* ($\Gamma \nVdash N$ for all Γ). In [153] the authors use the same statement as a lemma to prove that the *faulty expressions are untypable*. We use induction on the definition of the runtime errors, and have a proof case for each rule of Table 1.12.

Case 1.17 (E-SND). The rule says that $l[[a^{\Delta t}!\langle v\rangle.(P,Q)]]_\Gamma \overset{err}{\longmapsto}$ *if $\Gamma(l,a)$ is defined and $\Gamma(l,v) \not<: wobj(\Gamma(l,a))$*. Let us suppose that there is a type environment Δ such that the process generating a runtime error is well-typed under this environment, i.e., $\Delta \Vdash l[[a^{\Delta t}!\langle v\rangle.(P,Q)]]_\Gamma$. This means that $\Delta <: \Gamma$ and $\Gamma \vdash_l a^{\Delta t}!\langle v\rangle.(P,Q)$. Therefore there are two typing rules which can be applied, depending on the type of the output channel. If $a : - \notin \Gamma(l)$, then we have a contradiction with the fact that $\Gamma(l,a)$ *must be defined* from the definition of the rule. Otherwise we have to use (T-W), obtaining $\Gamma \vdash_l a : res\{w\langle T\rangle\}\Delta t$ and $\Gamma \vdash_l v : T$. Statement $\Gamma \vdash_l v : T$ implies that $\Gamma(l,v) <: T$. From $\Gamma \vdash_l a : res\{w\langle T\rangle\}\Delta t$ we get $\Gamma(l,a) = res\{w\langle T\rangle\}$ (by definition), which by application of the function $wobj$ leads to $wobj(\Gamma(l,a)) = T$. Together with $\Gamma(l,v) <: T$, this leads us to the contradiction $\Gamma(l,v) <: wobj(\Gamma(l,a))$.

Case 1.18 (E-GO). We have that $l[[go\,k.(P,Q)]]_\Gamma \overset{err}{\longmapsto}$ if $\Gamma(k)$ is defined and thus $\Gamma(k) \not<: loc\{go\}$. We suppose that there exists a type environment Δ such that we have $\Delta \Vdash l[[go\,k.\ (P,Q)]]_\Gamma$, and try to see if we can conclude a contradiction. If the location k is not defined in the type environment Γ, then we can use (T-GO1); however this would result in a contradiction. By (T-GO2), we have $\Gamma(k) : loc\{go\}$ which means that $\Gamma(k) <: loc\{go\}$; we get again a contradiction. Therefore we have the statement: *there is no type environment Δ such that $\Delta \Vdash l[[go\,k.(P,Q)]]_\Gamma$ and $l[[go\,k.(P,Q)]]_\Gamma \overset{err}{\longmapsto}$.*

Case 1.19 (E-RCV). We consider that there exists a type environment Δ such that $\Delta \Vdash l[[a^{\Delta t}?(X:T).(P,Q)]]_\Gamma$. Thus we have $\Delta <: \Gamma$ and $\Gamma \vdash_l a^{\Delta t}?(X:T).(P,Q)$. If we suppose our input channel to be *reading only*, then we apply the rule (T-RO) and we get $\Gamma \vdash_l a : res\{ro\langle T\rangle\}\Delta t$. Thus $\Gamma(l,a) = res\{ro\langle T\rangle\}$, and applying the function $roobj$ we get $roobj(\Gamma(l,a)) = T$, and thus $roobj(\Gamma(l,a)) <: T$, contradicting the definition.

Case 1.20 (E-COMM). We use the same method as before, and suppose a type environment Δ' such that $\Delta' \Vdash l[[a^{\Delta t}!\langle v\rangle.(P,Q)]]_\Gamma \mid l[[a^{\Delta t'}?(X:T).(P',Q')]]_\Delta$. By applying the rule (N-STR), and then (N-RUN), we get $\Delta' <: \Gamma$, $\Delta' <: \Delta$, and

$\Gamma \vdash_l a^{\Delta t}!\langle v\rangle.(P,Q)$, $\Delta \vdash_l a^{\Delta t'}?(X:T).(P',Q')$. By (T-W), we get $\Gamma \vdash_l a: res\{w\langle T\rangle\}$ which means that $\Gamma(l,a) = res\{w\langle T\rangle\}$. We apply the function $wobj$ and obtain that $wobj(\Gamma(l,a)) = T$ *(1)*. We suppose that the channel a under type environment Δ is an $r\langle\rangle$ channel, and infer from $\Delta \vdash_l a^{\Delta t'}?(X:T).(P',Q')$ that $\Delta \vdash_l a: res\{r\langle T\rangle\}$. As before, we can apply the function $robj$ and get $robj(\Delta(l,a)) = T$ *(2)*. From *(1)* and *(2)* we have the contradiction $wobj(\Gamma(l,a)) <: robj(\Delta(l,a))$.

Case 1.21 (E-SUBC, E-NEW, E-PAR *and* E-STR). These rules are the same as in $D\pi$, and the proofs are similar to those presented above. □

Exercise 1.4. Prove the other cases of Theorem 1.3.

Timed distributed pi-calculus uses a global clock which decrements all the timers. A simplified timed distributed pi-calculus called TiMo using local clocks and timing constraints to control migration and communication is defined in [58] and [59]. The authors provided an operational semantics for this new formalism of distributed systems with mobility, and succeeded in translating finite TiMo specifications into a class of high-level Petri nets with time.

PerTiMo extends TiMo by working with processes having appropriate access rights to communicate; moreover, the access permissions are dynamic (can change in time and space). PerTiMo uses local clocks and local maximal parallelism of actions. Operational semantics of PerTiMo and safety of communication and migration in terms of access permissions are presented in [60].

A software platform for timed mobility and timed interaction is presented in [56], and a high-level language for mobile agents with timers in [57].

1.3 Mobile Ambients

Another formalism able to express mobility is the *ambient calculus* [42]. The ambient calculus describes computation carried out on mobile devices (i.e. networks having a dynamic topology), and mobile computation (i.e. executable code able to move around the network). The primitive concept of the ambient calculus is the ambient defined as a bounded place in which computation can occur. Ambients can be nested inside other ambients. Each ambient has a name used to control access to it. Computation is represented as the movement of ambients: they can be moved as a whole, changing their location by consuming certain capabilities: in, out, open. These basic operations are expressive enough to simulate name-passing channels in the π-calculus. In certain conditions, the π-calculus is also able to simulate the ambient calculus [73].

We consider a variant of mobile ambients called safe ambients for which the movement of an ambient takes place only if both participants agree [112]. The mobility is provided by the consumption of *pairs* of capabilities. Safe ambients differ from ambients by co-actions: whereas in ambients a movement is initiated only by the moving ambient and the target ambient has no control over it, in safe ambients both participants must agree by using a matching between an action and its co-action. We present here a short description of pure safe ambients (SA); more information can be found in [112].

1.3.1 Syntax

Given an infinite set of names $\mathscr{N}$ (ranged over by $m, n, \ldots$), we define the set $\mathscr{A}$ of SA-processes (denoted by $A, A', B, B', \ldots$) together with their capabilities (denoted by $C, C', \ldots$) as follows:

Table 1.13 Safe Mobile Ambients Syntax

C	$::=$	capabilities	A	$::=$	processes
	$\mid in\ n \mid \overline{in}\ n$	in capabilities		$\mid \mathbf{0}$	inactive process
	$\mid out\ n \mid \overline{out}\ n$	out capabilities		$\mid C.A$	movement
	$\mid open\ n \mid \overline{open}\ n$	open capabilities		$\mid n[A]$	ambient
				$\mid A \mid B$	composition

Process $\mathbf{0}$ is an inactive mobile ambient. A movement $C.A$ is provided by the capability C, followed by the execution of A. An ambient $n[\,A\,]$ represents a bounded place labelled by n in which an SA-process A is executed. $A \mid B$ is a parallel composition of mobile ambients A and B.

1.3.2 Operational Semantics

The *structural congruence* $\equiv_{amb}$ over ambients is the least congruence such that $(\mathscr{A}, |, \mathbf{0})$ is a commutative monoid. The operational semantics of safe ambients is given in terms of a reduction relation $\Rightarrow_{amb}$ by the following axioms and rules:

Table 1.14 Safe Mobile Ambients Operational Semantics

Axioms

$$(In)\ \ n[\,in\ m.A \mid A'\,] \mid m[\,\overline{in}\ m.B \mid B'\,] \Rightarrow_{amb} m[\,n[\,A \mid A'\,] \mid B \mid B'\,];$$

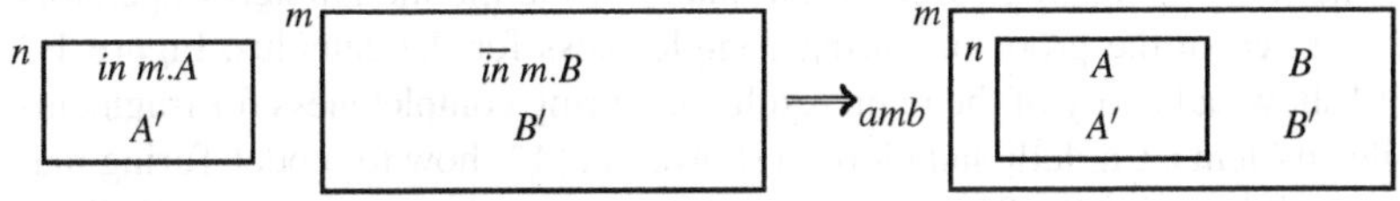

$$(Out)\quad m[\, n[\, out\ m.A \mid A' \,] \mid \overline{out}\ m.B \mid B' \,] \Rightarrow_{amb} n[\, A \mid A' \,] \mid m[\, B \mid B' \,];$$

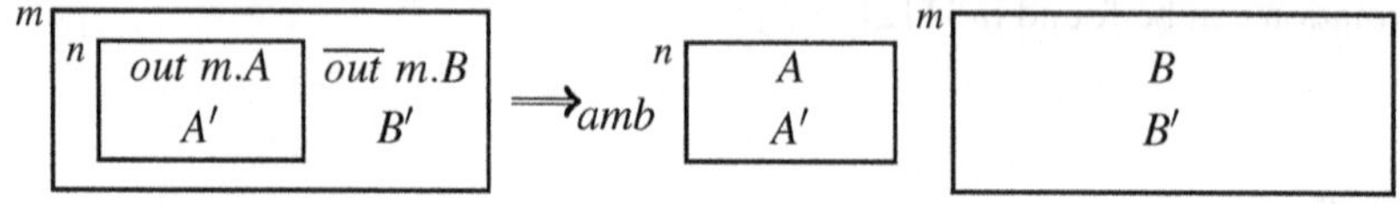

$$(Open)\quad open\ n.A \mid n[\, \overline{open}\ n.B \mid B' \,] \Rightarrow_{amb} A \mid B \mid B'.$$

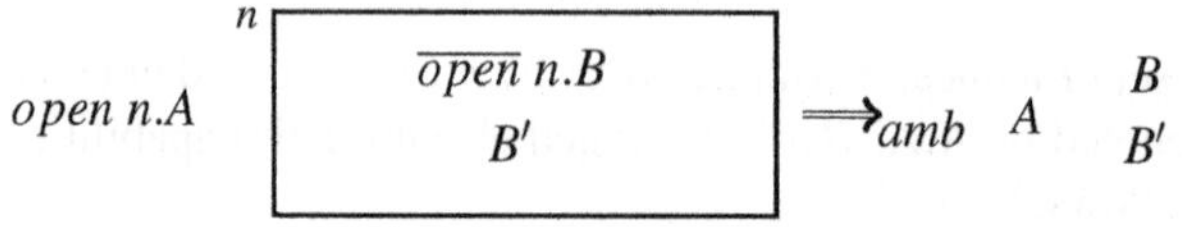

Rules:

$$(Comp1)\ \frac{A \Rightarrow_{amb} A'}{A \mid B \Rightarrow_{amb} A' \mid B};\ (Comp2)\ \frac{A \Rightarrow_{amb} A' \quad B \Rightarrow_{amb} B'}{A \mid B \Rightarrow_{amb} A' \mid B'};$$

$$(Amb)\ \frac{A \Rightarrow_{amb} A'}{n[\, A \,] \Rightarrow_{amb} n[\, A' \,]};\quad (Struc)\ \frac{A \equiv A',\ A' \Rightarrow_{amb} B',\ B' \equiv B}{A \Rightarrow_{amb} B}.$$

$\Rightarrow^{*}_{amb}$ denotes a reflexive and transitive closure of the binary relation $\Rightarrow_{amb}$.

1.3.3 Computability and Decidability

Over the years, many variants and dialects have been proposed; among them, we mention mobile safe ambients [112] (used to investigate security issues in mobile systems), push and pull ambient calculus [94, 134] (formalises objective rather than subjective mobility), boxed ambients [34] (used to model systems in which ambient boundaries cannot be dissolved and direct communication between parent and child ambients is permitted) and BioAmbients [138] (defined to model the behaviour of biological systems).

Following in the tradition of process calculi, mobile ambients and its variants have been equipped with a rich variety of formal tools (e.g., behavioural semantics [116], type system [41], logics [43]) for reasoning about and verifying properties of systems specified with these calculi (e.g., reasoning about both the behaviour and spatial structure of ambients). Another line of research looks at the expressiveness of these calculi to distinguish between necessary and redundant features.

The computational strength of mobile ambients has been investigated in several papers. The most interesting result is that many of the mobile ambients operators are not required in the proof of Turing completeness for the calculus. Figure 1.2 from [39] shows a history of the main results on Turing completeness for fragments of mobile ambients. Cardelli and Gordon showed in [42] how to model Turing machines in mobile ambients. This encoding of Turing machines made use of all the capabilities of mobile ambients as well as the restriction operator. Subsequently, in [38] it was proved that restriction is unnecessary. This result was obtained by showing how to model Random Access Machines [149] (which are a well-known register-based Turing-complete formalism) in mobile ambients without using the

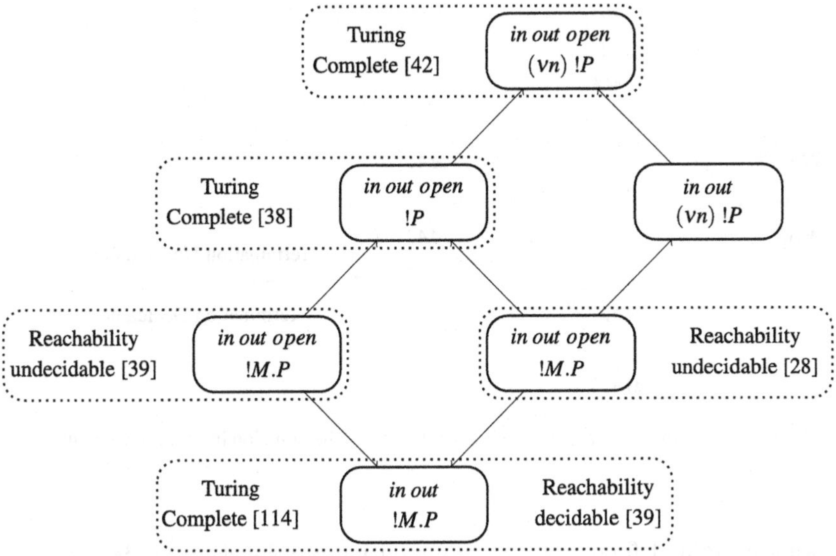

Fig. 1.2 Overview of the Results for Turing Completeness and the (Un)Decidability of Reachability in Pure Mobile Ambients (Arrows Represent the Sublanguage Relation) [39]

restriction operator. More recently, in [114] it was proved that the open capability is unnecessary by presenting an improved modelling of Random Access Machines without using open; moreover, the replication operator $!P$ is only applied to prefixed processes of the form $M.P$.

The proofs of Turing completeness mentioned above imply, as a direct consequence, that termination is not decidable in the fragments of mobile ambients considered. Another property of processes, which is in some cases even more interesting than termination, is process reachability: the reachability problem consists of verifying whether a target process can be reached from a source process.

The first work devoted to the investigation of reachability in mobile ambients was [28], which proved that reachability is undecidable even in a minimal fragment of pure mobile ambients where both the restriction operator and open capability are removed. Figure 1.2 indicates the known results of Turing completeness and the undecidability of reachability for fragments of mobile ambients considered in the papers mentioned above. The decidability of reachability in the fragment with *in*, *out*, $!M.P$ may appear surprising in light of the result on Turing completeness proved in [114]; it follows from the following monotonicity property deriving from the absence of the open capability (and from the impossibility of applying the replication operator to ambients): during a computation, the number of active ambients cannot increase. The existence of this bound allows the modeling of all possible computations as computations in finite Petri nets, a formalism in which the reachability problem is decidable.

The (un)decidability of the termination problem is presented in Figure 1.3. Even if the calculi with replication are not syntactically subsets of the corresponding cal-

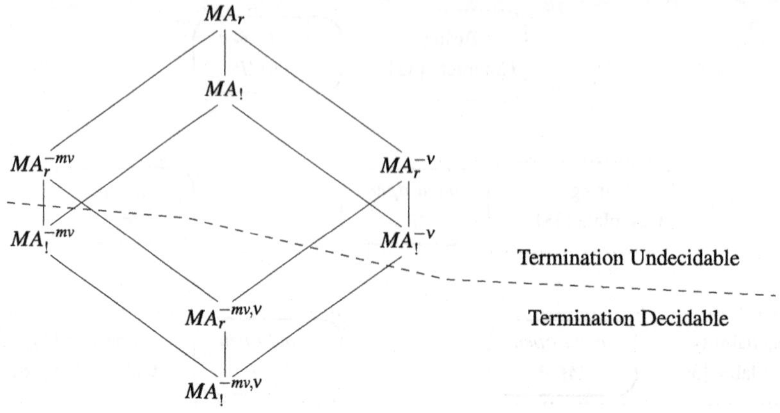

Fig. 1.3 Overview of the Results for the (Un)Decidability of Termination in Mobile Ambients [38]

culi with recursive definition, these are considered as sub-calculi in [38] because recursive definition is more general than replication. In Figure 1.3 the following notations are used: $MA_!$ (resp. MA_r) for the fragment of mobile ambients with replication (resp. recursion); $-mv$ (resp. $-\nu$) stands for the fragment of mobile ambients without movement (resp. name restriction).

1.4 Mobile Ambients with Timers

TTL value and strategies for retransmission in TCP/IP protocols provide a good motivation to add timers to ambients. In [9, 10, 15] we associate timers not only to ambients, but also to capabilities and communication channels. The resulting formalism is called mobile ambients with timers (tMA), and represents a conservative extension of the ambient calculus. Inspired by [41] we introduce types for ambients in tMA. The type system associates to each ambient a set of types in order to control its communication by allowing only well-typed messages. For instance, if a process inside an ambient sends a message of a type which is not included in the type system of the ambient, then the process fails. In tMA, by using timers, the process may continue its execution after the timer of the corresponding output communication expires.

1.4.1 Syntax

In tMA communication channels (input and output channels), capabilities and ambients are used as temporal resources. A timer Δt of each temporal resource makes the resource available only for a period of time t.

The novelty of this approach results from the fact that a location, represented by an ambient, can disappear. We denote by $n^{\Delta t}[P]^{\mu}$ the fact that an ambient n has a timer Δt, while the tag μ is a neutral tag that indicates whether an ambient is active or passive. If $t > 0$, the ambient $n^{\Delta t}[P]^{\mu}$ behaves exactly like the untimed ambient $n[P]$. Since the timer Δt can expire (i.e., $t = 0$) we use a pair $(n^{\Delta t}[P]^{\mu}, Q)$, where Q is a *safety process*. If no *open n* capability appears in t units of time, the ambient $n^{\Delta t}[P]^{\mu}$ is dissolved, the process P is cancelled, and the safety process Q is executed. If $Q = \mathbf{0}$ we can simply write $n^{\Delta t}[P]^{\mu}$ instead of $(n^{\Delta t}[P]^{\mu}, Q)$. If we want to simulate the behaviour of an untimed mobile ambient, then we use ∞ instead of Δt, i.e., $n^{\infty}[P]^{\mu}$.

Similarly, we add a safety process for the input and output communications and the movement processes. The process $open^{\Delta t} n.(P,Q)$ evolves to process P if before the timer Δt expires, the capability $open^{\Delta t} n$ is consumed; otherwise it evolves to process Q. The process $c^{\Delta t}!\langle m\rangle.(P,Q)$ evolves to process P if before the timer Δt expires, a process captures name m from channel c; otherwise it evolves to process Q.

Since messages are undirected, it is possible for a process $c^{\Delta t}!\langle m\rangle.(P,Q)$ to send a message which is not appropriate for any receiver. To restrict the communication and be sure that m reaches an appropriate receiver, we add types expressed by $Amb[\Gamma]$ and write $c^{\Delta t}!\langle m : Amb[\Gamma]\rangle.(P,Q)$. We use types inspired by [41]; the set of types is defined in Table 1.15. We use types for communication in order to validate the exchange of messages, namely that if we expect to communicate integers, then we cannot communicate boolean values. $\mathcal{B}$ represents a set of *base types*. The intuitive meaning of the subtyping relation is that $<:$ represents the inverse of the set inclusion relation ($\Gamma <: \Gamma'$ for types means $\Gamma \supset \Gamma'$ for sets and $\Gamma \sqcap \Gamma'$ for types means $\Gamma \cup \Gamma'$ for sets).

Table 1.15 Types in tMA

Set of types:	
$\Gamma ::= \mathcal{B} \mid Amb[\Gamma] \mid \Gamma \sqcap \Gamma'$	
$Amb[\Gamma]$	ambient name allowing Γ exchanges

If an appropriate message is received before the timer Δt expires, then the process $c^{\Delta t}?(x : Amb[\Gamma]).(P,Q)$ evolves to process P; otherwise it evolves to process Q. According to the syntax of tMA presented in Table 1.16, $Amb[\Gamma]$ can be used in a restriction process $(\nu n : Amb[\Gamma])P$, which means that n of type $Amb[\Gamma]$ is new in process P. A variable x is bound only in process P when we consider the process $c^{\Delta t}?(x : Amb[\Gamma]).(P,Q)$.

Table 1.16 Syntax of tMA

a, p	ambient tags		$P, Q ::=$	processes
c	channel name		$\mathbf{0}$	inactivity
n, m	ambient names		$M^{\Delta t}.(P, Q)$	movement
x	variable		$(n^{\Delta t}[P]^{\mu}, Q)$	ambient
M ::=	capabilities		$P \mid Q$	composition
	in n	can enter n	$(\nu n : Amb[\Gamma])P$	restriction
	out n	can exit n	$c^{\Delta t}!\langle m : Amb[\Gamma]\rangle.(P, Q)$	output action
	open n	can open n	$c^{\Delta t}?(x : Amb[\Gamma]).(P, Q)$	input action
			$*P$	replication

If it does not matter if an ambient is passive or active, we simple use μ as the tag of the ambient. When we initially describe the ambients, we consider that all ambients are active, and we associate the tag a to them.

1.4.2 Operational Semantics

The passage of time is described by a discrete global time progress function ϕ_{Δ} defined over the set $\mathscr{P}$ of mobile ambients with timers. The actions are performed at every tick of a universal clock. The opened ambients, the channels involved in a communication and the consumed capabilities disappear together with their timers. If a channel, capability or ambient has the timer ∞ we use $\infty - 1 = \infty$ when applying the function ϕ_{Δ}. This function modifies a process accordingly with the global passage of time. Another property of the function ϕ_{Δ} is that passive ambients can become active in the next unit of time in order to participate in other reductions.

Definition 1.5 (Global Time Progress Function). We define $\phi_\Delta : \mathscr{P} \to \mathscr{P}$ by:

$$\phi_\Delta(P) = \begin{cases} M^{\Delta(t-1)}.(R,Q) & \text{if } P = M^{\Delta t}.(R,Q),\, t>0 \\ Q & \text{if } P = M^{\Delta t}.(R,Q),\, t=0 \\ c^{\Delta(t-1)}!\langle m:Amb[\Gamma]\rangle.(R,Q) & \text{if } P = c^{\Delta t}!\langle m:Amb[\Gamma]\rangle.(R,Q),\, t>0 \\ Q & \text{if } P = c^{\Delta t}!\langle m:Amb[\Gamma]\rangle.(R,Q),\, t=0 \\ c^{\Delta(t-1)}?(x:Amb[\Gamma]).(R,Q) & \text{if } P = c^{\Delta t}?(x:Amb[\Gamma]).(R,Q),\, t>0 \\ Q & \text{if } P = c^{\Delta t}?(x:Amb[\Gamma]).(R,Q),\, t=0 \\ \phi_\Delta(R) \mid \phi_\Delta(Q) & \text{if } P = R \mid Q \\ (\nu n:Amb[\Gamma])\phi_\Delta(R) & \text{if } P = (\nu n:Amb[\Gamma])R \\ (n^{\Delta(t-1)}[\phi_\Delta(R)]^a, Q) & \text{if } P = (n^{\Delta t}[R]^\mu, Q),\, t>0 \\ Q & \text{if } P = (n^{\Delta t}[R]^\mu, Q),\, t=0 \\ P & \text{if } P = *R \text{ or } P = \mathbf{0} \end{cases}$$

For the processes $c^{\Delta t}!\langle m:Amb[\Gamma]\rangle.(P,Q)$, $c^{\Delta t}?(x:Amb[\Gamma]).(P,Q)$ and $M^{\Delta t}.(P,Q)$, the timer of process P is activated only after the consumption of $c^{\Delta t}!\langle m:Amb[\Gamma]\rangle$, $c^{\Delta t}?(x:Amb[\Gamma])$ and $M^{\Delta t}$ (in at most t units of time). Reduction rules (Table 1.18) show how the time function ϕ_Δ is used.

Processes are grouped into equivalence classes by the equivalence relation $\equiv$ called structural congruence. This relation provides a way of rearranging expressions such that interacting parts can be brought together.

Table 1.17 Structural Congruence in tMA

(S-Refl)	$P \equiv P$
(S-Sym)	$P \equiv Q$ implies $Q \equiv P$
(S-Trans)	$P \equiv R$, $R \equiv Q$ implies $P \equiv Q$
(S-Res)	$P \equiv Q$ implies $(\nu n:Amb[\Gamma])P \equiv (\nu n:Amb[\Gamma])Q$
(S-Par)	$P \equiv Q$ implies $P \mid R \equiv Q \mid R$
(S-Repl)	$P \equiv Q$ implies $*P \equiv *Q$
(S-Amb)	$P \equiv Q$ and $R \equiv S$ implies $(n^{\Delta t}[P]^\mu, R) \equiv (n^{\Delta t}[Q]^\mu, S)$
(S-Cap)	$P \equiv Q$ and $R \equiv S$ implies $M^{\Delta t}.(P,R) \equiv M^{\Delta t}.(Q,S)$
(S-Input)	$P \equiv Q$ and $R \equiv S$ implies $c^{\Delta t}?(x:Amb[\Gamma]).(P,R) \equiv c^{\Delta t}?(x:Amb[\Gamma]).(Q,S)$
(S-Output)	$P \equiv Q$ and $R \equiv S$ implies $c^{\Delta t}!\langle m:Amb[\Gamma]\rangle.(P,R) \equiv c^{\Delta t}!\langle m:Amb[\Gamma]\rangle.(Q,S)$
(S-Par Com)	$P \mid Q \equiv Q \mid P$
(S-Par Assoc)	$(P \mid Q) \mid R \equiv P \mid (Q \mid R)$
(S-Res Res)	$(\nu n:Amb[\Gamma])(\nu m:Amb[\Gamma'])P \equiv (\nu m:Amb[\Gamma'])(\nu n:Amb[\Gamma])P$ if $\Gamma \neq \Gamma'$
(S-Res Par)	$(\nu n:Amb[\Gamma])(P \mid Q) \equiv P \mid (\nu n:Amb[\Gamma])Q$ if $(n:Amb[\Gamma]) \notin fn(P)$
(S-Res Amb Dif)	$(\nu n:Amb[\Gamma])(m^{\Delta t}[P]^\mu, Q) \equiv (m^{\Delta t}[(\nu n:Amb[\Gamma])P]^\mu, Q)$ if $m \neq n$ and $n \notin fn(Q)$
(S-Res Amb Eq)	$(\nu n:Amb[\Gamma])(m^{\Delta t}[P]^\mu, Q) \equiv (m^{\Delta t}[(\nu n:Amb[\Gamma])P]^\mu, Q)$ if $m = n$, $n \notin fn(Q)$ and $\Gamma \neq \Gamma'$ where $m:Amb[\Gamma']$
(S-Zero Par)	$P \mid \mathbf{0} \equiv P$
(S-Repl Par)	$*P \equiv P \mid *P$
(S-Zero Res)	$(\nu n:Amb[\Gamma])\mathbf{0} \equiv \mathbf{0}$
(S-Zero Repl)	$*\mathbf{0} \equiv \mathbf{0}$

The rule **(S-Res Amb Eq)** states that if an ambient $n:Amb[\Gamma']$ is in the scope of a restriction $(\nu n:Amb[\Gamma])$ and $\Gamma \neq \Gamma'$, then the scope of $(\nu n:Amb[\Gamma])$ is restricted to the process running inside ambient $n : Amb[\Gamma']$. This rule is able to distinguish between two ambients having the same name ($m = n$), but different types.

We denote by $\not\to$ the fact that none of the rules **(R-In)**, **(R-Out)**, **(R-Open)** and **(R-Com)** can be applied. The behaviour of processes is given by the rules from Table 1.18.

In the rules **(R-In)**, **(R-Out)**, **(R-Open)** the ambient m can be *passive* or *active*, while in the rules **(R-In)**, **(R-Out)** the ambient n is *active*. The difference between *passive* and *active* ambients is that *passive* ambients can be used in several reductions in a unit of time, while *active* ambients can be used in at most one reduction in a unit of time, by consuming their capabilities. In the rules **(R-In)**, **(R-Out)** the *active* ambient n becomes *passive*, forcing it to consume only one capability in one unit of time. In **(R-Open)** we imposed the condition $\Gamma <: \Gamma'$ to avoid releasing an unwanted set of types inside the surrounding ambient m. The ambients which are tagged as *passive*, become *active* again by applying the global time-stepping function **(R-GTProgress)**.

Table 1.18 Reduction Rules in tMA

(R-GTProgress)	$\dfrac{P \not\to}{P \to \phi_\Delta(P)}$		
(R-In)	$(n^{\Delta t'}[in^{\Delta t} m.(P,P') \mid Q]^a, S') \mid (m^{\Delta t''}[R]^\mu, S'') \to (m^{\Delta t''}[(n^{\Delta t'}[P \mid Q]^p, S') \mid R]^\mu, S'')$		
(R-Out)	$(m^{\Delta t'}[(n^{\Delta t''}[out^{\Delta t} m.(P,P') \mid Q]^a, S'') \mid R]^\mu, S') \to (n^{\Delta t''}[P \mid Q]^p, S'') \mid (m^{\Delta t'}[R]^\mu, S')$		
(R-Com)	$c^{\Delta t}!\langle m : Amb[\Gamma]\rangle.(P,Q) \mid c^{\Delta t}?(x : Amb[\Gamma]).(P',Q') \to P \mid P'\{m/x\}$		
(R-Open)	$\dfrac{n : Amb[\Gamma'],\ m : Amb[\Gamma],\ \Gamma <: \Gamma'}{(m^{\Delta t'}[open^{\Delta t} n.\,(P,P') \mid (n^{\Delta t''}[Q]^\mu, S'')], S') \to (m^{\Delta t'}[P \mid Q]^\mu, S')}$		
(R-Res)	$\dfrac{P \to Q}{(\nu n : Amb[\Gamma])P \to (\nu n : Amb[\Gamma])Q}$		
(R-Amb)	$\dfrac{P \to Q}{(n^{\Delta t}[P]^\mu, R) \to (n^{\Delta t}[Q]^\mu, R)}$	**(R-Par1)**	$\dfrac{P \to Q}{R \mid P \to R \mid Q}$
(R-Par2)	$\dfrac{P \to P',\ Q \to Q'}{P \mid Q \to P' \mid Q'}$	**(R-Struct)**	$\dfrac{P' \equiv P,\ P \to Q,\ Q \equiv Q'}{P' \to Q'}$

In tMA, if one process evolves by one of the rules **(R-In)**, **(R-Out)**, **(R-Open)**, **(R-Com)**, while another one does not perform any reduction, then the rule **(R-Par1)** is applied. We define only the left composition **(R-Par1)**, because the right composition results from **(R-Struct)** and **(R-Par1)**. If more than one process evolves in parallel by applying one of the rules **(R-In)**, **(R-Out)**, **(R-Open)**, **(R-Com)** then the rule **(R-Par2)** is applied. The rule **(R-GTProgress)** is applied to simulate the global passage of time, changing all the p tags to a, and so permitting the ambients to participate in new reductions in the next unit of time.

Even if we consider types for ambients as in [41], we do not take into account the environment parameter. Instead, we consider that each ambient has its own set of types Γ, which control the communication of processes inside that ambient as it results from Table 1.18.

1.4.3 Subject Reduction

Well-typedness of a process is defined by a set of rules regarding only the communication inside ambients. The typing rules of Table 1.19 express the conditions which must be satisfied by each syntactic construction of a process in order for it to be well-typed. These rules describe the relationship of a process to its types, providing the static semantics of tMA. We write $P : \Gamma$ and say that a *process P is well-typed with respect to the set of types* Γ, meaning that process P can exchange only messages of types from set Γ; usually Γ represents the set of types valid in the ambient containing process P.

Table 1.19 Typing Rules in tMA

$$\textbf{(T-Null)}\ \mathbf{0}:\Gamma \quad \textbf{(T-Write)}\ \frac{P:\Gamma,\ Q:\Gamma,\ \Gamma <: Amb[\Gamma']}{c^{\Delta t}!\langle m : Amb[\Gamma']\rangle.(P,Q):\Gamma} \quad \textbf{(T-Par)}\ \frac{P:\Gamma,\ Q:\Gamma}{P\,|\,Q:\Gamma}$$

$$\textbf{(T-Read)}\ \frac{P:\Gamma\sqcap Amb[\Gamma'],\ Q:\Gamma}{c^{\Delta t}?(x : Amb[\Gamma']).(P,Q):\Gamma\sqcap Amb[\Gamma']} \quad \textbf{(T-New)}\ \frac{P:\Gamma\sqcap Amb[\Gamma']}{(\nu n : Amb[\Gamma'])P:\Gamma}$$

$$\textbf{(T-Amb)}\ \frac{n : Amb[\Gamma],\ P:\Gamma,\ Q:\Gamma'}{(n^{\Delta t}[P]^{\mu},Q):\Gamma'} \quad \textbf{(T-Cap)}\ \frac{P:\Gamma,\ Q:\Gamma}{M^{\Delta t}.(P,Q):\Gamma} \quad \textbf{(T-Repl)}\ \frac{P:\Gamma}{*P:\Gamma}$$

Since process **0** cannot communicate, **0** is well-typed under any set of types, this being expressed in rule **(T-Null)**. Rule **(T-Write)** states that only messages of types from the set Γ can be sent. Similar reasoning is expressed in **(T-Read)**. An ambient has only internal communication, meaning that it cannot send messages to sibling processes; therefore an ambient is well-typed under any set of types, and this is expressed in rule **(T-Amb)**. If P and Q are sibling processes which can exchange messages of types from the set Γ, then $P \mid Q$ is also such a process. Rule **(T-New)** states that if a process can exchange messages of types from the set $\Gamma \sqcap Amb[\Gamma']$, then the restricted process $(\nu n : Amb[\Gamma'])P$ cannot exchange messages of type $Amb[\Gamma']$ with sibling processes. By adding a capability to a process we do not affect the well-typedness of that process as it results from rule **(T-Cap)**.

Lemma 1.3. *If* $(\nu n : Amb[\Gamma'])P : \Gamma$ *and* $n \notin fn(P)$ *then* $P : \Gamma$.

Lemma 1.4. *If* $P : \Gamma \sqcap Amb[\Gamma']$, $x, m : Amb[\Gamma']$, $x \in fn(P)$ *and* $m \notin fn(P)$ *then* $P\{m/x\} : \Gamma \sqcap Amb[\Gamma']$.

In order to say that $c^{\Delta t}!\langle m : amb[\Gamma']\rangle.(P,Q)$ is well-typed with respect to the set of types Γ, the following statements should hold:

(i) $m : Amb[\Gamma']$, which means that ambient m contains the set of types Γ';

(ii) $\Gamma <: Amb[\Gamma']$, which means that Γ contains the type $Amb[\Gamma']$;

(iii) $P : \Gamma$; $Q : \Gamma$, which means that P and Q are well-typed with respect to the set of types Γ. If one of the statements is not true, the process $c^{\Delta t}!\langle m : Amb[\Gamma']\rangle.(P,Q)$ can still be well-typed, if the alternative process Q is well-typed, with respect to the same set of types Γ.

The following proposition states that the application of the global time progress function ϕ_Δ to a process P does not change its property of being well-typed.

Proposition 1.1 (Time Passage). *If* $P : \Gamma$ *then* $\phi_\Delta(P) : \Gamma$.

Proof. We take into account all the cases which enter in the definition of ϕ_Δ. We present only one case, the others being treated in a similar manner.

Case 1.22 $(P = M^{\Delta t}.(R,Q), t > 0)$. The syntax is a general notation to capture all the capabilities because their behaviour is the same in this context. As a consequence, the rule **(T-Cap)** is applied and the expected result $M^{\Delta(t-1)}.(R,Q) : \Gamma$ is obtained which is the same as $\phi_\Delta(P) : \Gamma$. □

Exercise 1.5. Prove the other cases of Proposition 1.1.

The following proposition states that if a process P is well-typed, then all the processes from its equivalence class are well-typed.

Proposition 1.2 (Subject Congruence). *If* $P \equiv Q$ *then* $P : \Gamma$ *iff* $Q : \Gamma$.

Proof. We proceed by structural induction. We present only one case, the others being treated in a similar manner.

Case 1.23 (**S-Res Amb Dif**). We have that $P = (\nu n : Amb[\Gamma'])(m^{\Delta t}[P']^{\mu}, P'')$ and $Q = (m^{\Delta t}[(\nu n : Amb[\Gamma'])P']^{\mu}, P'')$ with $n \neq m$. Assume $P : \Gamma$. This must have been derived from **(T-New)** and **(T-Amb)** with $P'' : \Gamma \sqcap Amb[\Gamma']$. Because n does not affect the process P'', by applying Lemma 1.3 we have that $P'' : \Gamma$. By applying **(T-Amb)** we obtain that $Q : \Gamma$. □

Exercise 1.6. Prove the other cases of Proposition 1.2.

The following proposition states that if a process P is well-typed, then the process obtained after applying a reduction rule is well-typed.

Proposition 1.3 (Subject Reduction). *If* $P \rightarrow Q$ *then* $P : \Gamma$ *iff* $Q : \Gamma$.

Proof. We proceed by induction on the derivation of $P \rightarrow Q$. We present only one case, the others being treated in a similar manner.

Case 1.24 (**R-Com**). We have that $P = c^{\Delta t}!\langle m\rangle.(P,Q) \mid c^{\Delta t'}?(x : Amb[\Gamma']).(P',Q')$ and $Q = P \mid P'\{m/x\}$. Assume $P : \Gamma$. This must have been derived from **(T-Par)** with $c^{\Delta t}!\langle m\rangle.(P,Q) : \Gamma$ and $c^{\Delta t'}?(x).(P',Q') : \Gamma$ and by applying the rules **(T-Write)** and **(T-Read)** we obtain that $P : \Gamma$, $P'\Gamma$ and $\Gamma <: Amb[\Gamma']$. By applying Lemma 1.4 and the rule **(T-Par)** we obtain that $P \mid P'\{m/x\} : \Gamma$. □

Exercise 1.7. Prove the other cases of Proposition 1.3.

Table 1.20 Error System in tMA

$$\textbf{(E-Com)}\ \frac{\Gamma \neq \Gamma'}{c^{\Delta t}!\langle m : Amb[\Gamma']\rangle.(P,Q) \mid c^{\Delta t}?(x : Amb[\Gamma]).(P',Q') \xrightarrow{err}}$$

$$\textbf{(E-Open)}\ \frac{n : Amb[\Gamma'],\, m : Amb[\Gamma],\, \Gamma \not<: \Gamma'}{(m^{\Delta t'}[open^{\Delta t} n.\,(P,P') \mid (n^{\Delta t''}[Q]^{\mu},S'')]^{\mu},S') \xrightarrow{err}} \qquad \textbf{(E-Par)}\ \frac{P \xrightarrow{err}}{P \mid Q \xrightarrow{err}}$$

$$\textbf{(E-Amb)}\ \frac{P \xrightarrow{err}}{(n^{\Delta t}[P]^{\mu},Q) \xrightarrow{err}} \qquad \textbf{(E-New)}\ \frac{P \xrightarrow{err}}{(\nu n : Amb[\Gamma])P \xrightarrow{err}} \qquad \textbf{(E-Str)}\ \frac{P \equiv Q \quad Q \xrightarrow{err}}{P \xrightarrow{err}}$$

In Table 1.20 we describe the error system of tMA, where by $\xrightarrow{err}$ we denote the fact that an error occurred. An error can occur only when a process tries to exchange a message of a wrong type. Note that if a process wants to communicate a message of a wrong type, it can still be well-typed if the alternative process Q is well-typed.

Rule **(E-Com)** states that a process can receive only messages of a certain type. In rule **(E-Open)** we express the fact that if messages of types from Γ' are exchanged in an ambient n, by opening the ambient, in order for the processes to exchange messages of types from Γ', the ambient m containing ambient n must allow exchange of messages of types from Γ'. The rest of the rules are obvious and state the fact that if a process generates an error then including it in another process, the error does not disappear.

Proposition 1.4. *If a process is well typed, then it does not generate errors:*
$$P : \Gamma \text{ implies } P \not\xrightarrow{err}.$$

Proof. The proof considers the opposite of the fact that if P gives rise to a run-time error ($P \xrightarrow{err}$), then P cannot be well-typed under any set of types Γ ($P \not: \Gamma$, *for all* Γ). We use induction on the structure of P and consider a proof case for each rule in Table 1.20. We present only one case, the others being treated in a similar manner.

Case 1.25 (**E-Com**). We suppose that there exists a set of types Γ such that $R : \Gamma$, where $R = c^{\Delta t}!\langle m\rangle.(P,Q) \mid c^{\Delta t'}?(x : Amb[\Gamma']).(P',Q')$. This must have been derived from **(T-Par)** with $c^{\Delta t}!\langle m\rangle.(P,Q) : \Gamma$ and $c^{\Delta t'}?(x : Amb[\Gamma']).(P',Q') : \Gamma$. Applying **(T-Write)**, **(T-Read)** we have that $\Gamma <: Amb[\Gamma']$ and $n : Amb[\Gamma']$, which is in contradiction with the hypothesis of the rule **(E-Com)**, and so we have that $R \not\xrightarrow{err}$. □

Exercise 1.8. Prove the other cases of Proposition 1.4.

We denote by $P \xrightarrow{t} Q$ the fact that process P evolves to process Q after applying the rule **(R-GTProgress)** $t \geq 0$ times, and by $t\phi_{\Delta}(R)$ the fact that function ϕ_{Δ} is applied t times to process R. We denote by $\cong$ the relation which respects all the rules of Table 1.17 except replication, namely rule **(S-Repl Par)**. The following result claims that if two processes are structurally congruent and both idle for t units of time, then the obtained processes are also structurally congruent.

Proposition 1.5. *Time passage cannot cause a nondeterministic behaviour:*

$$\text{if } P \cong Q,\ P \xrightarrow{t} P' \text{ and } Q \xrightarrow{t} Q' \text{ then } P' \cong Q'.$$

Proof. We proceed by structural induction and present only one case, the others being treated in a similar manner.

Case 1.26 (**S-Res**). We have $P = (\nu n : Amb[\Gamma'])P'$ and $Q = (\nu n : Amb[\Gamma'])Q'$ with $P \cong Q$. By induction, if $P' \cong Q'$, $P' \xrightarrow{t} P''$ and $Q' \xrightarrow{t} Q''$ then $P'' \cong Q''$. By applying **(R-Res)** to both $P' \xrightarrow{t} P''$ and $Q' \xrightarrow{t} Q''$ we obtain that $P \xrightarrow{t} (\nu n : Amb[\Gamma])P''$ and $Q \xrightarrow{t} (\nu n : Amb[\Gamma])Q''$. By applying **(S-Res)** to $P'' \cong Q''$ we obtain that $(\nu n : Amb[\Gamma])P'' \cong (\nu n : Amb[\Gamma])Q''$. □

Exercise 1.9. Prove the other cases of Proposition 1.5.

The following example motivates why we remove replication. Let $P = in^{\Delta 5} n$. Then we have $*P \equiv P \mid *P$. By applying the function ϕ_Δ, we obtain

$$\phi_\Delta(P \mid *P) = in^{\Delta 4} n \mid *P \not\equiv *P = \phi_\Delta(*P).$$

Example 1.3. We extend the cab protocol described in [94] by introducing new operations which describe a recall for a taxi when a certain period of time has passed, and the payment for the trip. Roughly speaking, the cab protocol is about a city with various sites, cabs and clients willing to go from one site to another. At http://www-sop.inria.fr/mimosa/ambicobjs/taxis.html, a graphical implementation of the cab protocol is presented. The implementation is written in Java, and presents the ambients as named and coloured circles, whose limits act as boundaries for what is inside. A capability *in c* is described by an anchor which remains in the ambient *a*, and an arrow outside which is linked to any ambient with name *c*. When such an arrow finds an ambient *c*, the ambient *a* is entirely moved inside *c*. A capability *out c* is described by an anchor pointing outside. A capability *open c* is represented as a small square trying to find an ambient with the same name. If it does, the boundaries are dissolved and the content of that ambient is released. A snapshot of the cab protocol is presented in the following figure:

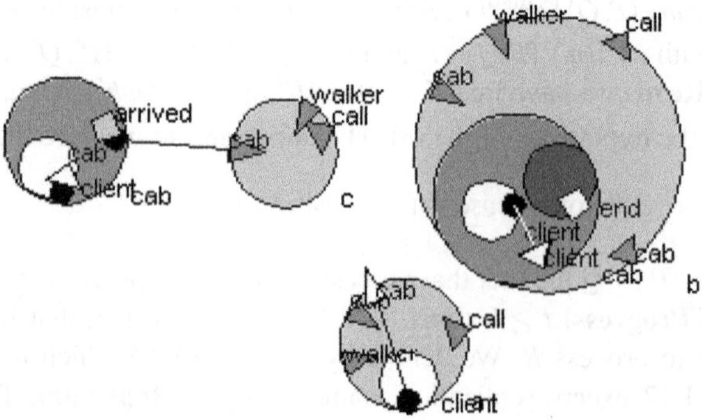

The whole system consists of one city, n sites, and several cabs and clients. The cabs can be empty waiting in a precise site or can have clients and be anywhere in the city, while the clients can be either at some sites waiting for a free cab to arrive, or already travelling with a cab. In order to initiate a trip a client must obtain a cab, and it does this by sending a request for an empty cab. In what follows we use this protocol to illustrate how mobile ambients with timers work, emphasizing the timing aspects. It is worth noting that each ambient of the system is well-typed because we do not consider communication. Considering the fact that we have only ambients with no internal communication, all the processes are well-typed under any set of types Γ.

A message emitted by a *client* located at a site *from* in order to call a *cab* is described by

$$load\ client = loading^{\Delta t_1}[out^{\Delta t_2}cab.\,in^{\Delta t_3}client]^{\mu}$$
$$call = call^{\Delta t_7}[out^{\Delta t_8}client.\,out^{\Delta t_9}from.\,in^{\Delta t_{10}}cab.\,in^{\Delta t_{11}}from.\,load\ client]^{\mu}$$
$$recall = recall^{\Delta t_{12}}[out^{\Delta t_{13}}cab.\,in^{\Delta t_{14}}from.\,in^{\Delta t_{15}}client]^{\mu}$$
$$call\ from\ client = (call, recall)$$

This ambient can enter a *cab*; here it gets opened and releases the process *load client*. After it exits the ambient *client* and successively the ambient *from* it looks for a *cab* to enter. If it finds a *cab* then it enters it by applying the **(R-In)** rule:

$$(call^{\Delta t_7}[in^{\Delta t_{10}}cab.in^{\Delta t_{11}}from.\ldots]^{a}, recall) \mid cab^{\infty}[\,]^{\mu}$$
$$\rightarrow cab^{\infty}[call^{\Delta t_7}[in^{\Delta t_{11}}from.\ldots]^{p}, recall]^{\mu}$$

If the timer Δt_7 of the ambient *call* expires before it enters a *cab*, then an ambient *recall* is released. This is possible if no *cab* ambient becomes sibling with the ambient *call* in the period of time represented by the timer Δt_7. To discard the ambient *call* with the expired timer we apply the **(R-GTProgress)** rule which launches the safety process *recall*:

$$(call^{\Delta t_7}[in^{\Delta t_{10}}cab.in^{\Delta t_{11}}from.\ldots]^{a}, recall) \rightarrow recall$$

The *recall* ambient enters the ambient *client*, and announces that he can make another call. This process of recalling is repeated until the process *load client* is released. The process *load client* is launched by opening the ambient *call* using the **(R-Open)** rule:

$$cab^{\infty}[(call^{\Delta t_7}[load\ client]^{\mu}, recall) \mid open^{\Delta t_{44}}call.open^{\Delta t_{45}}trip.\ldots]^{\mu}$$
$$\rightarrow cab^{\infty}[load\ client \mid open^{\Delta t_{45}}trip.\ldots]^{\mu}$$

As a consequence, the *cab* goes to *from* in order to meet its *client*, and it releases an ambient *loading*. All the steps necessary for a correct evolution of the trip are performed by applying the appropriate reduction rules. Once *loading* has been released, it enters the ambient *client*.

The address given to the *driver* by a *client* to go from the current location *from* to address *to*, as well as the payment for the trip are described by

$$trip\ from\ to\ c = trip^{\Delta t_{20}}[out^{\Delta t_{21}}client.\,out^{\Delta t_{22}}from.\,in^{\Delta t_{23}}to.\,pay\ driver]^{\mu}$$
$$pay\ driver = pay^{\Delta_{16}}[in^{\Delta t_{17}}c.\,in^{\Delta t_{18}}wallet.\,in^{\Delta t_{19}}money]^{\mu}$$

Whenever the *client* opens *loading* it means that the *cab* is present, and therefore the *client* may enter it. Consequently, the *client* enters the *cab* and releases an ambient *trip*, which the *cab* receives and opens. The process which is released moves the *cab* to its destination where it releases another synchronization ambient *pay* to inform the *client* to pay for the trip. An ambient *pay* enters the *client wallet* and moves an ambient *money* to the *driver wallet*.

$paid\ driver = paid^{\Delta t_{28}}[out^{\Delta t_{29}} money.out^{\Delta t_{30}} wallet.out^{\Delta t_{31}} driver.in^{\Delta t_{32}} c]^{\mu}$

$money\ client = money^{\infty}[open^{\Delta t_{33}} pay.out^{\Delta t_{34}} wallet.out^{\Delta t_{35}} c.in^{\Delta t_{36}} driver.$
$\quad in^{\Delta t_{37}} wallet.paid\ driver]^{\mu}$

$wallet\ client = wallet^{\infty}[money\ client \mid \ldots \mid money\ client]^{\mu}$

$bye\ cab = bye^{\Delta t_{24}}[out^{\Delta t_{25}} c.in^{\Delta t_{26}} cab.out^{\Delta t_{27}} to]^{\mu}$

$client\ from\ to = (\nu c)c^{\infty}[*(open^{\Delta t_{38}} recall.call\ from\ c) \mid recall^{\Delta t_{39}}[\,]^{\mu} \mid$
$\quad open^{\Delta t_{40}} loading.in^{\Delta t_{41}} cab.trip\ from\ to\ c \mid open^{\Delta t_{42}} paid.$
$\quad out^{\Delta t_{43}} cab.bye\ cab \mid wallet\ client]^{\mu}$

Once the ambient *money* enters the *driver wallet*, an ambient *paid* is released and sent to the *client* telling him to get out of the *cab*. The *client* opens it, leaves the *cab*, and sends the last synchronization ambient *bye* to the *cab*, instructing it to leave the current location *to*. The *cab* and the *city* are described by

$driver = driver^{\infty}[wallet^{\infty}[money^{\infty}[\,]^{\mu} \mid \ldots \mid money^{\infty}[\,]^{\mu}]^{\mu}]^{\mu}$

$cab = cab^{\infty}[rec\ X.open^{\Delta t_{44}} call.open^{\Delta t_{45}} trip.open^{\Delta t_{46}} bye.X \mid driver]^{\mu}$

$city = city^{\infty}[cab \mid \ldots \mid cab \mid site_1^{\infty}[client\ site_1\ site_i \mid client\ site_1\ site_j \mid \ldots]^{\mu}$
$\quad \mid \ldots \mid site_i^{\infty}[\ldots]^{\mu}]^{\mu}$

In the discussion above we have supposed that only the timer Δt_7 of the ambient *call* expires, and this may produce the execution of the safety process *recall*. This was made only for the sake of simplicity. In order to simulate other possible scenarios, we can suppose that other timers may also expire:

- Δt_1 - the *loading* ambient does not reach the ambient *client*, and a safety process should be released in order to instruct *cab* to create another *loading* ambient;
- Δt_{16} - the *pay* ambient does not reach the ambient *client*, and a safety process should be released in order to instruct *cab* to create another *pay* ambient;
- Δt_{20} - the *trip* ambient does not reach the ambient *cab*, and a safety process should be released in order to instruct the *client* to create another *trip* ambient;
- Δt_{28} - the *paid* ambient does not reach the ambient *client*, and a safety process should be released in order to instruct *cab* to create another *paid* ambient;
- Δt_{24} - the *bye* ambient does not reach the ambient *cab*, and a safety process should be released in order to instruct the *client* to create another *bye* ambient;
- various other scenarios can be simulated by introducing several other timers over capabilities and ambients.

1.5 Brane Calculi

Biological inspiration is predominant in the case of *brane calculi* [40]. The operations of the two basic brane calculi, namely *pino, exo, phago* (for the PEP fragment) and *mate, bud, drip* (for the MBD fragment) are directly inspired by the biological processes of *endocytosis, exocytosis* and *mitosis*. Since some proteins are embedded in cell membranes, and can act on both sides of the membrane simultaneously, brane calculi use both sides of the membrane, emphasizing that computation happens also on the membrane surface. We present here an overview of the PEP fragment of brane calculi without replication. Cardelli motivates that the replication operator is used to model the notion of a "multitude" of components of the same kind, which is in fact a standard situation in biology. We do not consider the replicator operator because we are not able to define a corresponding membrane system without knowing exactly the initial membrane structure. More details on brane calculi can be found in [40].

1.5.1 Syntax

A membrane structure consists of a collection of nested membranes as shown in Table 1.21. Membranes are formed of patches σ, where a patch can be composed from other patches $\rho \mid \tau$. A patch σ consists of an action a followed, after its consumption, by another patch σ_1; thus $\sigma = a.\sigma_1$. Actions often come in complementary pairs which cause interaction between membranes. The names n are used to pair-up actions and co-actions.

Table 1.21 Brane Calculi Syntax

Systems	$P,Q ::= P \circ Q \mid \sigma(\,) \mid \sigma(P)$
Branes	$\sigma,\tau ::= 0 \mid \sigma \mid \tau \mid a.\sigma$
Actions	$a,b ::= phago_n \mid \overline{phago}_n(\sigma) \mid exo_n \mid \overline{exo}_n \mid pino(\sigma)$
	$\mid mate_n \mid \overline{mate}_n \mid bud_n \mid \overline{bud}_n(\sigma) \mid drip(\sigma)$

We abbreviate $a.0$ as a, $0(P)$ as (P), and $0(\,)$ as $(\,)$.

1.5.2 Operational Semantics

The structural congruence relation is a way of rearranging the system such that the interacting parts come together; the structural congruence $\equiv_b$ is defined in Table 1.22.

Table 1.22 Brane Calculi Structural Congruence

$P \circ Q \equiv_b Q \circ P$	$\sigma \mid \tau \equiv_b \tau \mid \sigma$
$P \circ (Q \circ R) \equiv_b (P \circ Q) \circ R$	$\sigma \mid (\tau \mid \rho) \equiv_b (\sigma \mid \tau) \mid \rho$
	$\sigma \mid 0 \equiv_b \sigma$
$P \equiv_b Q$ implies $P \circ R \equiv_b Q \circ R$	$\sigma \equiv_b \tau$ implies $\sigma \mid \rho \equiv_b \tau \mid \rho$
$P \equiv_b Q$ and $\sigma \equiv_b \tau$ implies $\sigma(P) \equiv_b \tau(Q)$	$\sigma \equiv_b \tau$ implies $a.\sigma \equiv_b a.\tau$

Table 1.23 Brane Calculi Reduction Rules

$pino(\rho).\sigma \mid \sigma_0(P) \rightarrow_b \sigma \mid \sigma_0(\rho(\,) \circ P)$ — Pino

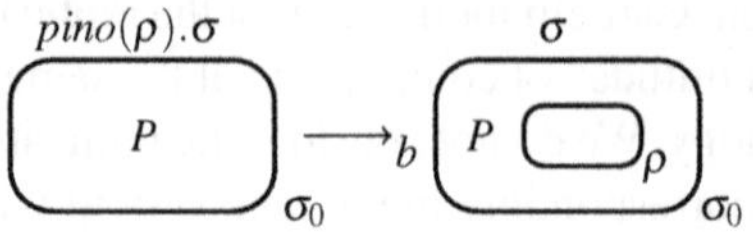

$\overline{exo}_n.\tau \mid \tau_0(exo_n.\sigma \mid \sigma_0(P) \circ Q) \rightarrow_b P \circ \sigma \mid \sigma_0 \mid \tau \mid \tau_0(Q)$ — Exo

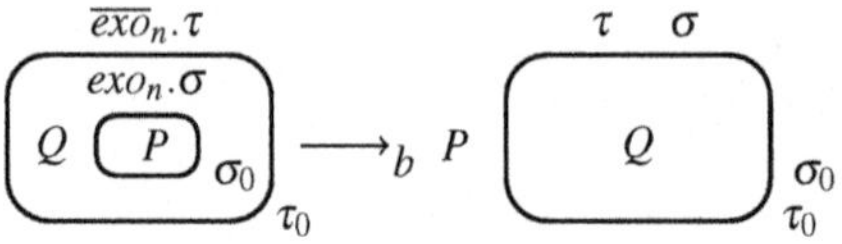

$phago_n.\sigma \mid \sigma_0(P) \circ \overline{phago}_n(\rho).\tau \mid \tau_0(Q) \rightarrow_b \tau \mid \tau_0(\rho(\sigma \mid \sigma_0(P)) \circ Q)$ — Phago

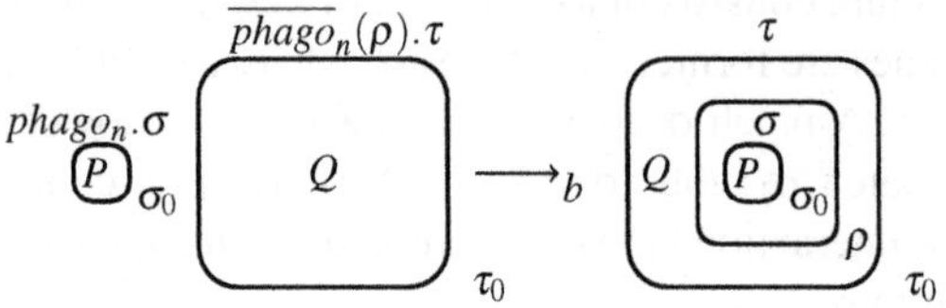

$mate_n.\sigma \mid \sigma_0(P) \circ \overline{mate}_n.\tau \mid \tau_0(Q) \rightarrow_b \tau \mid \tau_0 \mid \sigma \mid \sigma_0(P \circ Q)$ — Mate

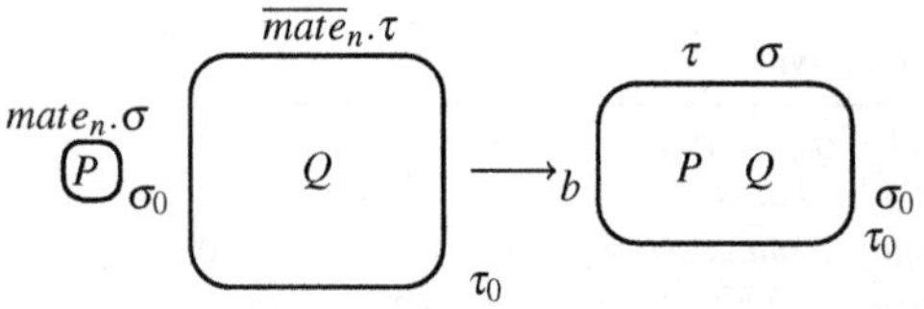

$\overline{bud}_n(\rho).\tau \mid \tau_0(bud_n.\sigma \mid \sigma_0(P) \circ Q) \rightarrow_b \rho(\sigma \mid \sigma_0(P)) \circ \tau \mid \tau_0(Q)$ — Bud

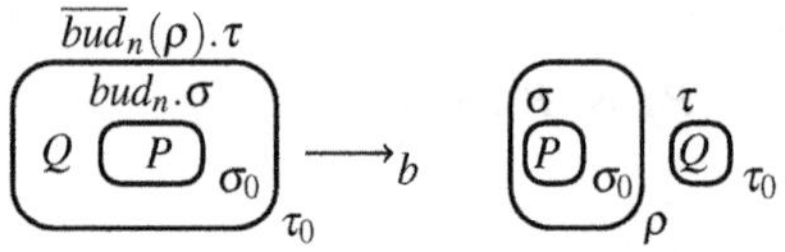

$drip(\rho).\sigma \mid \sigma_0(P) \rightarrow_b \rho(\,) \circ \sigma \mid \sigma_0(P)$ — Drip

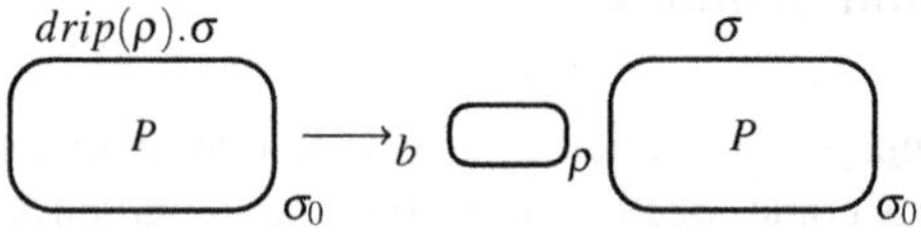

$P \rightarrow_b Q$ implies $P \circ R \rightarrow_b Q \circ R$ — Par
$P \rightarrow_b Q$ implies $\sigma(P) \rightarrow_b \sigma(Q)$ — Mem
$P \equiv_b P'$ and $P' \rightarrow_b Q'$ and $Q' \equiv_b Q$ implies $P \rightarrow_b Q$ — Struct

In what follows we explain the rules of Table 1.23. The action $pino(\sigma)$ creates an empty bubble within the membrane where the *pino* action resides. The original membrane buckles inwards and pinches off; the patch σ on the empty bubble is a parameter of *pino*. The exo action exo_n comes with a complementary co-action $\overline{exo}_n$; they model the merging of two nested membranes which starts with the membranes touching at a point. In this process (which is a smooth and continuous process), the subsystem P gets expelled to the outside, and all the residual patches of the two membranes become contiguous. Actions $mate_n$ and $\overline{mate_n}$ synchronize to obtain membrane fusion. Action bud_n permits splitting of an internal membrane, and synchronizes with the coaction $\overline{bud_n}$. Action $drip$ permits splitting off zero internal membranes. Actions *bud* and $drip$ are equipped with a process σ, that will be associated to the new membrane created by the brane performing the action.

The phago action $phago_n$ comes with a complementary co-action $\overline{phago_n}(\rho)$; they model a membrane (the one with Q) "eating" another membrane (the one with P). Again, the process has to be smooth and continuous, i.e., biologically implementable. It proceeds by the Q membrane wrapping around the P membrane and joining itself on the other side. Thus an additional layer of membrane is created around the eaten membrane: the patch on that membrane is specified by the parameter ρ of the co-phago action (similar to the parameter of the pino action).

1.5.3 Computability and Decidability

According to [37], a fragment of PEP, namely, the calculus without the pino action, is Turing powerful. The result is proved by showing how to model Random Access Machines (RAM). A direct consequence of this result is the undecidability of universal termination for PEP. Such an encoding is deterministic and enjoys the following property: the RAM terminates if and only if its encoding terminates. As a consequence, both the universal termination property (i.e., checking if the system has a divergent computation) and the existential termination property (i.e., checking if the system has a terminating computation) turn out to be undecidable for PEP.

In [37] it is shown that universal termination is a decidable property for the MBD fragment; the proof is based on the theory of well-structured transition systems [85]. In [36] it is stated that the decidability of universal termination provides an expressiveness gap between MBD and PEP, as a deterministic encoding of Random Access Machines can be provided in the second calculus, but not in the first calculus. Thus it is impossible to provide an encoding of PEP in MBD that preserves the universal termination property.

In [35] a non-deterministic encoding of RAMs in MBD is provided which preserves the existence of a terminating computation. The encoding is non-deterministic because it introduces additional computations which do not follow the expected behaviour of the modelled RAM. Since all the computations are infinite, given a RAM, its modelling has a terminating computation if and only if the RAM terminates. A direct consequence of this result is the undecidability of existential termination for MBD.

The decidability of universal termination for MBD in [37] ensures that it is impossible to provide a deterministic encoding of RAMs in MBD. If it is required that the RAM terminates if and only if all the computations of the encoding terminate, it is also impossible to provide a (non-deterministic) encoding of RAMs in MBD.

The computational power of MBD is increased if the interleaving semantics is replaced with a maximal parallelism semantics as in membrane computing [130]. According to the maximal parallelism semantics, at each computational step a maximal set of independent reductions is simultaneously executed. Hence, all the membranes that can evolve have to do so. Using such a semantics, in [35] a deterministic encoding of RAMs in MBD with maximal parallelism is provided that preserves the existence of a terminated computation (hence also the existence of a divergent computation). In this case is obtained the undecidability of both existential and universal termination for MBD with maximal parallelism. This result confirms the intuition emerging from [89], where the interleaving (sequential) and the maximal parallelism semantics of many variants of P systems are compared: in most cases, the computational power increases when moving from interleaving to maximal parallelism.

Chapter 2
Mobility in Membrane Computing

Abstract Membrane computing is part of natural computing, being a rule-based formalism inspired by biological cells. The basic model of membrane computing is usually called transition membrane systems. When membrane systems are considered as computing devices, two main research directions are considered: their computational power in comparison with the classical notion of Turing computability, and their efficiency in algorithmically solving hard problems (e.g., NP-problems) in polynomial time. In this chapter we define mobile membrane systems which are both powerful (equivalent to Turing machines) and efficient (algorithms have been developed which provide efficient solutions to NP-complete problems).

2.1 Mobility in Cell Biology

The cell is the functional basic unit of life, and is often called the building block of life [7]. The cells of living organisms are categorised into prokaryotic and eukaryotic cells. Any organism contains either prokaryotic or eukaryotic cells. Prokaryotic are very simple cells that are smaller than the more complex eukaryotic cells. Bacteria are made up of one or more prokaryotic cells. The cell membrane in a prokaryotic cell consists of a fluid phospholipid bilayer without carbohydrates. The prokaryotic cell is incapable of endocytosis or exocytosis unlike the eukaryotic cell.

Eukaryotic cells are found inside plant and animal life and are more advanced and larger, and differ fundamentally from their prokaryotic counter-parts by their possession of membrane-bounded internal compartments. The emergence of an endomembrane system was a crucial stage in the prokaryote-to-eukaryote evolutionary transition [76]. A feature that distinguishes prokaryote cells from eukaryote cells is the presence in eukaryote cells of a cytoskeleton that maintains their structural integrity. Thus the cell can afford to have a membrane that consists of a fluid phospholipid bilayer that includes carbohydrates. The increased fluidity of the outer membrane allowed the development of two mechanisms, called endocytosis and exocytosis. Endocytosis and exocytosis are complementary operation that allow substances

B. Aman, G. Ciobanu, *Mobility in Process Calculi and Natural Computing*, Natural Computing Series, DOI 10.1007/978-3-642-24867-2_2,

to enter (endocytosis) or exit (exocytosis) the cell through membrane-bounded vesicles. In endocytosis, the vesicle is formed by the invagination of a small segment of the outer membrane that contains substances from outside the cell. In the reverse process of exocytosis, the vesicle fuses with the outer cell membrane releasing its content into the extracellular space.

The development of endocytosis and exocytosis prepared the way for all subsequent steps of eukaryotic evolution [100]. According to [76], several innovations appeared during this evolution from prokaryote to eukaryote cells (see Figure 2.1 and Table 2.1).

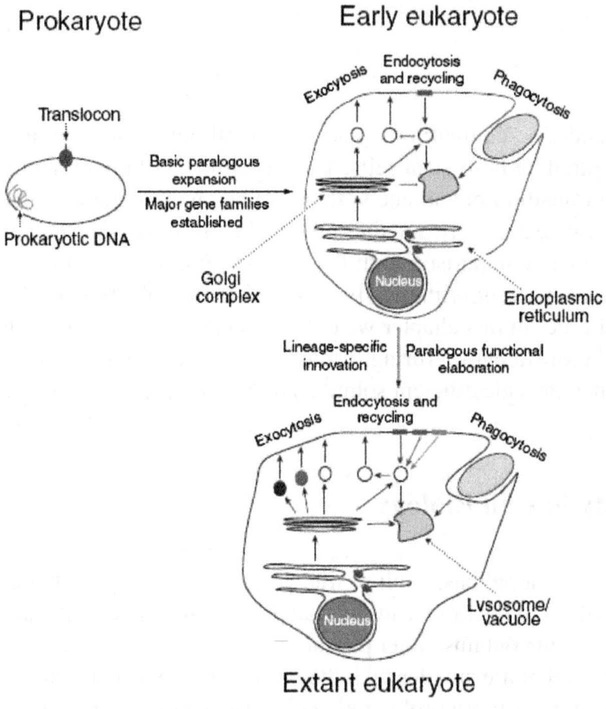

Fig. 2.1 Prokaryote to Eukaryote Evolutionary Transition

Table 2.1 Major Innovations in the Evolution of the Endomembrane System

Prokaryote	Early Eukaryote	Extant Eukaryote
Unfolded substrate exported	Folded substrate exported	Multiple modes of endocytosis
Membrane translocation	Vesicular transport	Multiple modes of exocytosis
	Limited modes of endocytosis	Tissue-specific pathways
	Phagocytosis	Lineage-specific pathways
	Lysosomal degradation	Secondary losses

Endocytosis of eukaryotic cells also enhances communication between cells [150], a feature that enables eukaryote cells to form multicellular organisms. Cells sense the environment and communicate with each other through activation of signalling receptors on the cell surface. Endocytosis regulates cell signalling by physically reducing the number of signalling receptors available for activation. In some cases, a reduction in the number of surface receptors does not attenuate the maximal signalling response, but instead shifts the dose response relationship so that a higher concentration of ligand is required to trigger a response of the same magnitude. This is of physiological importance in settings of limited ligand concentration. In this way, endocytosis and cell signaling are intertwined in many biological processes, such as cell motility and cell fate determination [150].

We use endocytosis and exocytosis as abstract operations in membrane computing, a new biologically inspired model of computation.

2.2 Membrane Computing

Membrane computing is part of natural computing, being a rule-based formalism inspired by biological cells. The basic model of membrane computing is usually referred to as transition membrane systems. In this model, objects are represented using symbols from a given alphabet, and each symbol from this alphabet can appear inside a region in many different copies. That is, the content of a region is represented as a multiset over a given alphabet. Rules are essentially multiset rewriting rules with some extra features: targets appear in the right hand side of the rules and are used to specify where to move the produced objects.

Definition 2.1. A transition membrane system of degree $n \geq 1$ is a construct

$$\Pi = (V, H, T, C, \mu, w_1, \ldots, w_n, (R_1, \rho_1), \ldots, (R_n, \rho_n), i_O)$$

where:

1. n represents the number of membranes;
2. V is an alphabet of symbols; its elements are called objects;
3. $T \subseteq V$ is the terminal (or output) alphabet;
4. $C \subseteq V$, $C \cap T = \emptyset$ is the alphabet of catalysts;
5. H is a finite set of labels for membranes;
6. $\mu \subset H \times H$ describes the membrane structure, such that $(i, j) \in \mu$ denotes that the membrane labelled by j is contained within the membrane labelled by i; we distinguish the external membrane (usually called the "skin" membrane) and several internal membranes; a membrane without any other membrane inside it is said to be elementary;
7. $w_i \in V^*$, for each $1 \leq i \leq n$ is a multiset of objects assigned initially to membrane i;
8. R_i, for all $1 \leq i \leq n$, is a finite set of evolution rules which is associated with membrane i; an evolution rule is a multiset rewriting rule of the form $u \to v$, with $u \in V^+$, $v = v'$ or $v = v'\delta$, $v' \in ((V \times \{here, out\}) \cup (V \times \{in_j \mid 1 \leq j \leq n\}))^*$, and δ a special symbol not appearing in V;

9. ρ_i, for all $1 \leq i \leq n$, is a partial order relationship defined over the rules in R_i specifying a priority relation between these rules;
10. i_O is the label of an elementary membrane of μ which identifies the output region.

$$R_1 = \{r_1 : a \to (a, in_2)\}$$
$$\cup \{r_2 : b \to (a, in_2)\}$$
$$R_2 = \{r_3 : a \to (b, out)(a, here)\}$$
$$\cup \{r_4 : b \to (b, out)\}$$

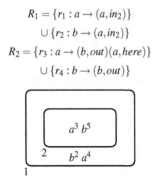

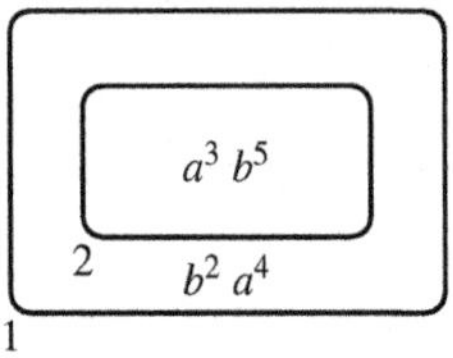

Fig. 2.2 A Transition Membrane System

As an example, we consider a membrane system with two nested membranes (the inner membrane labelled by 2, the outer membrane labelled by 1), two sets of evolution rules R_2 and R_1 and two symbols (a and b). Initially, membrane 1 contains the multiset $b^2\ a^4$, and membrane 2 contains the multiset $a^3\ b^5$.

Therefore, a transition membrane system of degree $n \geq 1$ consists of a membrane structure μ containing n membranes where each membrane i is assigned a finite multiset of objects w_i and a finite set of evolution rules R_i, with an associated priority relation ρ_i. An evolution rule is a multiset rewriting rule which consumes a multiset of objects from V and produces a multiset of pairs (a,t), with $a \in V$ and $t \in \{here, out, in\}$ a *target* specifying where to move the objects after the application of the rule. As well as this, an evolution rule can produce the special object δ to specify that, after the application of the rule, the membrane where the rule has been applied has to be dissolved. After dissolving a membrane, all objects and membranes previously contained in it become contained in the membrane containing it, while the rules of the dissolved membrane are removed.

2.3 Mobile Membranes

Mobile membranes represent a formalism describing the movement of membranes inside a spatial structure by applying rules from a given set. Mobile membranes represent a variant of membrane computing [128]. Several systems of mobile membranes are studied in [17], and their computational universality are proved by using a small number of membranes [18]. The model is characterized by two essential features:

- A spatial structure consisting of a hierarchy of membranes (which do not intersect). The membranes produce a delimitation between regions. For each membrane there is a unique associated region. To each membrane we associate a multiset of objects placed on its surface and a multiset of objects placed inside the

corresponding region. A membrane without any other membranes inside is called elementary, while a membrane containing other membranes is called composite.
- The general rules describing the evolution of the structure: endocytosis (moving an elementary membrane inside a neighbouring membrane) and exocytosis (moving an elementary membrane outside the membrane where it is placed). More specific rules are given by pinocytosis (creating an elementary membrane) and phagocytosis (engulfing just one sibling elementary membrane). A movement rule consists in fact of two steps: rewriting the objects that initiated the movement to multisets of objects and changing the membrane structure.

The computations are performed in the following way: starting from an initial structure, the system evolves by applying the rules in a nondeterministic and maximally parallel manner. A rule is applicable when all the involved objects and membranes appearing in its left hand side are available. The maximally parallel way of using the rules means that in each step we apply a maximal multiset of rules, namely a multiset of rules such that no further rule can be added to the set. A halting configuration is reached when no rule is applicable. The result is represented by the number of objects associated to a specified membrane.

Let $\mathbb{N}$ be the set of positive integers, and consider a finite alphabet V of symbols. A multiset over V is a mapping $u : V \to \mathbb{N}$. The empty multiset is represented by λ. We use the string representation of multisets that is widely used in the field of membrane systems. An example of such a representation is the multiset $u = aabca$, where $u(a) = 3$, $u(b) = 1$, $u(c) = 1$. Using such a representation, the operations over multisets are defined as operations over strings. Given two multisets u, v over V, for any $a \in V$, we have $(u \uplus v)(a) = u(a) + v(a)$ as the multiset union, and $(u \backslash v)(a) = max\{0, u(a) - v(a)\}$ as the multiset difference.

2.3.1 Simple Mobile Membranes

A first definition of mobile P systems is given in [133] with rules coming from mobile ambients [42]. Inspired by the operations of endocytosis and exocytosis, namely moving a membrane inside a neighbouring membrane (endocytosis) and moving a membrane outside the membrane where it is placed (exocytosis), P systems with mobile membranes are introduced in [108] as a variant of P systems with active membranes [128]. We use the name *simple mobile membrane system* instead of P systems with mobile membranes.

Definition 2.2. A simple mobile membrane system is a construct

$$\Pi = (V, H, \mu, w_1, \ldots, w_n, R, i_O)$$

where:

1. n, V, H, μ, w_i, i_O are as in Definition 2.1;
2. R is a finite set of developmental rules, of the following forms:

   (a) $[[a \to v]_m]_k$, for $k, m \in H$, $a \in V$, $v \in V^*$; local object evolution

$$M_3 \left[M_2 \left(a\, M_1 \right)_m \right]_k \longrightarrow M_3 \left[M_2 \left(v\, M_1 \right)_m \right]_k$$

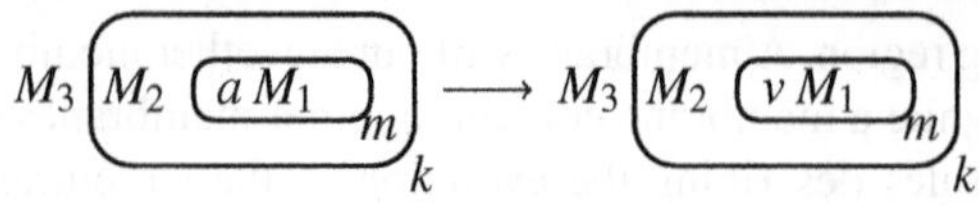

These rules are called *local* because the evolution of an object a of membrane m is possible only when membrane m is inside membrane k. By M_1, M_2 and M_3 we denote (possible empty) multisets of objects, elementary and composite membranes.

$$M_2 \left(a\, M_1 \right)_m \longrightarrow M_2 \left(v\, M_1 \right)_m$$

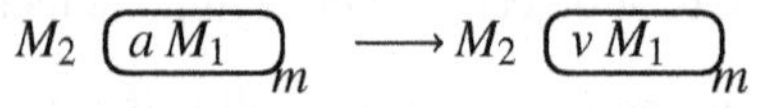

If the restriction of nested membranes is not imposed, that is, the evolution of the object a in membrane m is allowed irrespective of where membrane m is placed, then we say that we have a global evolution rule, and write it simply as $[a \to v]_m$. By M_1 and M_2 we denote (possible empty) multisets of objects, elementary and composite membranes.

(b) $[a]_h[\ ]_m \to [[b]_h]_m$, for $h, m \in H$, $a, b \in V$;   endocytosis

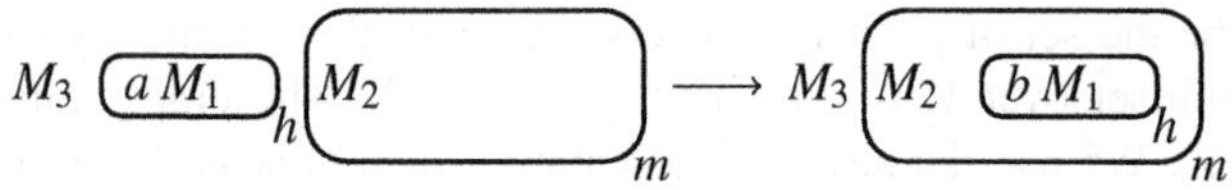

$$M_3 \left(a\, M_1 \right)_h \left[M_2 \right]_m \longrightarrow M_3 \left[M_2 \left(b\, M_1 \right)_h \right]_m$$

An elementary membrane labelled h enters the adjacent membrane labelled m, under the control of object a; the labels h and m remain unchanged during the process; however the object a is modified to b during the operation; m is not necessarily an elementary membrane. By M_1 we denote a (possibly empty) multiset of objects, and by M_2 and M_3 we denote (possibly empty) multisets of objects, elementary and composite membranes.

(c) $[[a]_h]_m \to [b]_h[\ ]_m$, for $h, m \in H$, $a, b \in V$;   exocytosis

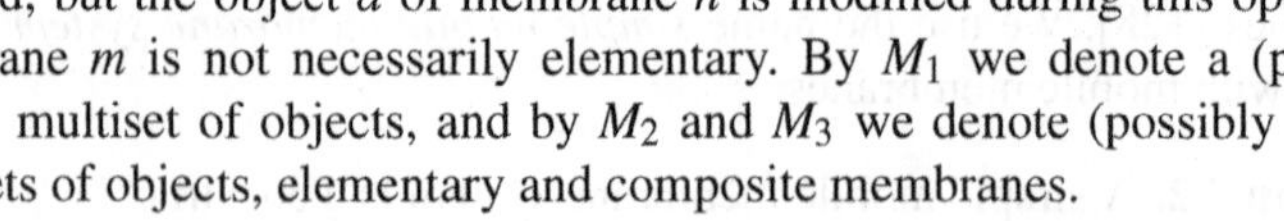

$$M_3 \left[M_2 \left(a\, M_1 \right)_h \right]_m \longrightarrow M_3 \left(b\, M_1 \right)_h \left[M_2 \right]_m$$

An elementary membrane labelled h is sent out of a membrane labelled m, under the control of object a; the labels of the two membranes remain unchanged, but the object a of membrane h is modified during this operation; membrane m is not necessarily elementary. By M_1 we denote a (possibly empty) multiset of objects, and by M_2 and M_3 we denote (possibly empty) multisets of objects, elementary and composite membranes.

The rules are applied according to the following principles:

1. Rules are applied in parallel, non-deterministically choosing the rules, the membranes, and the objects in such a way that the parallelism is maximal; this means that in each step we apply a certain set of rules such that no further rule can be added to the set.

2. The membrane m from the rules of type $(a)-(c)$ is said to be *passive*, while the membrane h is said to be *active*. In any step of a computation, any object and any active membrane can be involved in at most one rule. However, the passive membranes can be used by several rules at the same time. In a rule $[a \rightarrow v]_m$ of type (a), object a is active, while membrane m is passive.
3. When a membrane is moved across another membrane, by endocytosis or exocytosis, its whole contents (its objects) are moved; the inner objects evolve first (if rules are applicable for them), and then any membrane is moved with the contents as obtained after its internal evolution.
4. If a membrane exits the system (by exocytosis), then its internal evolution stops, even if there are rules of type (a) which could be applied.
5. The objects and membranes which do not evolve at a given step are passed unchanged to the next configuration of the system.

By using the rules in this way, we get transitions among the configurations of the system. A sequence of transitions is a computation, and a computation is successful if, starting from the initial configuration, it halts (it reaches a configuration where no rule can be applied). The multiplicity vector of the multiset from a special membrane called the output membrane is considered to be the result of the computation. Thus, the result of a halting computation consists of all the vectors describing the multiplicity of objects from the output membrane; a non-halting computation provides no output. The set of vectors of natural numbers produced in this way by a system Π is denoted by $Ps(\Pi)$. A computation can produce several vectors, all of them considered in the set $Ps(\Pi)$.

Hence a computation is structured as follows: it starts with an initial configuration of the system (the initial membrane structure and the initial distribution of objects within regions), then the computation proceeds, and when it stops the result is to be found in a specific output membrane.

2.3.2 Enhanced Mobile Membranes

Enhanced mobile membranes were introduced in [16] for describing some biological mechanisms of the immune system. The presentation of the immune system is taken from [98], a book which is revised every few years to keep pace with the new discoveries in this field. The cells of the immune system work together with different proteins to seek out and destroy anything foreign or dangerous which enters our body. It takes some time for the immune cell to be activated, but once this happens very few hostile organisms have a chance. There are several types of immune cells, each of them with its own strengths and weaknesses. Some seek out and engulf invaders, while others destroy infected or mutated body cells. One type of immune cells are the B cells which have the ability to release special proteins called antibodies which mark intruders so that they may be destroyed by macrophages. The immune system also has the ability to produce some cells able to remember enemies which it fought in the past. In this way, once the immune system recognizes an invader it attacks more quickly and strongly against it.

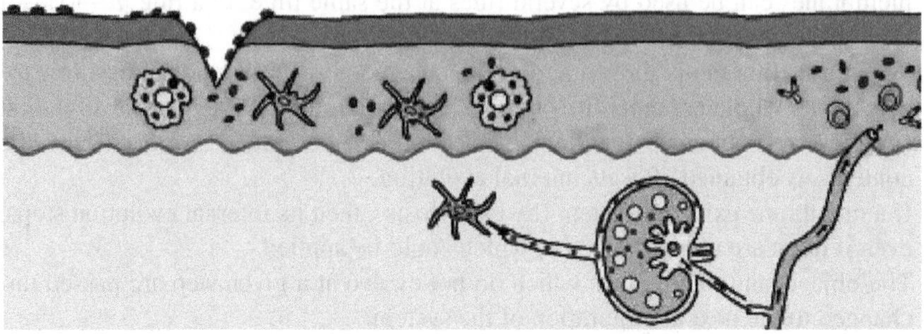

Fig. 2.3 Immune System Mechanisms

Dendritic cells can engulf bacteria, viruses, and other cells. Once a dendritic cells engulfs a bacterium, it dissolves this bacterium and places portions of bacterium proteins on its surface (see Figure 2.3). These surface markers serve as an alarm to other immune cells, namely helper T cells, which then infer the form of the invader. This mechanism makes the T cells sensitive to recognize the antigens or other foreign agents which trigger a reaction of the immune system. Antigens are often found on the surface of bacteria and viruses.

New rules are introduced following this biological example. We define a new variant of mobile membranes, namely the enhanced mobile membranes, originally introduced in [16]. The object a is the one indicating the membrane which initializes the move in the rules of type $(a) - (d)$.

Definition 2.3. *An enhanced mobile membrane system* is a construct

$$\Pi = (V, H, \mu, w_1, \ldots, w_n, R, i_O), \text{ where:}$$

1. n, V, H, μ, w_i, i_O are as in Definition 2.1;
2. R is a finite set of *developmental rules* of the following forms:

(a) $[a]_h[\;]_m \to [[w]_h]_m$, for $h, m \in H, a \in V, w \in V^*$; endo

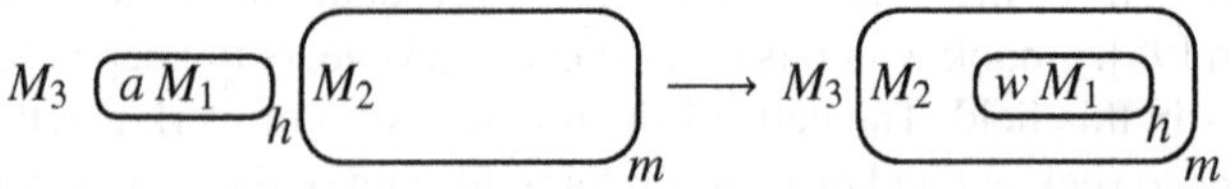

An elementary membrane labelled h enters the adjacent membrane labelled m, under the control of object a; the labels h and m remain unchanged during this process, however, the object a is modified to w during the operation; m is not necessarily an elementary membrane.

(b) $[[a]_h]_m \to [w]_h[\;]_m$, for $h, m \in H, a \in V, w \in V^*$; exo

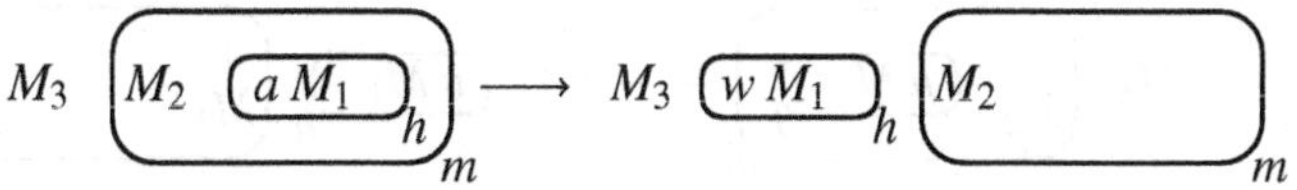

$$M_3 \; \boxed{M_2 \; \boxed{a\,M_1}_{h}}_{m} \longrightarrow M_3 \; \boxed{w\,M_1}_{h} \; \boxed{M_2}_{m}$$

An elementary membrane labelled h exits a membrane labelled m, under the control of object a; the labels of the two membranes remain unchanged, but the object a from membrane h is modified during this operation; membrane m is not necessarily elementary. By M_1 we denote a (possibly empty) multiset of objects, and by M_2 and M_3 we denote (possibly empty) multisets of objects, elementary and composite membranes.

(c) $[\,]_h[a]_m \rightarrow [[\,]_h w]_m$, for $h, m \in H, a \in V, w \in V^*$;   fendo

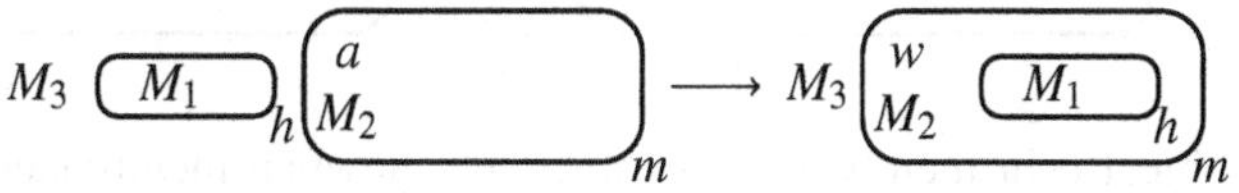

$$M_3 \; \boxed{M_1}_{h} \; \boxed{\begin{array}{c} a \\ M_2 \end{array}}_{m} \longrightarrow M_3 \; \boxed{\begin{array}{c} w \\ M_2 \quad \boxed{M_1}_{h} \end{array}}_{m}$$

An elementary membrane labelled h is engulfed by the adjacent membrane labelled m, under the control of object a of m; the labels h and m remain unchanged during this process, however, the object a is modified to w during the operation; m is not necessarily an elementary membrane. By M_1 we denote a (possibly empty) multiset of objects, and by M_2 and M_3 we denote (possibly empty) multisets of objects, elementary and composite membranes.

(d) $[a[\,]_h]_m \rightarrow [\,]_h[w]_m$, for $h, m \in H, a \in V, w \in V^*$;   fexo

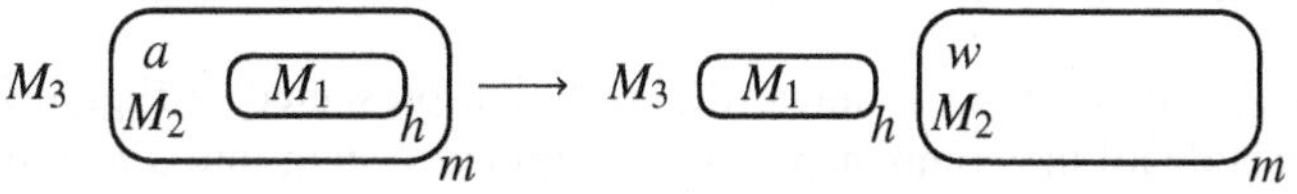

$$M_3 \; \boxed{\begin{array}{c} a \\ M_2 \quad \boxed{M_1}_{h} \end{array}}_{m} \longrightarrow M_3 \; \boxed{M_1}_{h} \; \boxed{\begin{array}{c} w \\ M_2 \end{array}}_{m}$$

An elementary membrane labelled h is expelled by a membrane labelled m, under the control of object a of m; the labels of the two membranes remain unchanged, but the object a from membrane m is modified during this operation; membrane m is not necessarily elementary. By M_1 we denote a (possibly empty) multiset of objects, and by M_2 and M_3 we denote (possibly empty) multisets of objects, elementary and composite membranes.

(e) $[u]_h[v]_m \rightarrow [[u]_h v]_m$, for $h, m \in H, u, v \in V^*$;   pendo

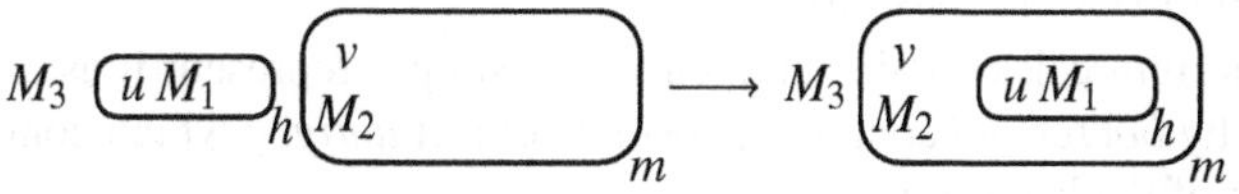

$$M_3 \; \boxed{u\,M_1}_{h} \; \boxed{\begin{array}{c} v \\ M_2 \end{array}}_{m} \longrightarrow M_3 \; \boxed{\begin{array}{c} v \\ M_2 \quad \boxed{u\,M_1}_{h} \end{array}}_{m}$$

An elementary membrane labelled h containing u enters the adjacent membrane containing v; the objects do not evolve in the process. By M_1 we denote a (possibly empty) multiset of objects, and by M_2 and M_3 we denote (possibly empty) multisets of objects, elementary and composite membranes.

(f) $[v[u]_h]_m \rightarrow [u]_h[v]_m$, for $h, m \in H, u, v \in V^*$;   pexo

$$M_3 \left[\begin{array}{c} v \\ M_2 \end{array} \boxed{u\, M_1}_h \right]_m \longrightarrow M_3 \boxed{u\, M_1}_h \left[\begin{array}{c} v \\ M_2 \end{array} \right]_m$$

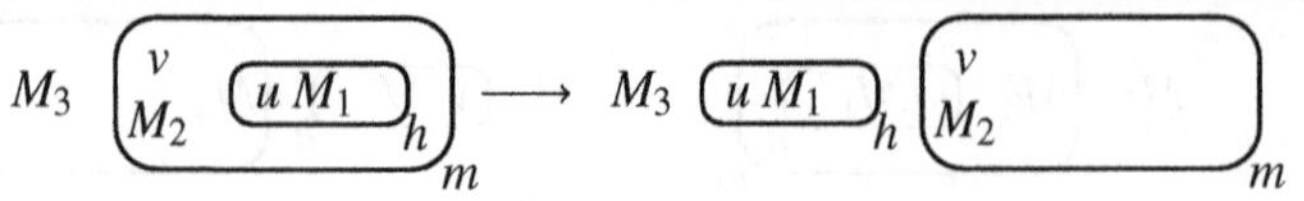

An elementary membrane labelled h containing u comes out of the membrane labelled m containing v. The objects do not evolve in the process. By M_1 we denote a (possibly empty) multiset of objects, and by M_2 and M_3 we denote (possibly empty) multisets of objects, elementary and composite membranes.

(g) $[[a]_j[b]_h]_k \rightarrow [[w]_j[b]_h]_k$ for $h, j, k \in H, a, b \in V, w \in V^*$;   cevol

$$M_4 \left[M_3 \boxed{a\, M_1}_j \boxed{b\, M_2}_h \right]_k \longrightarrow M_4 \left[M_3 \boxed{w\, M_1}_j \boxed{b\, M_2}_h \right]_k$$

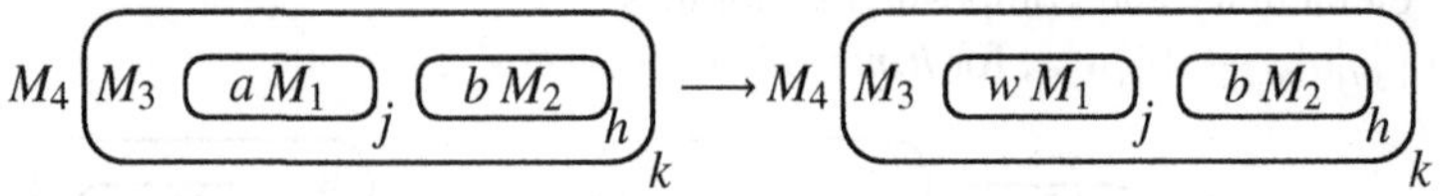

An object a in membrane m evolves into w when membranes h and m are adjacent to each other inside membrane k. By M_1, M_2, M_3 and M_4 we denote (possibly empty) multisets of objects, elementary and composite membranes.

The rules of enhanced mobile membranes are applied according to the principles of simple mobile membranes. Here *endo* and *exo* represent endocytosis and exocytosis, *fendo* and *fexo* represent forced endocytosis and forced exocytosis, *pendo* and *pexo* represent pure endocytosis and pure exocytosis, while *cevol* represents contextual evolution. When we restrict $|w| = 1$ in rules (a) - (d), we call the operations *rendo*, *rexo*, *rfendo* and *rfexo* where r stands for "restricted".

Example 2.1. In order to simulate the evolution presented in Figure 2.3, we need first to encode all the components of the immune system into a membrane system. This can be realized by associating a membrane to each component, and objects to the signals, states and parts of molecules. For the steps done by the dendritic cells presented in Figure 1 we use the following encodings:

- dendritic cell - $[eat]_{DC}$

  An immature dendritic cell is willing to eat any bacterium it encounters, so we translate it into a membrane labelled by DC which has inside an object *eat* used to engulf the bacterium. Once the dendritic cell matures, the object *eat* is consumed.
- bacterium cell - $[antigen]_{bacterium}$

  A bacterium cell contains antigen so we simply represent it as a membrane labelled by *bacterium* containing a single object *antigen* which contains the information of the bacterium.
- lymph node - $[\ ]_{lymph\ node}$

  The lymph node is the place to which the mature dendritic cells migrate in order to start the immune response, so we translate it into a membrane labelled by *lymph node*.

Using the above membranes we can describe the membrane system as follows (here *skin* stands for the body skin):

$$[[eat]_{DC}[\,]_{lymph\ node}]_{skin}[antigen]_{bacterium}$$

with the following rules which describe its evolution:

- $[antigen]_{bacterium}[\,]_{skin} \rightarrow [[antigen]_{bacterium}]_{skin}$
  A bacterium enters through the skin by performing an endocytosis rule in order to infect the body. The bacterium contains an object *antigen* which represent its signature.
- $[eat]_{DC}[\,]_{bacterium} \rightarrow [eat[\,]_{bacterium}]_{DC}$
  Once an immature dendritic cell becomes sibling to a bacterium, it "eats" the bacterium by performing a forced endocytosis rule. Until this moment the bacterium has controlled its own movement; in this step of the evolution the movement becomes controlled by the dendritic cell which eats the bacterium.
- $[[antigen]_{bacterium}]_{DC} \rightarrow [[antigen\ \delta]_{bacterium}]_{DC}$
  Once the bacterium has entered the dendritic cell, an object δ is created in order to dissolve the membrane *bacterium*, and the content of the bacterium is released into the dendritic cell.
- $[antigen]_{DC}[\,]_{lymph\ node} \rightarrow [[antigen]_{DC}]_{lymph\ node}$
  Once the dendritic cell contains antigen, it enters the lymph node in order to activate a special class of T cells, namely the helper T cells.
- $[[eat]_{DC}]_{lymph\ node} \rightarrow [[\,]_{DC}]_{lymph\ node}$
  Once the dendritic cell enters the lymph node it matures and the capacity to engulf bacteria disappears, namely the *eat* object is consumed.

Using only these few rules we can simulate the way a bacterium is engulfed and its content is displayed by the eater cell. The proteins produced by helper T cells activate the B cells.

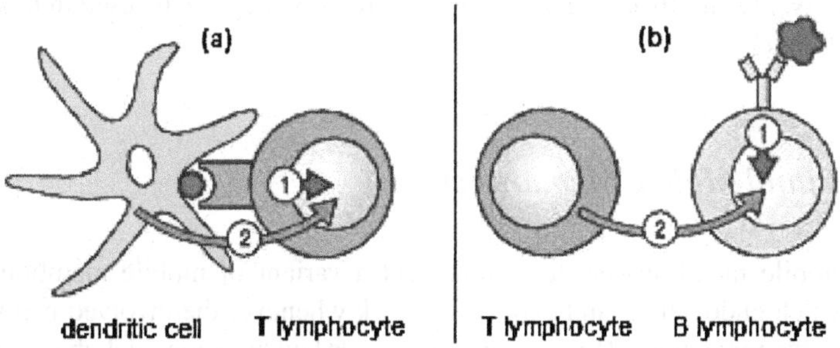

Fig. 2.4 Activation of T cells (a) and B cells (b)

For the process of activating the helper T cells and B cells we use the following encodings:

- helper T cell - $[passive]_{helper\ T\ cell}$

A helper T cell is initially passive, so we represent it as a membrane labelled *helper T cell* in which the object *passive* is placed. When the cell is activated the object *passive* is transformed into *active*.

- B cell - $[passive]_{B\ cell}$

  A B cell is initially passive, so we represent it as a membrane labelled *B cell* in which the object *passive* is placed. When the cell is activated the object *passive* is transformed into *active*.

The activation of the helper T cells and B cells is conditioned by the presence in the lymph node of the dendritic cells, and that is why we use the following contextual evolution rules:

- $[[antigen]_{DC}[passive]_{helper\ T\ cell}]_{lymph\ node} \to$
  $[[antigen]_{DC}[active]_{helper\ T\ cell}]_{lymph\ node}$

  Once the dendritic cell containing antigen enters the lymph node, it activates a special class of T cells, namely the helper T cells. This is denoted by changing the object *passive* to *active* in helper T cells.
- $[[passive]_{B\ cell}[active]_{helper\ T\ cell}]_{lymph\ node} \to$
  $[[active]_{B\ cell}[active]_{helper\ T\ cell}]_{lymph\ node}$

  Once the helper T cells are activated, the B cells that are sibling with them are the next cells which are activated.

The B cell searches for antigen matching its receptors. If it finds such antigen, then inside the B cell a triggering signal is set off. Using the proteins produced by helper T cells, the B cell starts to divide and produce clones of itself. During this process, two new cell types are created: plasma cells which produce an antibody, and memory cells which are used to "remember" specific intruders.

These examples motivate the introduction of the new class of mobile membranes; more exactly, they motivate the new rules and the way they can be used in modelling some biological systems.

2.3.3 Mutual Mobile Membranes

Mutual mobile membrane systems represent a variant of mobile membrane systems in which endocytosis and exocytosis work whenever the involved membranes "agree" on the movement. This agreement is described by using dual objects a and $\overline{a}$ in the involved membranes, with $\overline{\overline{a}} = a$. The duality relation is distributive over a multiset, namely $\overline{u} = \overline{a_1} \ldots \overline{a_n}$ for $u = a_1 \ldots a_n$.

Definition 2.4. A mutual mobile membrane system is a construct

$$\Pi = (V, H, \mu, w_1, \ldots, w_n, R, i_O), \quad \text{where:}$$

1. n, V, H, μ, w_i, i_O are as in Definition 2.1;
2. R is a finite set of developmental rules of the following forms:

(a) $[uv]_h[\overline{u}v']_m \rightarrow [\,[w]_h w']_m$ for $h,m \in H, u,\overline{u} \in V^+, v,v',w,w' \in V^*$; mendo

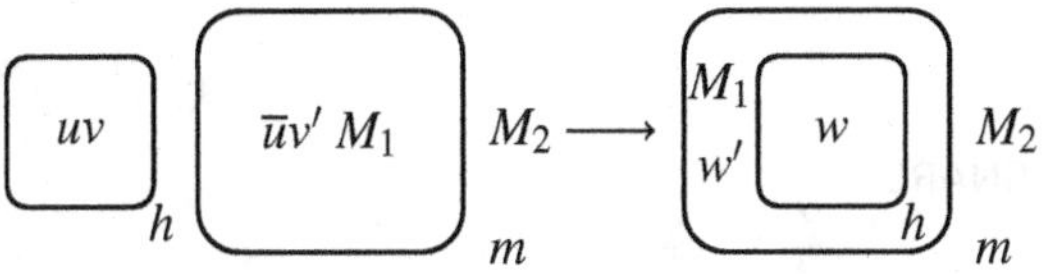

An elementary membrane labelled h enters the adjacent membrane labelled m under the control of the multisets of objects uv and $\overline{u}v'$. The labels h and m remain unchanged during this process; however the multisets of objects uv and $\overline{u}v'$ are replaced with the multisets of objects w and w', respectively. By M_1 and M_2 we denote (possibly empty) multisets of objects, elementary and composite membranes.

(b) $[\overline{u}v'[uv]_h]_m \rightarrow [w]_h[w']_m$ for $h,m \in H, u,\overline{u} \in V^+, v,v',w,w' \in V^*$; mexo

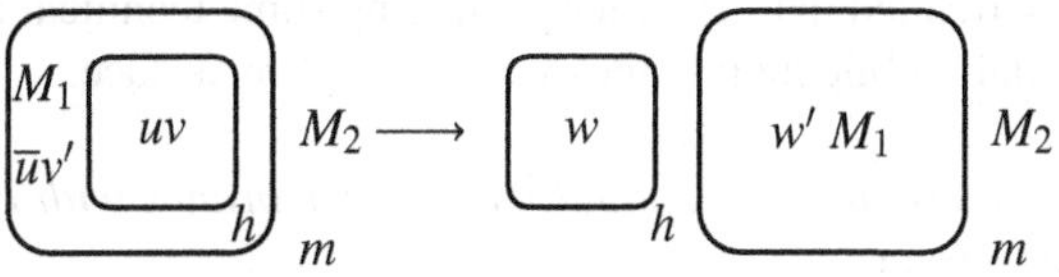

An elementary membrane labelled h exits a membrane labelled m, under the control of the multisets of objects uv and $\overline{u}v'$. The labels of the two membranes remain unchanged, but the multisets of objects uv and $\overline{u}v'$ are replaced with the multisets of objects w and w', respectively. By M_1 and M_2 we denote (possibly empty) multisets of objects, elementary and composite membranes.

The rules of the mutual mobile membranes are applied according to the principles of simple mobile membranes. Here *mendo* and *mexo* represent mutual endocytosis and mutual exocytosis. A multiset u indicates the membrane which initializes the move in the rules of type $(a)-(b)$, while a multiset $\overline{u}$ indicates the membrane which accepts the movement.

2.3.4 Mutual Mobile Membranes with Objects on Surface

Membrane fusion occurs when two separate membranes containing complementary proteins merge into a single membrane. The process described in Figure 2.5 is performed in several well-distinguished steps.

Initially, the two involved membranes mutually identify each other by means of complementary proteins: v-SNARES and t-SNARES. Then SNARES located on the vesicles (v-SNARES) and on the target membranes (t-SNARES) interact with one another to form a stable complex that brings the two membranes very close. Finally, the vesicle and target membranes distort and then fuse. Each vesicle must only fuse with the correct target membrane in order to avoid an unwanted mixing of proteins.

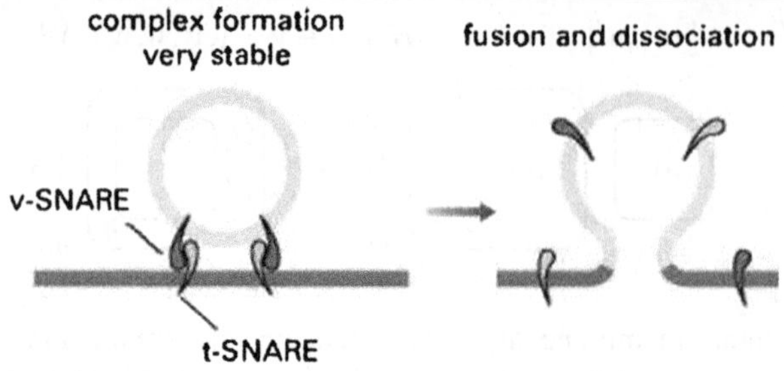

Fig. 2.5 Vesicle Fusion Mediated by Complex Formation of Complementary SNARES

These biological facts provide the motivation for using objects and co-objects for the pino, exo and phago rules as done in [20]. These rules are also related to the formal approach defined in [45]. After presenting some technical notions, we define systems of mutual mobile membranes with objects on surface.

Definition 2.5. A *system of n mutual mobile membranes with objects on surface* (M^3OS_n) is a construct

$$\Pi = (V, \mu, u_1, \ldots, u_n, R, i_O)$$

where:

1. n, V, i_O are as in Definition 2.1;
2. μ is a membrane hierarchical structure with $n \geq 2$ membranes;
3. $u_1, \ldots, u_n$ are multisets of proteins (represented by strings over V) bound to the n membranes at the beginning of the computation; the membranes are bijectively mapped to $\{1, \ldots, n\}$; the skin membrane is labelled with 1 and $u_1 = \varepsilon$;
4. R is a finite set of rules of the following forms:

(a) $[\,]_{au\bar{a}v} \rightarrow_m [[\,]_{cu}]_{dv}$, for $a, \bar{a} \in V, c, d, u, v \in V^*$ pino

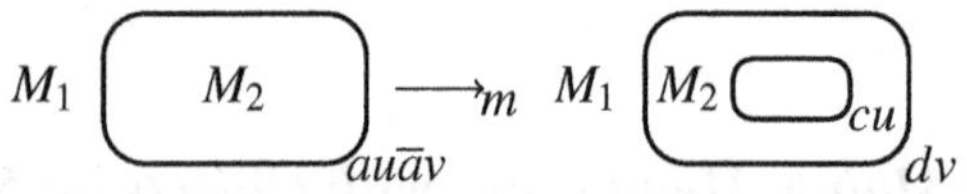

An object a together with a complementary object $\bar{a}$ model the creation of an empty membrane within the membrane on which are objects a and $\bar{a}$. We should imagine that the original membrane buckles towards the inside, and pinches off by breaking the connection between a and $\bar{a}$. The multiset of objects u on the newly created (empty) membrane is transferred from the initial membrane. The objects a and $\bar{a}$ are modified during this step to the multisets c and d, respectively. On the surface of the membrane appearing on the left hand side of the rule there are some objects (other than $au\bar{a}v$) which are ignored;

these objects are also not specified on the right hand side of the rule, being randomly distributed between the two resulting membranes. By M_1 and M_2 we denote (possibly empty) multisets of elementary and composite membranes.

(b) $[[\,]_{au}]_{\overline{a}v} \rightarrow_m [\,]_{cudv}$, for $a, \overline{a} \in V, c, d, u, v \in V^*$ **exo**

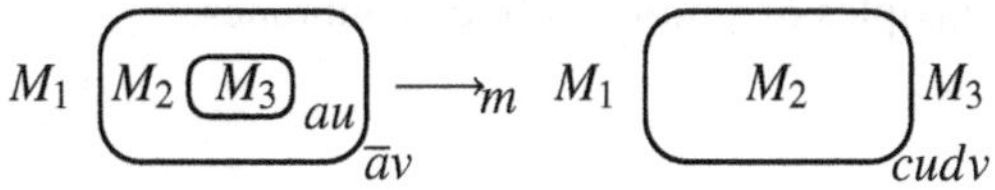

An object a together with a complementary object $\overline{a}$ model the merging of a nested membrane with its surrounding membrane. We should imagine that the connection between a and $\overline{a}$ represent the point where the membranes connect to each other. In this merging process (which is a smooth and continuous process), the membrane having the multiset au on its surface gets expelled to the outside, and all objects of the two membranes are united into a multiset on the membrane which initially contained v. The objects a and $\overline{a}$ are modified during this evolution to the multisets c and d, respectively. If the membrane having on its surface the object a is composite, then its content is released near the newly merged membrane after applying the rule. On the surface of the membranes appearing on the left hand side of the rule there are some objects (other than au and $\overline{a}v$) which are ignored; these objects are also not specified on the right hand side of the rule, being moved onto the resulting membrane. By M_1, M_2 and M_3 we denote (possibly empty) multisets of elementary and composite membranes.

(c) $[\,]_{au}[\,]_{\overline{a}v} \rightarrow_m [[[\,]_{cu}]_d]_v$, for $a, \overline{a} \in V, c, d, u, v \in V^*$ **phago**

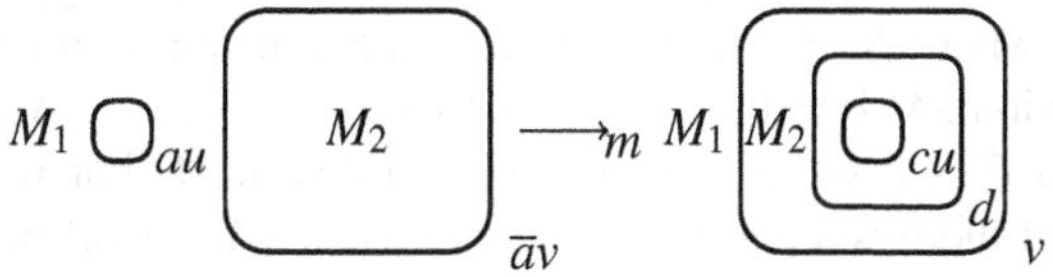

An object a together with its complementary object $\overline{a}$ model a membrane (the one with $\overline{a}$ on its surface) "eating" an elementary membrane (the one with a on its surface). The membrane having $\overline{a}$ and v on its surface wraps around the membrane having a and u on its surface. An additional membrane is created around the eaten membrane; the objects a and $\overline{a}$ are modified during this evolution to the multisets c and d (the multiset c corresponds to a and remains on the eaten membrane, while the multiset d corresponds to $\overline{a}$ and is placed on the newly created membrane). On the surface of the membranes appearing on the left hand side of the rule there are some objects (other than au and $\overline{a}v$) which are ignored; these objects are also not specified on the right hand side of the rule. The objects appearing on the membrane initially having the object a

on surface remain unchanged, while the objects appearing on the membrane initially having the object $\overline{a}$ on surface are randomly distributed between the two resulting membranes (the ones with d and v). By M_1 and M_2 we denote (possibly empty) multisets of elementary and composite membranes.

The rules of mutual mobile membranes with objects on surface are applied according to the principles of simple mobile membranes.

2.4 Computability Power of Mobile Membranes

Several notions and notations from the field of formal languages that are used here can be found in [80] and [142].

2.4.1 *Preliminaries*

For an alphabet $V = \{a_1, \ldots, a_n\}$, we denote by V^* the set of all strings over V; λ denotes the empty string. V^* is a monoid with λ as its unit element. For a string $x \in V^*$, $|x|_a$ denotes the number of occurrences of symbol a in x. A multiset over V is represented by a string over V (together with all its permutations), and each string precisely identifies a multiset. For an alphabet V, the *Parikh vector* is $\psi_V : V^* \rightarrow \mathbf{N}^n$ with $\psi_V(x) = (|x|_{a_1}, \ldots, |x|_{a_n})$, for all $x \in V^*$. For a language L, the Parikh vector is $\psi_V(L) = \{\psi_V(x) \mid x \in L\}$, while for a family FL of languages, the Parikh vector is $PsFL = \{\psi_V(L) \mid L \in FL\}$.

Definition 2.6. A *matrix grammar with appearance checking* $G = (N, T, S, M, F)$ is a construct where N, T are disjoint alphabets of non-terminals and terminals, $S \in N$ is the axiom, M is a finite set of matrices $(A_1 \rightarrow x_1, \ldots, A_n \rightarrow x_n)$ of context-free rules, and F is a set of occurrences of rules in M. For $w, z \in (N \cup T)^*$, we write $w \Rightarrow_m z$ if there is a matrix $m = (A_1 \rightarrow x_1, \ldots, A_n \rightarrow x_n)$ in M and the strings $w_i \in (N \cup T)^*, 1 \leq i \leq n+1$, such that $w = w_1, z = w_{n+1}$, and for all i, $1 \leq i \leq n$, either (1) $w_i = w_i' A_i w_i''$, $w_{i+1} = w_i' x_i w_i''$, for some w_i', $w_i'' \in (N \cup T)^*$, or (2) $w_i = w_{i+1}$, A_i does not appear in w_i, and the rule $A_i \rightarrow x_i$ appears in F. The language generated by G is $L(G) = \{x \in T^* \mid S \Rightarrow^* x\}$.

Definition 2.7. A *matrix grammar in the strong binary form* $G = (N, T, S, M, F)$ is a construct where $N = N_1 \cup N_2 \cup \{S, \#\}$, with these three sets mutually disjoint, two distinguished symbols $B^{(1)}, B^{(2)} \in N_2$, and the matrices in M of one of the following forms:

(1) $(S \rightarrow XA)$, with $X \in N_1, A \in N_2$;
(2) $(X \rightarrow Y, A \rightarrow x)$, with $X, Y \in N_1, A \in N_2, x \in (N_2 \cup T)^*$, $|x| \leq 2$;
(3) $(X \rightarrow Y, B^{(j)} \rightarrow \#)$, with $X, Y \in N_1, j = 1, 2$;

(4) $(X \to \lambda, A \to x)$, with $X \in N_1, A \in N_2, x \in T^*$, $|x| \leq 2$.

If we ignore the empty string when comparing languages, then the rules of type (4) are of the form $(X \to a, A \to x)$, with $X \in N_1$, $a \in T$, $A \in N_2$, $x \in T^*$.

We denote by *PsRE* the family of Turing computable sets of vectors generated by arbitrary grammars.

By $\mathbf{N}RCM(M, CF - \lambda, ac)$ and $\mathbf{N}MAT_{ac}$ we denote the families of sets of numbers computed by random context non-erasing matrix grammars with appearance checking, and non-erasing matrix grammars with appearance checking, respectively. These can also be looked at as the families of sets of numbers recognized by these languages. It is known that $\mathbf{N}RCM(M, CF - \lambda, ac) = \mathbf{N}MAT_{ac} \subset \mathbf{N}RE$ (see [80]).

Definition 2.8 (Left Quotient). The left quotient of a language L by a letter a is given by $\partial_a(L) = \{x \mid ax \in L\}$.

Definition 2.9 (Random Context Matrix Grammars). A random context matrix grammar is a construct $G = (N, T, M, S, F)$ where N, T, S are as in a usual matrix grammar and M is a finite set of triples $((A_1 \to x_1, A_2 \to x_2, \ldots, A_n \to x_n), Q, R)$ where $A_i \to x_i$ are context-free rules, $1 \leq i \leq n$, $Q, R \subseteq N$, $Q \cap R = \emptyset$. A matrix can be applied to a string $x = x_1X_1x_2X_2\ldots X_lx_{l+1}$ in order to rewrite effectively the symbols $X_1, \ldots, X_l$ only if $x_1, \ldots x_{l+1}$ contains all symbols of Q and no symbols of R. We denote by $RCM(M, \beta, \max(\alpha, \gamma))$ the family of languages generated by random context matrix grammars $G = (N, T, S, M, F)$ with rules of type β, with $\beta \in \{CF, CF - \lambda\}$. If $\gamma = ac$, then F is arbitrary, and if γ is empty, then $F = \emptyset$. If $\alpha = ac$, then R is arbitrary in $((r_1, \ldots, r_n), Q, R) \in M$ and if α is empty, no forbidding contexts are involved. max $(\alpha, \gamma) = ac$ if at least one of α, γ is ac. Thus, if no appearance checking is used, and if no forbidding contexts are used, we have the family $RCM(M, \beta, \emptyset)$.

Minsky introduced the concept of register machines in [125] by showing that the power of Turing machines can be achieved by such abstract machines using a finite number of registers for storing arbitrarily large non-negative integers. A register machine runs a program consisting of labelled instructions which encode simple operations for updating the content of the register.

Definition 2.10 (Register Machine). An n-register machine is $M = (n, B, l_0, l_h, I)$, where:

- n is the number of registers; B is a set of labels; l_0 and l_h are the labels of the initial and halting instructions;
- I is a set of labelled instructions of the form $l_i : (op(r), l_j, l_k)$, where $op(r)$ is an operation on register r of M, and l_i, l_j, l_k are labels from the set B.
- the machine is capable of the following instructions:

  1. $l_i : (ADD(r), l_j, l_k)$: Add one to the content of register r and proceed, in a non-deterministic way, to instruction with label l_j or to instruction with label l_k; in the deterministic variant, $l_j = l_k$ and then the instruction is written in the form $l_i : (ADD(r), l_j)$.

2. $l_i : (SUB(r), l_j, l_k)$: If register r is not empty, then subtract one from its contents and go to instruction with label l_j, otherwise proceed to instruction with label l_k.
3. l_h : *halt*: This instruction stops the machine and can only be assigned to the final label l_h.

Theorem 2.1 ([146]). *A 3-register machine can compute any partial recursive function of one variable. It starts with the argument in a counter, and (if it halts) leaves the answer in a counter.*

Definition 2.11 (E0L System). $G = (V, T, \omega, R)$ is a construct where V is the alphabet, $T \subseteq V$ is the terminal alphabet, $\omega \in V^*$ is the axiom, and R is a finite set of rules of the form $a \to v$ with $a \in V$ and $v \in V^*$ such that for each $a \in V$ there is at least one rule $a \to v$ in R. For $w_1, w_2 \in V^*$, we say that $w_1 \Rightarrow w_2$ if $w_1 = a_1 \dots a_n$, $w_2 = v_1 \dots v_n$ for $a_i \to v_i \in R$, $1 \le i \le n$. The generated language is $L(G) = \{x \in T^* \mid \omega \Rightarrow^* x\}$.

Definition 2.12 (ET0L System). $G = (V, T, \omega, R_1, \dots R_n)$ is a construct such that each (V, T, ω, R_i) is an E0L system; each R_i is called a table, $1 \le i \le n$. The generated language is defined as $L(G) = \{x \in T^* \mid \omega \Rightarrow_{R_{j_1}} \cdots \Rightarrow_{R_{j_m}} w_m = x\}$, with $m \ge 0$, $1 \le j_i \le n$, $1 \le i \le m$. We denote by $PsET0L$ the families of languages generated by extended table 0L grammars.

2.4.2 Simple Mobile Membranes

The computational power of simple mobile membranes is treated in [108].

The family of all sets $Ps(\Pi)$ generated by systems of degree at most n using rules $\alpha \in \{levol, gevol, endo, exo\}$ is denoted by $PsMM(\alpha)$ If the number of membranes is not bounded, this is denoted by $PsMM_*(levol, endo, exo)$. Here *levol* and *gevol* represent local and global evolution, *endo* and *exo* represent endocytosis and exocytosis. The number of membranes does not increase during the computation, but it can decrease by sending membranes out of the skin.

The following result establishes a universality result using nine membranes and the operations of endocytosis and exocytosis:

Theorem 2.2 ([108]). $PsMM_9(endo, exo) = PsRE$.

A strengthening of the previous universality result is:

Corollary 2.1 ([108]). $PsMM_*(endo, exo) = PsMM_n(endo, exo) = PsMM_n(gevol, endo, exo) = PsMM_n(levol, endo, exo) = PsRE$*, for all* $n \ge 9$.

An improvement of Theorem 2.2 is:

Theorem 2.3 ([103]). $PsMM_4(gevol, endo, exo) = PsRE$.

We improve the previous result by decreasing the number of membranes to three.

Theorem 2.4. $PsMM_3(levol, endo, exo) = PsRE$.

Proof. Consider a matrix grammar $G = (N, T, S, M, F)$ in the improved strong binary normal form (hence with $N = N_1 \cup N_2 \cup \{S; \#\}$), having n_1 matrices of types (2) and (4) (that is, not used in the appearance checking mode), and n_2 matrices of type (3) (with appearance checking rules). Let $B^{(1)}$ and $B^{(2)}$ be the two objects in N_2 for which we have rules $B^{(j)} \to \#$ in matrices of M. The matrices of the form $(X \to Y, B^{(j)} \to \#)$ are labelled by m_i', with $i \in lab_j$, for $j \in \{1,2\}$, such that lab_1, lab_2, and $lab_0 = \{1,\ldots,n_1\}$ are mutually disjoint sets.

We construct a mobile membrane system $\Pi = (V, H, \mu, w_1, w_2, w_3, R, 2)$ of degree three, where:

$V = N \cup \{X, X_{i,j} \mid X \in N_1, 1 \le i \le n_1, 0 \le j \le n_1\}$
$\cup \{a, a' \mid a \in T\} \cup \{x \mid x \in (N_2 \cup T)^*\}$
$\cup \{A, A_{i,j} \mid A \in N_2, 1 \le i \le n_1, 0 \le j \le n_1\}$
$H = \{1,2,3\}$
$\mu = \{(1,2);(1,3)\}$
$w_2 = XA$, where $(S \to XA)$ is the initial matrix of G
$w_h = \lambda$, for all $h \in \{1,3\}$

The set R of rules is constructed as follows:

(i) For each (nonterminal) matrix $m_i : (X \to Y, A \to x)$, $X, Y \in N_1$, $x \in (N_2 \cup T)^*$, $A \in N_2$, with $1 \le i \le n_1$, we consider the rules:

1. $[X]_2[\ ]_3 \to [[X_{i,0}]_2]_3$ (endo)
2. $[[A]_2]_3 \to [A_{j,0}]_2[\ ]_3$ (exo)
3. $[[X_{i,k} \to X_{i,k+1}]_2]_1$, $k < i$ (levol)
4. $[[A_{j,k} \to A_{j,k+1}]_2]_1$, $k < j$ (levol)
5. $[[A_{j,j}X_{i,i} \to xY]_2]_1$, $j = i$ (levol)
6. $[[A_{j,i}X_{i,i} \to \#]_2]_1$, $j > i$ (levol)
7. $[[A_{j,j}X_{i,j} \to \#]_2]_1$, $j < i$ (levol)

In the initial configuration, we have the objects X and A corresponding to the initial matrix in membrane 2. To simulate a matrix of the above type we start by applying the endocytosis rule 1, thus replacing X with $X_{i,0}$, followed by the exocytosis rule 2, thus replacing a single $A \in N_2$ with $A_{j,0}$. No other $A \in N_2$ can be replaced until membrane 2 enters membrane 3. Rule 3 (for X) and rule 4 (for A) are used to increment the second indices of X and A. This is done to check if the first indices of X and A are the same, and in this case to rewrite A according to the matrix m_i. Once the first indices are equal, rule 5 is applied to complete the simulation of matrix m_i. If the first indices of X and A are not the same, rule 6 (if the first indices of X is lower than the first indices of A) or rule 7 (if the first indices of X is bigger than the first indices of A) is applied, the computation is blocked without producing any output. We now illustrate the evolution of the configurations during one simulation of a type (2) matrix.

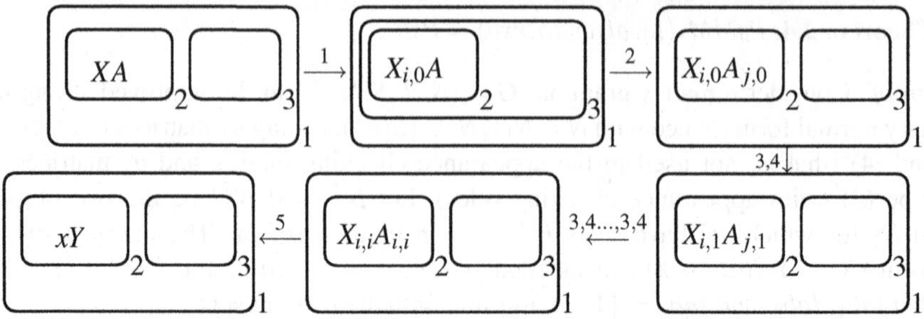

(ii) For a terminal matrix $m_i : (X \to a, A \to x), X \in N_1, a \in T, A \in N_2, x \in T^*$, where $1 \le i \le n_1$, we use rules 1-7, where rule 5 is replaced by the rules:

8. $[a_{i,i}X_{i,i} \to a'Y]_1$ (levol)
9. $[[a']_2]_1 \to [a]_2[\,]_1$ (exo)

Observe that simulation of a type (4) matrix follows similar steps, except that we have an a in place of Y. During the finishing stages of a type (4) simulation, we use rule 8 to replace $a_{i,i}$ by a', and then to rewrite it to a when sending the membrane 2 out of the skin membrane, namely membrane 1. We now illustrate the evolution of the configurations during one simulation of a type (4) (terminal) matrix.

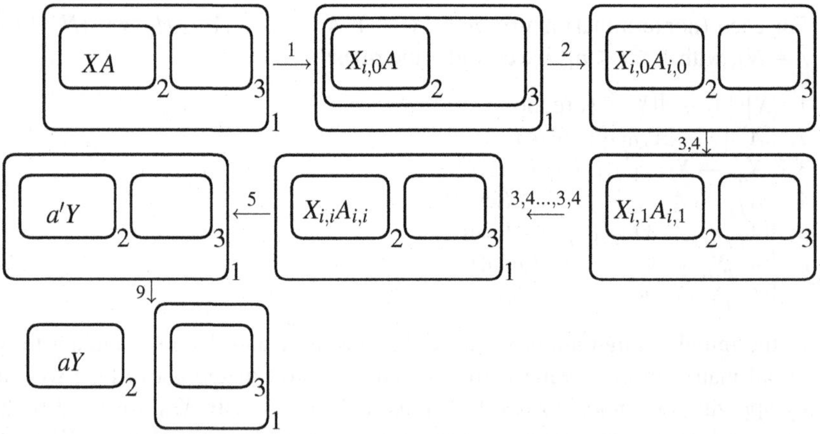

(iii) For each matrix $m'_i : (X \to Y, B^{(k)} \to \#), X, Y \in N_1$, where $n_1 + 1 \le j \le n_1 + n_2$, $j \in lab_k$, $k = 1, 2$, we consider the rules:

10. $[X]_2[\,]_3 \to [[X_k]_2]_3$, for $i \in lab_k$ (endo)
11. $[[X_kB^{(k)} \to \#]_2]_3$, $k = 1, 2$ (levol)
12. $[[X_k]_2]_3 \to [Y]_2[\,]_3$, $k = 1, 2$ (exo)

The simulation of matrices of type (3) begins with a rule of type 10. This is followed by a rule 11 in case $B^{(k)}$ exists, blocking membrane 2 inside membrane 3 and the computation stops without producing any output. If no $B^{(k)}$ exists, then rule 12 can be used to send out membrane 2, successfully completing the simulation. We now illustrate the evolution of the configurations during one simulation of a type (3) matrix.

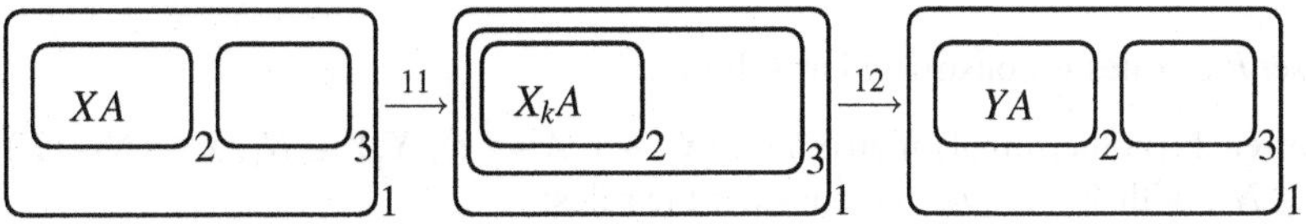

□

2.4.3 Enhanced Mobile Membranes

The operations governing the mobility of enhanced mobile membranes are endocytosis (endo), exocytosis (exo), enhanced endocytosis (fendo) and enhanced exocytosis (fexo). The interplay between these four operations is quite powerful, and the computational power of a Turing machine is obtained using twelve membranes without using the context-free evolution of objects [107].

The family of all sets $Ps(\Pi)$ generated by systems of degree at most n using rules $\alpha \subseteq \{exo, endo, fendo, fexo, pendo, pexo, rendo, rexo, rfendo, rfexo, cevol\}$ is denoted by $PsEMM_n(\alpha)$. Here *cevol* represents contextual evolution. The main results are the following.

Theorem 2.5 ([107]). $PsEMM_{12}(endo, exo, fendo, fexo) = PsRE$.

Theorem 2.6 ([107]). $PsEMM_3(cevol) = PsRE$.

Theorem 2.7 ([107]). $PsEMM_3(endo, exo) = PsEMM_3(fendo, fexo)$.

We improve the result of Theorem 2.5 as follows:

Theorem 2.8. $PsEMM_9(endo, exo, fendo, fexo) = PsRE$.

Proof. Consider a matrix grammar $G = (N, T, S, M, F)$ in the improved strong binary normal form (hence with $N = N_1 \cup N_2 \cup \{S; \#\}$), having n_1 matrices $m_1, \dots, m_{n_1}$ of types (2) and (4) (that is, not used in the appearance checking mode), and n_2 matrices of type (3) (with appearance checking rules). The initial matrix is m_0, with $m_0 : (S \to XA)$. Let $B^{(1)}$ and $B^{(2)}$ be the two objects in N_2 for which we have rules $B^{(j)} \to \#$ in matrices of M. The matrices of the form $(X \to Y, B^{(j)} \to \#)$ are labelled by m'_i, $1 \le i \le n_2$ with $i \in lab_j$, for $j \in \{1,2\}$, such that lab_1, lab_2, and $lab_0 = \{1, 2, \dots, n_1\}$ are mutually disjoint sets.

We construct a mobile membrane system $\Pi = (V, H, \mu, w_1, \dots, w_9, R, 7)$ of degree nine, where:

$V = N \cup T \cup \{X'_{0i}, A'_{0i} \mid X \in N_1, A \in N_2, 1 \le i \le n_1\}$

$\quad \cup \{X_{ji}, A_{ji} \mid 0 \le i, j \le n_1\} \cup \{X_i^{(j)}, X_j \mid X \in N_1, j \in \{1,2\}, 1 \le i \le n_2\}$

$H = \{1, \dots, 9\}$

$\mu = \{(1,7); (1,8); (1,9); (1,2); (2,3); (2,4); 2,5); (2,6)\}$

$w_7 = XA$, where $(S \to XA)$ is the initial matrix of G
$w_h = \lambda$, for all $h \in \{1,\ldots,9\} \setminus \{7\}$

The set R of rules is constructed as follows:

(i) For each (nonterminal) matrix $m_i : (X \to Y, A \to x)$, $X,Y \in N_1$, $x \in (N_2 \cup T)^*$, $A \in N_2$, with $1 \leq i \leq n_1$, we consider the rules:

1. $[X]_7[\,]_8 \to [[X_{i,i}]_7]_8$ (endo)
2. $[[A]_7]_8 \to [A_{j,j}]_7[\,]_8$ (exo)
3. $[X_{k,i}]_7[\,]_9 \to [[X_{k-1,i}]_7]_9$, $k > 0$ (endo)
4. $[[A_{k,j}]_7]_9 \to [A_{k-1,j}]_7[\,]_9$, $K > 0$ (exo)
5. $[\,]_8[X_{0,i}]_7 \to [X'_{0,i}[\,]_8]_7$ (fendo)
6. $[\,]_9[A_{0,j}]_7 \to [A'_{0,j}[\,]_9]_7$ (fendo)
7. $[\,]_8[X_{0,j}]_7 \to [\#[\,]_8]_7$ (fendo)
8. $[[A_{0,j}]_7]_9 \to [\#]_7[\,]_9$ (exo)
9. $[X'_{0,j}[\,]_8]_7 \to [\,]_8[Y]_7$ (fexo)
10. $[A'_{0,j}[\,]_9]_7 \to [\,]_9[x]_7$ (fexo)

We now illustrate the evolution of the configurations during one simulation of a type (2) matrix, when $i = j$.

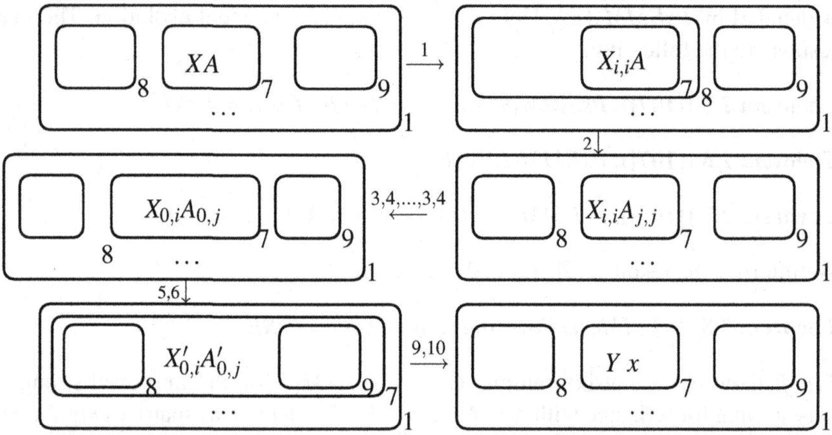

In the initial configuration, we have the objects X, A corresponding to the initial matrix in membrane 7. To simulate a matrix of type (2), we start by applying the endocytosis rule 1, thus replacing X with $X_{i,i}$, followed by the exocytosis rule 2, thus replacing a single $A \in N_2$ with $A_{j,j}$. Rule 3 (for X) and rule 4 (for A) are used to decrement the first indices of X and A. This is done to check if the indices of X and A are the same, and in this case to rewrite A according to the matrix m_i. By using fendo rules 5 and 6, membranes 8 and 9 enter membrane 7 replacing $X_{0,i}$ and $A_{0,j}$ with $X'_{0,i}$ and $A'_{0,j}$, respectively. This is then followed by rules 9 and 10, when membranes 8 and 9 exit membrane 7 by fexo rules replacing $X'_{0,i}$ and $A'_{0,i}$ with Y and x, respectively. If $i > j$, then we obtain $A_{0,j}$ before $X_{0,i}$. In this case, we have a configuration where membrane 7 is inside membrane 9 containing $A_{0,j}$. Then rule 8 is used, replacing $A_{0,j}$ with #, and an infinite computation is obtained (rule 17). If $j > i$, then we obtain $X_{0,i}$ before $A_{0,j}$. In this case, we reach a configuration with $X_{0,i}A_{k,j}$, $k > 0$ in membrane 7, and membrane 7 placed inside membrane 1. Rule 3 cannot be used now, and the only possibility is to use rule 7, which leads to an infinite computation. Thus, if $i = j$, then we can correctly simulate a matrix of type (2).

(ii) For each matrix $m_i' : (X \to Y, B^{(k)} \to \#), X, Y \in N_1, A \in N_2, n_1 + 1 \leq j \leq n_1 + n_2$, $j \in lab_k$, $k = 1,2$, we consider the rules:

11. $[X]_7[\,]_2 \to [[X_i^{(j)}]_7]_2$, $j = 1,2$ (endo)
12. $[\,]_{j+2}[X_i^{(j)}]_7 \to [X_i^{(j)}[\,]_{j+2}]_7$, $j = 1,2$ (fendo)
13. $[\,]_{j+4}[B^{(j)}]_7 \to [\#[\,]_{j+4}]_7$, $j = 1,2$ (fendo)
14. $[X_i^{(j)}[\,]_{j+2}]_7 \to [\,]_{j+2}[Y_j]_7$, $j = 1,2$ (fexo)
15. $[[Y_j]_7]_2 \to [Y]_7[\,]_2$, $j = 1,2$ (exo)

The simulation of matrices of type (3) begins with a rule of type 11. Inside membrane 2, rules 12 and 13 are used, and so membrane $(j+2)$ enters membrane 7, and membrane $(j+4)$ enters membrane 7 if the symbol $B^{(j)}$ is present. In this case, $B^{(j)}$ is replaced with #. Otherwise, membrane $(j+2)$ comes out of the membrane 7 replacing $X_i^{(j)}$ with Y_j. Then membrane 7 exits membrane 2, by replacing Y_j with Y thus successfully simulating a matrix of type (3). We now illustrate (only the membranes appearing in the rules 11-15 and $j = 1$) the evolution of the configurations during one simulation of a type (3) matrix.

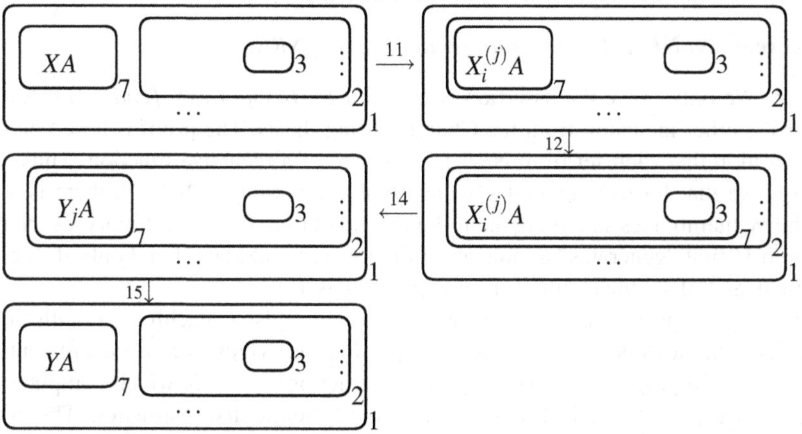

(iii) For a terminal matrix $m_i : (X \to a, A \to x), X \in N_1, a \in T, A \in N_2, x \in T^*$, where $1 \leq i \leq n_1$:

16. $[[a']_7]_1 \to [a]_7[\,]_1$ (exo)
17. $[\,]_8[\#]_7 \to [\#[\,]_8]_7$ (fendo)
$[\#[\,]_8]_7 \to [\,]_8[\#]_7$ (fexo)

Observe that simulation of a matrix of type (4) is similar to that of a matrix of type (2), except that we have an a' in place of Y in rule 9. During the finishing stages of a matrix of type (4) simulation, we use rule 16 to replace a' with a when sending the membrane 7 out of the skin membrane. We now illustrate (only the membranes appearing in the rules 1-8,16-17) the evolution of the configurations during one simulation of a type (4) (terminal) matrix.

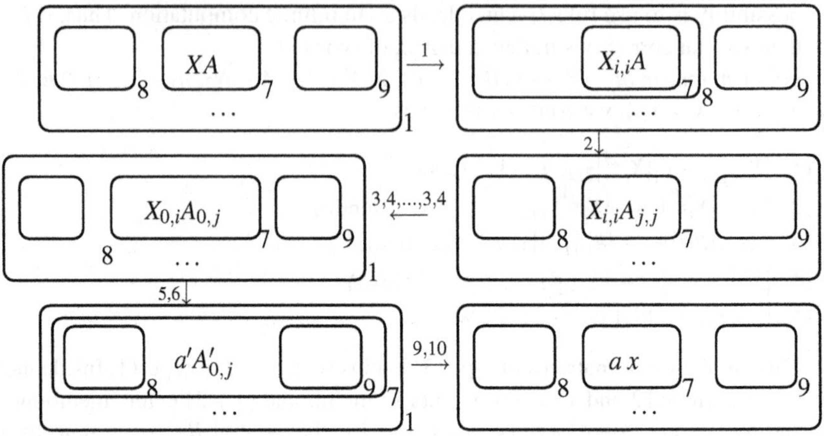

□

The family of all sets of numbers $\mathbf{N}(\Pi)$ which are obtained as a result of a halting computation by a P system Π with enhanced mobile membranes of degree at most n using rules $\alpha \subseteq \{exo, endo, fendo, fexo, pendo, pexo\}$, is denoted by $\mathbf{N}EMM_n(\alpha)$. In what follows we present the results obtained in [61].

Theorem 2.9. $\mathbf{N}EMM_5(endo, exo, fendo, fexo) = \mathbf{N}RE$.

Proof. We only prove the assertion $\mathbf{N}RE \subseteq \mathbf{N}EMM_5(endo, exo, fendo, fexo)$, and infer the other inclusion from the Church-Turing thesis. The proof is based on the observation that each set from $\mathbf{N}RE$ is the range of a recursive function. Thus, we will prove that for each recursively enumerable function $f : \mathbf{N} \to \mathbf{N}$, there is a Π with five membranes satisfying the following condition: For any arbitrary $x \in \mathbf{N}$, the system Π first "generates" a multiset of the form c^x and halts if and only if $f(x)$ is defined, and, if so, the result of the computation is $f(x)$.

In order to prove this assertion, we consider a register machine $\mathcal{M}$ with three registers, the last one being a special output register which is never decremented. Let there be a program P consisting of h instructions $l_1, \ldots, l_h$ which computes f. Let l_h correspond to the instruction HALT and l_1 be the first instruction. The input value x is expected to be in register 1 and the output value in register 3. Without loss of generality, we can assume that all registers other than the first one are empty at the beginning of a computation. We construct the membrane system

$\Pi = (V, \{0,1,2,3,4\}, \{(0,1);(0,2);(0,3);(0,4)\}, \emptyset, \{a_0\}, \{a_1\}, \{K_0\}, \emptyset, R, 3)$ with $V = \{l_i, l_i', l_i'', L_i, L_i' \mid 1 \le i \le h\} \cup \{K_0, a_0, a_1, c\}$. The rules R are:

1. $[K_0]_3[\,]_1 \to [[K_0]_3]_1$ (endo)
   $[[\,]_3 a_0]_1 \to [\,]_3[a_0 c]_1$ (fexo)
   $[[K_0]_3]_1 \to [l_1]_3[\,]_1$ (exo)

Generation of c^x, the initial contents of register 1: Membrane 3 with K_0 enters membrane 1, and comes out each time adding a c to membrane 1. To terminate, K_0 is changed to l_1.

2. $[l_i]_3[\,]_1 \rightarrow [[l_j]_3]_1$ (endo)
   $[[\,]_3 a_0]_1 \rightarrow [\,]_3[a_0 c]_1$ (fexo)
3. $[l_i]_3[\,]_2 \rightarrow [\,[l_j]_3]_2$ (endo)
   $[[\,]_3 a_1]_2 \rightarrow [\,]_3[a_1 c]_2$ (fexo)
4. $[l_i]_3[\,]_4 \rightarrow [[l_j c]_3]_4$ (endo)
   $[[l_j]_3]_4 \rightarrow [l_j]_3[\,]_4$ (exo)

   Simulation of an increment instruction: $l_i : (inc(k), l_j)$. An endo and fexo rule given by rule 2 is used to increment counter 1: membrane 3 enters membrane 1 changing the instruction label, and comes out after adding a c to membrane 1. There is a similar rule (rule 3) between membranes 2, 3 for incrementing counter 2, with a_1 playing the role of a_0 for increment. To increment counter 3, we use the rules (rule 4) between membranes 3 and 4.
5. $[\,]_1[l_i]_3 \rightarrow [[\,]_1 L_i]_3$ (fendo)
6. $[[c]_1]_3 \rightarrow [\,]_1[\,]_3$ (exo)
   $[\,]_4[L_i]_3 \rightarrow [[\,]_4 L_i']_3$ (fendo)
7. $[[\,]_1 L_i']_3 \rightarrow [\,]_1[l_i'']_3$ (fexo)
8. $[[\,]_4 l_i'']_3 \rightarrow [\,]_4[l_i'']_3$ (fexo)
9. $[[\,]_4 L_i']_3 \rightarrow [\,]_4[L_i']_3$ (fexo)
10. $[L_i']_3[\,]_4 \rightarrow [[l_i']_3]_4$ (endo)
11. $[[l_i']_3]_4 \rightarrow [l_i']_3[\,]_4$ (exo)

   Simulation of a decrement instruction $l_i : (dec(1), l_i', l_i'')$. The simulation is initiated by rule 5: when membrane 1 enters membrane 3 by a fendo rule, l_i is replaced with L_i. If there is a c in membrane 1, then membrane 1 exits membrane 3 using rule 6; in parallel, membrane 4 enters membrane 3 using a fendo rule, replacing L_i with L_i'. If there were no c's in membrane 1, then membrane 1 will still be inside membrane 3, hence rule 7 is used, replacing L_i' with l_i'', a fexo rule. Membrane 4 exits membrane 3 irrespective of when membrane 1 exits membrane 3. If the symbol L_i' is present in membrane 3 after both membranes 1 and 4 exit it, then it means that there was a c in membrane 1; this L_i' is now replaced with l_i' using the endo, exo rules 10, 11. Rules for decrementing counter 2 are similar, with membrane 2 playing the role of membrane 1.

If $\mathcal{M}$ halts, then eventually we will have the instruction l_h in membrane 3 and membranes 1, 2 will have the final contents of counters 1, 2. Using the rule $[l_h]_3[\,]_4 \rightarrow [[\,]_3]_4$, the label l_h is erased. If we assign 3 as the output membrane, then its contents will be same as the contents of the output counter 3 at the end of a halting computation. □

Theorem 2.10. $\mathbf{N}EMM_{10}(fendo, fexo) = \mathbf{N}RE$.

Proof. The proof is done by simulating a matrix grammar $G = (N, T, S, M, F)$ with appearance checking in the strong binary normal form. We construct the P system

$$\Pi = (V, \{0, 1, 1', 2, 2', 3, 4, 5, 6, 7\}, \mu, \{XA\}, \emptyset, \ldots, \emptyset, R, 0)$$

with $\mu = \{(7,0);(7,3);(7,4);(7,5);(7,1);(7,1');(7,2);(7,2');(7,6)\}$.

$V = N \cup T \cup \{X_{ij}, A_{ij} \mid 0 \le i,j \le n_1, X \in N_1, A \in N_2\}$
$\cup \{X'_j, X''_j \mid X \in N_1, n_1+1 \le j \le n_1+n_2\} \cup \{Z, \dagger\}$.

Here, XA in membrane 0 corresponds to the initial matrix $(S \to XA)$. Membrane 0 is the output membrane. Let there be n_1 matrices of types (2), (4) in G labelled $1, \ldots n_1$ and n_2 matrices of type (3) in G labelled $n_1+1, \ldots, n_1+n_2$. The rules are

1. $[X]_0[\,]_3 \to [X_{ii}[\,]_3]_0$ (fendo)
   $[A[\,]_3]_0 \to [A_{jj}]_0[\,]_3$ (fexo)
2. $[X_{ki}]_0[\,]_4 \to [X_{k-1i}[\,]_4]_0$ (fendo)
   $[A_{kj}[\,]_4]_0 \to [A_{k-1j}]_0[\,]_4$, $k > 0$ (fexo)
3. $[X_{0,i}]_0[\,]_5 \to [Y[\,]_5]_0$ (fendo)
   $[A_{0,j}[\,]_5]_0 \to [x]_0[\,]_5$ (fexo)
4. $[A_{kj}[\,]_5]_0 \to [\dagger]_0[\,]_5$, $k > 0$ (fexo)
5. $[A_{0j}[\,]_4]_0 \to [\dagger]_0[\,]_4$ (fexo)
6. $[\dagger]_0[\,]_4 \to [\dagger[\,]_4]_0$ (fendo)
   $[\dagger[\,]_4]_0 \to [\dagger]_0[\,]_4$ (fexo)
   Simulation of a type (2) matrix $m_i : (X \to Y, A \to x)$. Rules 1 are used to remember the matrix m_i to be simulated. If X, A belong to the same matrix, then we obtain X_{ii} and A_{ii}. Rules 2 are then used to check if both X, A belong to the same matrix. If yes, then A_{0i} is generated in membrane 0 in the immediate next step after X_{0i}. This is followed by rule 3, by which X_{0i} and A_{0i} are replaced. In case rule 1 gives rise to X_{ii} and A_{jj} with $i \neq j$, then an infinite computation is triggered by rules 4, 5 and 6.
   For $i \in \{1,2\}$, and a matrix $m_j : (X \to Y, B^{(i)} \to \dagger)$ of type (3),
7. $[X]_0[\,]_i \to [X'_j[\,]_i]_0$ (fendo)
8. $[X'_j]_0[\,]_{i'} \to [X''_j[\,]_{i'}]_0$ (fendo)
   $[B^{(i)}[\,]_i]_0 \to [\dagger]_0[\,]_i$ (fexo)
9. $[X''_j[\,]_{i'}]_0 \to [Y]_0[\,]_{i'}$ (fexo)
10. $[Y[\,]_i]_0 \to [Y]_0[\,]_i$ (fexo)
11. $[\dagger]_0[\,]_i \to [\dagger[\,]_i]_0$ (fendo)
    $[\dagger[\,]_i]_0 \to [\dagger]_0[\,]_i$ (fexo)
    Simulation of a type (3) matrix $m_j : (X \to Y, B^{(i)} \to \dagger)$. The membrane labelled i enters membrane 0 replacing X with X'_j. This is followed by two parallel rules: membrane i' entering membrane 0 replacing X'_j with X''_j, and membrane i exiting membrane 0 in the presence of $B^{(i)}$. If $B^{(i)}$ is present, an infinite computation is triggered by rule 11. Membrane i' exits membrane 0 replacing X''_j with Y. If $B^{(i)}$ is absent, then membrane i will be inside membrane 0. In this case, it exits membrane 0 replacing Y with Y.
12. $[Z]_0[\,]_6 \to [[\,]_6]_0$ (fendo)
    $[A[\,]_6]_0 \to [\dagger]_0[\,]_6, A \in N_2$(fexo)
    Simulation of a type (4) matrix $m_j : (X \to \lambda, A \to x)$. This is done using the rules 1-6, replacing X with a new symbol Z. After this, we check if membrane 0 contains any non-terminals, and if so, an infinite computation is triggered by

rule 6. Otherwise, a halting computation is obtained, with membrane 0 containing the output. □

Theorem 2.11. $\mathbf{N}EMM_9(endo, exo, pendo) = \mathbf{N}RE$.

Proof. The proof is done by simulating a matrix grammar $G = (N, T, S, M, F)$ with appearance checking in the strong binary normal form. As in Theorem 2.10, let there be n_1 matrices of types 2, 4 and n_2 matrices of type 3. We construct the P system

$$\Pi = (V, \{0,1,2,3,4,5,6,7,8\}, \mu, w_0, \ldots, w_8, R, 0)$$

with $\mu = \{(8,0); (8,3); (8,4); (8,5); (8,6); (8,7); (6,1); (6,2)\}$
$w_0 = XA$, $w_1 = \alpha$, $w_2 = B^{(1)}B^{(2)}$, $w_3 = \ldots = w_8 = \emptyset$
$V = N \cup T \cup \{X_{ij}, A_{ij} \mid 0 \le i, j \le n_1, X \in N_1, A \in N_2\}$
$\cup \{X_j \mid n_1 + 1 \le j \le n_2\} \cup \{\alpha, \beta\} \cup \{Z, \dagger\}$.
Membrane 0 is the output membrane and XA corresponds to the initial matrix $(S \to XA)$. The rules are

1. $[X]_0[\,]_3 \to [[X_{ii}]_0]_3$ (endo)
   $[[A]_0]_3 \to [A_{jj}]_0[\,]_3$ (exo)
2. $[X_{il}]_0[\,]_4 \to [[X_{i-1l}]_0]_4$ (endo)
   $[[A_{jk}]_0]_4 \to [A_{j-1k}]_0[\,]_4$ for $i, j > 0$ (exo)
3. $[X_{0i}]_0[\,]_5 \to [[Y]_0]_5$ (endo)
   $[[A_{0i}]_0]_5 \to [x]_0[\,]_5$ (exo)
4. $[[A_{jk}]_0]_5 \to [\dagger]_0[\,]_5$ if $j > 0$ (exo)
   $[[A_{0k}]_0]_4 \to [\dagger]_0[\,]_4$ (exo)
5. $[\dagger]_0[\,]_5 \to [[\dagger]_0]_5$ (endo)
   $[[\dagger]_0]_5 \to [\dagger]_0[\,]_5$ (exo)
   Simulation of a type (2) matrix $m_i : (X \to Y, A \to x)$. Similar to Theorem 2.10.
6. $[X]_0[\,]_6 \to [[X_j]_0]_6$ (endo)
   $[X_j B^{(i)}]_0[B^{(i)}]_2 \to [[B^{(i)}]_2 X_j B^{(i)}]_0$ (pendo)
   $[\alpha]_1[]_0 \to [[\alpha]_1]_0$ (endo)
7. $[[B^{(i)}]_2]_0 \to [\dagger]_2[]_0$ (exo)
   $[[\alpha]_1]_0 \to [\beta]_1[\,]_0$ (exo)
8. $[X_j]_0[\beta]_1 \to [[X_j]_0\beta]_1$ (endo)
   $[[X_j]_0]_1 \to [Y]_0[\,]_1$ (exo)
9. $[\beta]_1[Y]_0 \to [[\beta]_1 Y]_0$ (pendo)
   $[[\beta]_1]_0 \to [\alpha]_1[\,]_0$ (exo)
10. $[[Y]_0]_6 \to [Y]_0[\,]_6$ (exo)
11. $[[\beta]_1]_6 \to [\dagger]_1[\,]_6$ (exo)
12. $[[\dagger]_i]_6 \to [\dagger]_i[\,]_6$ (exo)
    $[\dagger]_i[\,]_6 \to [[\dagger]_i]_6$ for $i = 1, 2$ (endo)

Simulation of a type (3) matrix $m_j : (X \to Y, B^{(i)} \to \dagger)$. Membrane 0 enters membrane 6 replacing X with X_j. This is followed by the *pendo*, *endo* rules 6, by which membranes 1, 2 enter 0. Of course, membrane 2 enters only if there is a $B^{(i)}$ in membrane 0. The presence of a $B^{(i)}$ in membrane 0 triggers an infinite computation. Membrane 1 exits membrane 0 replacing α with β. This is followed by a *pendo* rule in 8, by which membrane 0 enters membrane 1. This helps in replacing X_j with Y. Next, β is replaced with α by rule 9. If rule 10 is used before rule 9, we get an infinite computation. Membrane 0 comes out using rule 10, and another matrix can be simulated.

13. $[Z]_0[\,]_7 \rightarrow [[\,]_0]_7$ (endo)
$[[A]_0]_7 \rightarrow [\dagger]_0[\,]_7, A \in N_2$ (exo)
Simulation of a type (4) matrix $m_j : (X \rightarrow \lambda, A \rightarrow x)$. This is similar to Theorem 2.10. □

Theorem 2.12. $NEMM_8(fendo, fexo, pendo) = \mathbf{N}RE$.

Theorem 2.13. $NEMM_{12}(endo, exo, fendo) = \mathbf{N}RE$.

Exercise 2.1. Prove Theorems 2.12 and 2.13 using the techniques from Theorems 2.9 and 2.10.

The following result can be observed from the above Theorems.

Theorem 2.14. $\mathbf{N}MAT \subseteq \mathbf{N}EM_5(endo, exo)$.

Theorem 2.15. *For all $n \in \mathbf{N}$, we have*

$$\mathbf{N}EM_n(rendo, rexo, rfendo, rfexo) \subseteq \mathbf{N}MAT_{ac} \subset \mathbf{N}RE.$$

Proof. Consider the membrane system $\Pi = (V, H, \mu, w_1, \ldots, w_n, i_0)$. We construct a random context matrix grammar $G = (N, T, S, M, F)$ with appearance checking but without λ-rules, and use the result $\mathbf{N}RCM(M, CF - \lambda, ac) = \mathbf{N}MAT_{ac} \subset \mathbf{N}RE$. In our construction, $F = \emptyset$, the appearance checking comes from the forbidden sets R used in the matrices.

Let $V_i = \{a_i \mid a \in V\}$, $V_i' = \{a_i' \mid a \in V\}, V_i'' = \{a_i'' \mid a \in V\}$ for $1 \le i \le n$, and $H = \{(i,j) \mid 1 \le i,j \le n, i \ne j\}$, and $Q = \{E_{j,\emptyset}, E'_{j,\emptyset}, N_{j,list}, N'_{j,list}, N'_{j,list\overline{w}} \mid 1 \le j \le n\}$. By *list*, w, and $list\overline{w}$ we denote strings of length at most n over the symbols $1, \ldots, n$, with no repetition of symbols. Then, $N = Q \cup \{C, E, U, Z, X\} \cup \mathscr{P}(H) \cup V_j \cup V_j' \cup V_j''$ for $1 \le j \le n$ is the set of non-terminals of G, and $T = V \cup \{Y\}$, is the set of terminal symbols, where Y is a new symbol. $\mathscr{P}(H)$ is the power set of all distinct pairs of labels of membranes.

The idea is to construct a grammar that not only simulates Π, but also keeps track of the membrane structure at each step. Let $E_{i,\emptyset}$ denote that membrane i is elementary, and $N_{j,list}$ denote that membrane j is non-elementary, and *list* is the list of its children. One step of Π is simulated by G in several steps: G selects pairs of membranes one after the other in random fashion and checks if there is any applicable rule between them; if so, the rule is used, else the next pair of membranes is selected.

We start with the initial matrix

$$(S \rightarrow CZE_{1,\emptyset} \ldots N_{i,list} \ldots E_{n,\emptyset} w_1 w_j \ldots w_{i_0}, \emptyset, \emptyset)$$

where

1. C is a symbol to choose randomly a pair (i, j) of membranes whose interaction we are going to simulate,

2. $E_{i,\emptyset}, N_{j,list}$ gives the status of membranes in the initial configuration,
3. w_i is the content of membrane i in the initial configuration,
4. Z is a set that keeps track of the pairs of membranes that have interacted so far in the simulation of a step of Π. To simulate one step of Π, we have to check if all rules applicable to all pairs of membranes $(i,j), i \neq j$ have indeed been applied.

The content of the output membrane w_{i_0} is kept at the tail of the string. Z is initialized to $\emptyset$. We randomly keep selecting pairs of membranes to simulate a rule for a step of Π; this is continued until we have examined all possible pairs. The proof idea is essentially to update the symbols $E_{i,\emptyset}, N_{j,list}$ to reflect (i) whether i has been active in a step : we replace $E_{i,\emptyset}$ with $E'_{i,\emptyset}$ if i is elementary; (ii) if i is elementary and some membrane k entered i, we replace $E_{i,\emptyset}$ with $N'_{i,\bar{k}}$; (iii) if i is non-elementary and some membrane k entered i, we replace one of $N_{i,list}, N'_{i,list}$ with $N'_{i,list\bar{k}}$; (iv) if i is non-elementary, and some membrane k left membrane i, we replace one of $N_{i,list}, N'_{i,list}$ with $N'_{i,list-\{k\}}$. The prime on a symbol just tells us that the corresponding membrane has already been a part of a rule. In the case of $E'_{i,\emptyset}$, this means we can no longer utilize i, in the case of $N'_{j,list}$, since j is passive in the current step, j can be part of more than one rule. Let k_i be the number of symbols in membrane i in the initial configuration. This number remains constant, since the rules do not change the length of symbols evolving. We begin to write matrices that simulate Π and keep track of the membrane structure at every step. Assume that we decide to pick the pair of membranes $(1,i)$ first for simulation. Then we have the following matrices:

1. $((C \to C_{1,i}, Z \to Z \cup \{(1,i)\}, a_1 \to b'_1, E_{1,\emptyset} \to E'_{1,\emptyset}, N_{i,list} \to N'_{i,list\bar{1}},$
   $N_{j,list'} \to N'_{j,list'-\{1\}}), \emptyset, \{(1,i)\})$ if there is a *rendo* rule $[a]_1[\ ]_i \to [[b]_1]_i$, and $list'$ contains 1 and i (hence j is the parent of $1, i$). Here we assume i is non-elementary. In case i is elementary, a similar matrix with $N_{i,list}$ replaced with $E_{i,\emptyset}$ can be considered. In that case, we will have
   $((C \to C_{1,i}, Z \to Z \cup \{(1,i)\}, a_1 \to b'_1, E_{1,\emptyset} \to E'_{1,\emptyset}, E_{i,\emptyset} \to N'_{i,\bar{1}},$
   $N_{j,list'} \to N'_{j,list'-\{1\}}), \emptyset, \{(1,i)\})$.
2. $((C \to C_{1,i}, Z \to Z \cup \{(1,i)\}, a_1 \to b'_1, E_{i,\emptyset} \to E'_{i,\emptyset}, E_{1,\emptyset} \to N'_{1,\bar{i}},$
   $N_{j,list'} \to N'_{j,list'-\{i\}}), \emptyset, \{(1,i)\})$ if there is a $rfendo$ rule $[a]_1[\ ]_i \to [b[\ ]_i]_1$, and $list'$ contains 1 and i. Similarly, we can consider a matrix when membrane 1 is non-elementary.
3. $((C \to C_{1,i}, Z \to Z \cup \{(1,i)\}, a_1 \to b'_1, E_{1,\emptyset} \to E'_{1,\emptyset}, N_{i,list} \to N'_{i,list-\{1\}},$
   $N_{j,list'} \to N'_{j,list'\bar{1}}), \emptyset, \{(1,i)\})$ if there is a *rexo* rule $[[a]_1]_i \to [b]_1[\ ]_i$ and j is the parent of i; i.e, $list'$ contains i.
4. $((C \to C_{1,i}, Z \to Z \cup \{(1,i)\}, a_i \to b'_i, E_{1,\emptyset} \to E'_{1,\emptyset}, N_{i,list} \to N'_{i,list-\{1\}},$
   $N_{j,list'} \to N'_{j,list'\bar{1}}), \emptyset, \{(1,i)\})$ if there is a $rfexo$ rule $[a[\ ]_1]_i \to [\ ]_1[b]_i$ and j is the parent of i; i.e, $list'$ contains i.
5. We can consider $rexo, rfexo$ when i is a child of 1. This is similar to the above and we do not give details.

Now we consider cases when any rule cannot be used for a pair of membranes $(1,i)$. Recall that the number of objects in any membrane remains constant and there are k_i objects in membrane i. In the case when 1, i are adjacent, first check all symbols in membrane 1 to check that they are not involved in any rule. This is followed by checking all symbols of membrane i. A total of $(k_1+k_i)\cdot|R|$ steps are required where R is the set of rules of Π. Assume 1, i are siblings, 1 is elementary and i is non-elementary.

6. $((C \to C^1_{1,i}, Z \to Z\cup\{(1,i)\}, a_1 \to a''_1), \{N_{j,list'}, E_{1,\emptyset}, N_{i,list}\}, \emptyset)$ if *list'* contains both 1 and i, and there is no *endo* rule involving a in membrane 1 between $1,i$.

   Continue incrementing the superscript $C^l_{1,i}$ to $C^{l+1}_{1,i}$ until we finish checking all symbols of membranes 1 and i.
7. $((C^l_{1,i} \to C^{l+1}_{1,i}, b_1 \to b''_1), \{N_{j,list'}, E_{1,\emptyset}, N_{i,list}\}, \emptyset)$ if *list'* contains both 1 and i, and there is no *endo* rule involving b in membrane 1 between $1,i$, and $l<k_1$.
8. $((C^{k_1}_{1,i} \to C^{k_1+1}_{1,i}, a_i \to a''_i), \{N_{j,list'}, E_{1,\emptyset}, N_{i,list}\}, \emptyset)$ if *list'* contains both 1 and i, and there is no *fendo* rule involving a in membrane i between $1,i$.

   Once the superscript reaches k_1+k_i, we are done in checking. As $(1,i)$ is added to Z, we remember that this pair is checked already. Now, to make the symbols of membranes $1,i$ available for rules with other membranes, we unprime them.
9. $((C^{k_1+k_2}_{1,i} \to D_{1,i}, a''_1 \to a_1), \emptyset, \emptyset)$
10. $((D_{1,i} \to D_{1,i}, a''_1 \to a_1), \emptyset, \emptyset)$

    Rules 6-10 consider the case when membranes $1,i$ are siblings, but cannot participate in a rule with each other. We run through all elements in membranes $1,i$ to make sure of this. A similar set of rules can be written when 1 is non-elementary and i is elementary. If $1,i$ are siblings, and both are non-elementary, then we cannot use them in any mutual rule. In this case, it is enough to have a matrix that directly produces $D_{1,i}$ checking the presence of $N_{1,list}, N_{i,list'}$ (or their primed versions) and $N_{j,list''}$ (or its primed version) such that *list''* contains both $1,i$. Another possibility to consider is when $1,i$ are such that one is a child of the other and we cannot employ any rules between them. This is done in a similar manner to rules 6-10. The third possibility is that $1,i$ are not adjacent. Then we have
11. $((C \to D_{1,i}, Z \to Z\cup\{(1,i)\}), \emptyset, \{N_{j,list}, N_{1,list'}, N_{i,list''} \mid j \neq 1,i\})$

    We check all the $N_{\alpha,list}$ symbols to ensure that $1,i$ are not adjacent. Thus, we need to put these in the forbidden symbols. Here *list* contains membranes $1,i$, while *list'* contains membrane i; *list''* contains membrane 1.

The next pair of membranes will be chosen by replacing $D_{1,i}$ with an appropriate symbol (either $D_{r,s}$ or $C_{r,s}$ or $C^1_{r,s}$ as is the case); we must check that all double primed symbols are replaced - this is achieved by placing $V''_1 \cup V''_i$ in the forbidden list.

The above rules represent the first choice of a pair of membranes to begin a simulation. In general, for a pair of membranes r,s,

(a) For the case when a rule is applicable between the pair r,s, we will have a matrix

$$((D_{i,j} \rightarrow C_{r,s}, Z \rightarrow Z \cup \{(r,s)\}, rules\ \ as\ \ appropriate), \emptyset, \{(r,s)\} \cup V_i'' \cup V_j'')$$

or

$$((C_{i,j} \rightarrow C_{r,s}, Z \rightarrow Z \cup \{(r,s)\}, rules\ \ as\ \ appropriate), \emptyset, \{(r,s)\})$$

or

$$((C \rightarrow C_{r,s}, Z \rightarrow Z \cup \{(r,s)\}, rules\ \ as\ \ appropriate), \emptyset, \{(r,s)\})$$

The third matrix represents the case when we begin a round of simulation of Π. In all matrices, we will choose rule depending on the current relationship between r,s. The rules are based on the following guidelines:

- When r,s are siblings and not both are non-elementary: A matrix each for the cases when there is a $rendo, rfendo$ rule between r,s. Things to look out for: replace a_r or a_s as is the case by b_r' or b_s' (the prime on the symbols is so that they do not get used by another pair involving r or s); change the status of r, s: (i) if we have $E_{r,\emptyset}$, and if r is active in that step, indicate it by changing it into $E_{r,\emptyset}'$; update the parent information of r, and also the child list of the former parent of r, s; similarly if s is active - this covers the case when r,s are both elementary (ii) if r is passive and s is active, then we may have (a) $N_{r,list}$ (this indicates that so far nothing has entered r or left r) or (b) $N_{r,list}'$ (this indicates that something left r) or (c) $N_{r,list\bar{w}}'$ (this indicates that everything in w entered r previously in this step). Update these symbols appropriately: In cases (a), (b) we get $N_{r,list\bar{s}}'$, and in case (c), we get $N_{r,list\bar{w}\bar{s}}'$. The case when s is active and r is passive is similar. Update the parent j of r,s by deleting from j's list r or s: if we had $N_{j,list}$, it becomes $N_{j,list-\{r\}}'$ or $N_{j,list-\{s\}}'$, if we had $N_{j,list}'$, it becomes $N_{j,list-\{r\}}'$ or $N_{j,list-\{s\}}'$ and if we had $N_{j,list\bar{w}}'$, it becomes $N_{j,list-\{r\}\bar{w}}'$ or $N_{j,list-\{s\}\bar{w}}'$.
- When r is a parent of s and s is elementary: A matrix each for the cases of $rfexo, rexo$. Updates to be done similarly to the above case.
- When s is a parent of r and r elementary: A matrix each for the cases of $rfexo, rexo$. Updates to be done similarly to the above case.

(b) When there is no applicable rule between r,s:

Case (i): r,s are adjacent, and both are non-elementary. In this case, we have a matrix which directly gives $D_{r,s}$; Z is updated to contain (r,s). The permitting context is either $\{N_{r,list}, N_{s,list'}, N_{k,list''}\}$ such that $list''$ contains r,s, or is $\{N_{r,list}, N_{s,list'}\}$ such that $list'$ contains r or $list$ contains s. The permitting context can also be a set consisting of primed versions of these symbols; the only point to note is that we do not have a symbol α and its primed version α' together. The forbidding context contains (r,s).

Case (ii) r,s are adjacent but not both are non-elementary. We have a matrix of one of the forms

$$((C_{i,j} \rightarrow C_{r,s}^1, Z \rightarrow Z \cup \{(r,s)\}, rules\ \ as\ \ appropriate), A, \{(r,s)\})$$

or

$$((D_{i,j} \to C^1_{r,s}, Z \to Z \cup \{(r,s)\}, rules\ as\ appropriate), A, \{(r,s)\} \cup V_i'' \cup V_j'')$$

or

$$((C \to C^1_{r,s}, Z \to Z \cup \{(r,s)\}, rules\ as\ appropriate), A, \{(r,s)\})$$

The last matrix represents the beginning of a round of simulation of Π. We are checking all forms of adjacency : siblings, parent or child. Here we consider $A = \{E_{r,\emptyset}, N_{s,list'}, N_{j,list''}\}$ or $\{N_{r,\emptyset}, E_{s,list'}, N_{j,list''}\}$ where $list''$ contains both r, s or $A = \{E_{r,\emptyset}, N_{s,list'}\}$ or $\{E_{s,\emptyset}, N_{r,list''}\}$ where $list'$ contains r and $list''$ contains s. For brevity, we explain only these cases for A here: the case when the members of A are primed also should be included - the only point to notice is that a symbol and its prime are not both together in A. The rules are chosen by the following guidelines: the permitting context A checks that r,s are indeed adjacent. For each symbol a_r, we check that there is no applicable rule with respect to s, and increment the counter till k_r. We have matrices for each of these, until we reach counter k_r. Note that since the number of rules in Π and the number of symbols in membrane r are both finite, this can be done. Then we check for each symbol of s that no rule is applicable till the counter reaches $k_r + k_s$. At this point, we switch to $D_{r,s}$ and unprime the symbols of r, s, and finally $D_{r,s}$ is replaced with some symbol that kick starts the simulation for another pair.

Case (iii) r,s are not adjacent. We check this is the case by having a matrix of the form

$$((C_{i,j} \to D_{r,s}, Z \to Z \cup \{(r,s)\}), \emptyset, \{(r,s)\} \cup B)$$

or

$$((D_{i,j} \to D_{r,s}, Z \to Z \cup \{(r,s)\}), \emptyset, \{(r,s)\} \cup V_i'' \cup V_j'' \cup B)$$

or

$$((C \to D_{r,s}, Z \to Z \cup \{(r,s)\}), \emptyset, \{(r,s)\} \cup B)$$

The last matrix represents the beginning of a round of simulation of Π. Let

$$B = \{N_{r,list}, N'_{r,list}, N'_{r,list\bar{w}}, N_{s,list'}, N'_{s,list'}, N'_{s,list'\bar{y}}, N_{j,list''}, N'_{j,list''}, N'_{j,list''\bar{x}}\}$$

where j is a membrane different from r,s, and $list$ contains s, $list'$ contains r, and $list''$ contains r,s. The matrix says it all; the absence of symbols of B, which spans all possibilities of r,s being adjacent, is enough. We add (r,s) to Z, so that we know that this pair has already been considered.

To terminate simulation of one step of Π, we do the following:

1. $((C_{p,q} \to E), \{Z \mid |Z| = \binom{n}{2} - n\}, \emptyset)$ or

   $((D_{p,q} \to E), \{Z \mid |Z| = \binom{n}{2} - n\}, V_p'' \cup V_q'')$

   At some point, we check that all pairs of membranes have been examined. This is the case if the size of Z is $\binom{n}{2} - n$.

2. $((Z \to \emptyset), \{E\}, \emptyset)$, $((a'_i \to a_i), \{E\}, \emptyset)$ for all i,
3. $((E'_{i,\emptyset} \to E_{i,\emptyset}), \{E\}, \emptyset, \emptyset)$,
   In the presence of E, unprime all the primed symbols, replace Z with $\emptyset$
4. $((N'_{j,list\bar{w}} \to N_{j,list'}), \{E\}, \emptyset)$, where $list'$ is $list$ union w,
5. $((N'_{j,list} \to N_{j,list}), \{E\}, \emptyset)$ if $list \neq \emptyset$, $((N'_{j,list} \to E_{j,\emptyset}), \{E\}, \emptyset)$ if $list = \emptyset$
   Update the lists appropriately after a round of simulation
6. $((E \to C), \emptyset, \{$primed symbols like $a'_i, N'_{j,list}, E'_{j,\emptyset}, N'_{j,list\bar{w}}\})$
   Once all updates are completed, replace E with C to start the next round of simulation. The absence of primed symbols tells us that all updates are complete.

To terminate the simulation of G, we guess that Π has reached a halting configuration and validate our guess.

1. $((C \to U), \emptyset, \emptyset)$,
2. For $i \neq i_0$ and i is elementary: $((a_i \to X), W \cup \{U\}, \emptyset)$ where W is the set of all symbols $E_{i,\emptyset}$, $N_{j,list}$ such that $list$ contains i, and there are no *rexo* rules in Π between a in membrane i and membrane j, and for all $l \in list$, there are no $rendo, rfendo$ rules in Π between a in membrane i and membrane l. This can be checked in finite time: Π has a finite collection of rules, and the number of symbols in the current sentential form of G is finite : $(k_1 + \cdots + k_n) + n + 2$. Comparing each symbol a_i with the rules of Π pertaining to the appropriate membrane l in $list$ and membrane j is enough to apply this matrix.
3. i_0 is elementary: $((a_{i_0} \to a), W \cup \{U\}, \emptyset)$ where W is similar to above,
4. For $i \neq i_0$ and i is non-elementary: $((a_i \to X), W' \cup \{U\}, \emptyset)$ where W' is the set of all symbols $N_{i,list}, N_{k,list''}$ such that $list''$ contains i, and there are no $rfendo$ rules in Π between a in membrane i and membranes in $list''$, there are no $rfexo$ rules between a in membrane i and membranes in $list$. As in rule 1, this can be checked in finite time.
5. i_0 is non-elementary: $((a_{i_0} \to a), W' \cup \{U\}, \emptyset)$ where W' is similar to above,
6. $((U \to X), \emptyset, V_i)$,
   After all symbols a_i are replaced with X or a, we replace U with X.
7. $((N_{j,list} \to X), \{X\}, V_i \cup \{U\})$, $((E_{j,\emptyset} \to X), \{X\}, V_i \cup \{U\})$,
   $((Z \to X), \{X\}, V_i \cup \{U\})$
   After U has also been replaced, replace all the symbols $N_{j,list}$ and $E_{j,\emptyset}, Z$ with X.
8. $((X \to Y), \emptyset, N \setminus \{X\})$
   Replace all occurrences of X in the absence of all non-terminals different from X.

To terminate, we have to make sure that we have reached a halting computation. For this, at some non-deterministic point of time, we replace C with U. Then for all membranes other than the output, we replace the contents with symbols X after making sure that they have no rule to participate in given the current configuration. We replace symbols a_{i_0} of i_0 with a after ensuring the same. We are then left with a string of the form $XZN_{1,list} \ldots E_{n,\emptyset} X \ldots Xw$, where w is the contents of i_0. Then we also replace the symbols $N_{j,list}, E_{i,\emptyset}, Z$ with X. Then the number of symbols X we have is $\kappa = n + 2 + (k_1 + \cdots + k_n) - k_{i_0}$ which is finite. When we get a string of the form $X^\kappa w$, made over X and V, we replace all occurrences of X with Y. Since the family MAT_{ac} is closed under quotient by letters, we do the quotient operation κ times on $L(G)$. Then we have $\mathbf{N}(\partial_Y^{\kappa}(L)) = \mathbf{N}(\Pi)$.

Note that if our guess is incorrect, i.e, if we replace C with U before Π has reached a halting configuration, we will not be able to proceed further, since the remaining matrices will not be applicable. In this case, we do not obtain a terminal string.

Note: In the above matrices, for ease of notation, we have kept the symbol Z in the rules and updated it, in an algorithmic style. In pure grammar notation, this can be thought of as replacing the current set with a larger set, and in the last step of a simulation, replacing a set of size $\binom{n}{2} - n$ with $\emptyset$. □

Theorem 2.16. *For all $n \in \mathbf{N}$, we have*
$\mathbf{N}EM_n(rendo, rexo, rfendo, rfexo, pendo, pexo) \subseteq \mathbf{N}MAT_{ac} \subset \mathbf{N}RE.$

Proof. A minor change to the proof of Theorem 2.15 will do. The only extra book-keeping that needs to be done is to keep track of a *pendo*, *pexo* rule that has happened. After choosing a pair (r,s) of membranes, we check if there are any *pendo*, *pexo* rules which could be simulated as follows:

1. If r,s are siblings: If there exist symbols in membranes r,s such that r is entering s (*pendo*), then prime these symbols, prime the symbol $E_{r,\emptyset}$, change $N_{s,list}$ to $N_{s,list\bar{r}}$ or $E_{s,\emptyset}$ to $N_{s,\bar{r}}$, and continue as usual. The case of a *pexo* is similar.
2. In case we have chosen r,s and there are no *pendo*, *pexo* rules applicable to them, after checking that there are no *rendo*, *rexo*, *rfendo*, *rfexo* rules applicable to them, do a check for non-applicability of *pendo*, *pexo* rules. We have the symbol $D_{r,s}$ after checking for non-applicability of *rendo*, *rexo*, *rfendo*, *rfexo*. In case r, s are adjacent, this will look like

$$((D_{r,s} \to E_{r,s}, a''_r \to a'''_r, N_{j,list} \text{ or } N'_{j,list} \to \text{ itself}), \emptyset, B)$$

where $B = \{b''_r, c''_s \mid$ there is a pendo rule involving a,b of membrane r and symbol c of membrane $s\}$. Continue this process by replacing a double primed symbol with a triple primed symbol, provided it is not part of a context that will enable a *pendo*, *pexo* rule. $E_{r,s}$ is retained until this finishes. The exact description of B depends on the rules of Π. After replacing all symbols of r,s with triple primed symbols, replace $E_{r,s}$ with the next choice of membranes. Similarly check to ensure non-applicability of *pexo* rules. □

2.4.4 Mutual Mobile Membranes

The family of all sets $Ps(\Pi)$ generated by systems of n mobile membranes using the mutual endocytosis rule *mendo* and the mutual exocytosis rule *mexo* is denoted by $PsMM_n(mendo, mexo)$. We denote by $PsRE$ the family of Turing computable sets of vectors generated by arbitrary grammars.

We prove that it is enough to consider only systems with three mobile membranes together with the operations of mutual endocytosis and mutual exocytosis to get the full computational power of a Turing machine. The proof is done in a similar manner to the proof of the computational universality of enhanced mobile membranes [107].

Theorem 2.17. *$PsMM_3(mendo, mexo) = PsRE$.*

Proof. It is proved in [90] that each recursively enumerable language can be generated by a matrix grammar in the strong binary normal form. We consider a matrix grammar $G = (N,T,S,M,F)$ in the strong binary normal form, having n_1 matrices $m_1,\dots,m_{n_1}$ of types (2) and (4), and n_2 matrices of type (3) labelled by m'_i, $n_1+1 \le i \le n_1+n_2$ with $i \in lab_j$, for $j \in \{1,2\}$, such that lab_1, lab_2, and $lab_0 = \{1,2,\dots,n_1\}$ are mutually disjoint sets. The initial matrix is $m_0 : (S \to XA)$.

We construct a system $\Pi = (V,H,\mu,w_1,w_2,w_3,R,2)$ with three mutual mobile membranes, where

$V = N \cup T \cup \{\alpha,\overline{\alpha},\beta,\overline{\beta}\}$
$\quad \cup\{X'_{0i}, A'_{0i} \mid X \in N_1, A \in N_2, 1 \le i \le n_1\}$
$\quad \cup\{X_{ji}, A_{ji} \mid 0 \le i,j \le n_1\}$
$\quad \cup\{\beta_j, \overline{\beta}_j, X_i^{(j)}, Y_j \mid X \in N_1, j \in \{1,2\}, 1 \le i \le n_2\}$
$H = \{1,2,3\}$; $\quad \mu = \{(1,2);(1,3)\}$
$w_1 = \emptyset$; $\quad w_2 = \overline{\alpha}\,\overline{\alpha}_1\,\overline{\alpha}_2\beta\beta_1\beta_2XA$; $\quad w_3 = \alpha\alpha_1\alpha_2\overline{\beta}\,\overline{\beta}_1\,\overline{\beta}_2$

where $(S \to XA)$ is the initial matrix and the set R is constructed as follows:

(i) For each (nonterminal) matrix $m_i : (X \to Y, A \to x)$, $X,Y \in N_1$, $x \in (N_2 \cup T)^*$, $A \in N_2$, with $1 \le i \le n_1$, we consider the rules:

1. $[X\beta]_2[\overline{\beta}]_3 \to [[X_{ii}\beta]_2\overline{\beta}]_3$ (mendo)
   $[[A\beta]_2\overline{\beta}]_3 \to [A_{jj}\beta]_2[\overline{\beta}]_3$ (mexo)
2. $[X_{ki}\beta]_2[\overline{\beta}]_3 \to [[X_{k-1i}\beta]_2\overline{\beta}]_3$, $k > 0$ (mendo)
   $[[A_{kj}\beta]_2]\overline{\beta}]_3 \to [A_{k-1j}\beta]_2[\overline{\beta}]_3$, $k > 0$ (mexo)
3. $[X_{0i}A_{0j}\beta]_2[\overline{\beta}]_3 \to [[X'_{0i}A'_{0j}\beta]_2\overline{\beta}]_3$ (mendo)
   $[[X'_{0i}A'_{0j}\beta]_2\overline{\beta}]_3 \to [Yx\beta]_2[\overline{\beta}]_3$ (mexo)
4. $[[X_{ki}A_{0j}\beta]_2\overline{\beta}]_3 \to [\#\beta]_2[\overline{\beta}]_3$, $k > 0$ (mexo)
5. $[X_{0i}A_{kj}\beta]_2[\overline{\beta}]_3 \to [[\#\beta]_2\overline{\beta}]_3$, $k > 0$ (mendo)

By rule 1, membrane 2 enters membrane 3, replacing $X \in N_1$ with X_{ii}. A symbol $A \in N_2$ is replaced with A_{jj}, and membrane 2 comes out of membrane 3. The subscripts represent the matrices $m_i(m_j)$, $1 \le i,j \le n_1$ corresponding to which X, A have a rule. Next, rule 2 is used until X_{ii} and A_{jj} become X_{0i} and A_{0j}, respectively. If $i = j$, then we have X_{0i} and A_{0j} simultaneously in membrane 2. Then rule 3 is used, by which membrane 2 enters membrane 3 replacing X_{0i} and A_{0j} with X'_{0i} and A'_{0j}, while X'_{0i} and A'_{0j} are replaced with Y and x when membrane 2 exits membrane 3. If $i > j$, then we obtain A_{0j} before X_{0i}. In this case, we have a configuration where membrane 2 is inside membrane 3 containing A_{0j}. Rule 2 cannot be used now, and the only possibility is to use rule 4, replacing X_{ki} and A_{0j} with #, which leads to an infinite computation (due to rule 12). If $j > i$, then we obtain X_{0i} before A_{0j}. In this case, we reach a configuration with $X_{0i}A_{kj}$, $k > 0$ in membrane 2, and membrane 2 is in the skin membrane. Rule 2 cannot be used now, and the only possibility is to use rule 5, replacing X_{0i} and A_{kj} with #, which leads to an infinite computation. In this way, we correctly simulate a type (2) matrix whenever $i = j$. We now illustrate the evolution of the configurations during one simulation of a type (2) matrix, for $i = j$.

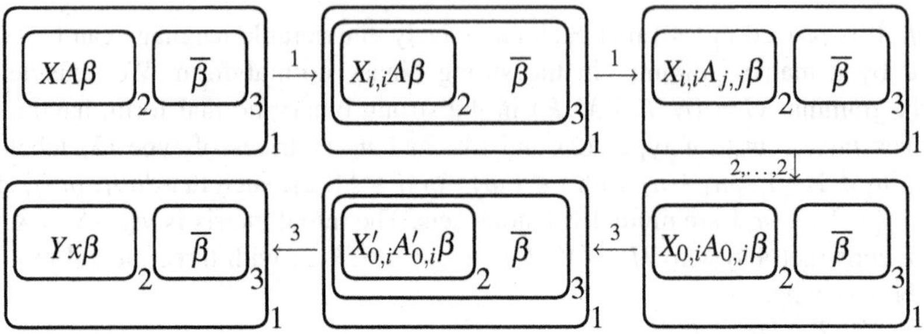

(ii) For each matrix $m'_i : (X \to Y, B^{(j)} \to \#), X, Y \in N_1, B^{(j)} \in N_2, n_1 + 1 \le i \le n_1 + n_2$, $i \in lab_j$, $j = 1, 2$, we consider the rules:

6. $[X\beta_j]_2[\overline{\beta}_j]_3 \to [[X_i^{(j)}\beta_j]_2\overline{\beta}_j]_3$ (mendo)
   $[[X_i^{(j)}\beta_j]_2\overline{\beta}_j]_3 \to [X_i^{(j)}\beta_j]_2[\overline{\beta}_j]_3$ (mexo)
7. $[B^{(j)}\beta_j]_2[\overline{\beta}_j]_3 \to [[\#\beta_j]_2\overline{\beta}_j]_3$ (mendo)
8. $[X_i^{(j)}\beta_j]_2[\overline{\beta}_j]_3 \to [[Y_j\beta_j]_2\overline{\beta}_j]_3$ (mendo)
9. $[[Y_j\beta_j]_2\overline{\beta}_j]_3 \to [Y\beta_j]_2[\overline{\beta}_j]_3$ (mexo)

Membrane 2 enters membrane 3 by rule 6, and creates an object $X_i^{(j)}$ depending on whether it has the symbol $B^{(j)}$, $j = 1, 2$ associated with it, and then exits with the newly created object. Next, by rule 7, membrane 2 enters membrane 3 if the object $B^{(j)}$ is present, replacing it with #. If this rule is applied, membrane 2 exits membrane 3 by applying rule 12. Regardless of the existence of object $B^{(j)}$, membrane 2 enters membrane 3 replacing $X_i^{(j)}$ with Y_j. Membrane 2 exits membrane 3, replacing Y_j with Y, successfully simulating a matrix of type (3). We now illustrate the evolution of the configurations during one simulation of a type (3) matrix.

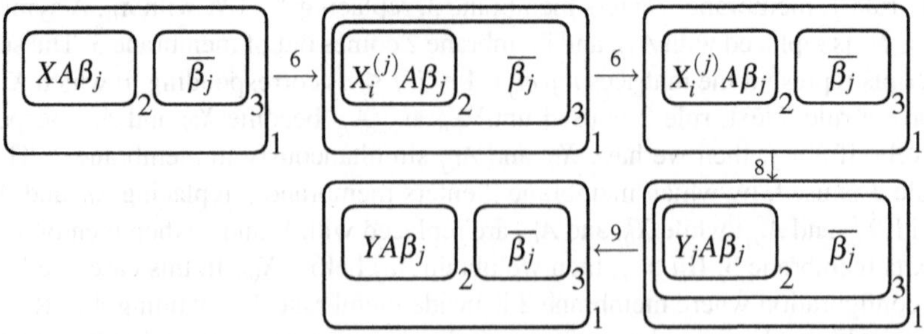

(iii) For a terminal matrix $m_i : (X \to \lambda, A \to x)$, $X \in N_1$, $A \in N_2$, $x \in T^*$, $1 \le i \le n_1$. We begin with rules 1-5 as before and simulate the matrix $(X \to Z, A \to x)$ in place of m_i, where Z is a new symbol.

10. $[\alpha]_3[Z\overline{\alpha}]_2 \to [\lambda\overline{\alpha}[\alpha]_3]_2$ (mendo)
11. $[A\overline{\alpha}[\alpha]_3]_2 \to [\alpha]_3[\#\overline{\alpha}]$ (mexo)
12. $[\#\beta]_2[\overline{\beta}]_3 \to [[\#\beta]_2\overline{\beta}]_3$ (mendo)
    $[[\#\beta]_2\overline{\beta}]_3 \to [\#\beta]_2[\overline{\beta}]_3$ (mexo)

Now we use rule 10 to erase the symbol Z while membrane 3 enters membrane 2. This is followed by rule 11 if there are any more symbols $A \in N_2$ remaining, in which case they are replaced with the trap symbol #. An infinite computation is obtained if the symbol # is present in membrane 2. It is clear that if the computation proceeds correctly, then membrane 2 contains a multiset of terminal symbols $x \in T^*$. In this way we can conclude that $Ps(\Pi)$ equals $Ps(L(G))$. We now illustrate the evolution of the configurations during one simulation of a type (4) (terminal) matrix.

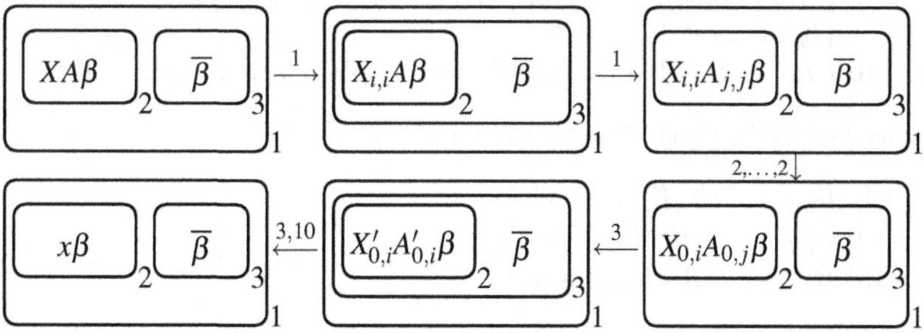

□

It is worth noting that three is the smallest number of membranes when using effectively the movement of membranes given by endocytosis and exocytosis.

It is reasonable to investigate whether we can obtain new computability results using parallel mechanisms instead of sequential mechanisms. For systems of mobile membranes using mutual endocytosis and mutual exocytosis, we get the same computation power, but the results can be obtained more efficiently using parallel mechanisms. The following proof links parallel systems of mutual mobile membranes to sequential register machines. The register machines work in a slow and biologically unrealistic way; the results show that it is possible to get similar results with parallel mechanisms (based on the Church-Turing thesis).

Considering the number of objects and reduction to a register machine, we prove that the family **N***RE* of all sets of natural numbers generated by arbitrary grammars is the same as the family $\mathbf{N}MM_3(mendo, mexo)$ of all sets of natural numbers generated by systems with three mobile membranes using *mendo* and *mexo* rules. This is calculated by looking at the cardinality of the objects in a specified *output membrane* of the mutual mobile membrane system at the end of a halting configuration.

Theorem 2.18. $\mathbf{N}MM_3(mendo, mexo) = \mathbf{N}RE$.

Proof. In view of the Church-Turing thesis, only $\mathbf{N}RE \subseteq \mathbf{N}MM_3(mendo, mexo)$ has to be proved. The proof is based on the observation that each set from $\mathbf{N}RE$ is the range of a partial recursive function. Thus, we prove that for each partial recursive function $f : \mathbf{N} \to \mathbf{N}$, there is a mutual mobile membrane system Π with three membranes satisfying the following condition: for any arbitrary $x \in \mathbf{N}$, the system first "generates" a multiset of the form o_1^x and halts if and only if $f(x)$ is defined, and, if so, the result is $f(x)$.

In order to prove the assertion, using similar arguments as in Ibarra et al [96], we can assume that the output register is never decremented during computation. This happens without loss of generality. Let there be a program P consisting of h instructions $P_1, \ldots, P_h$ which computes f. Let P_h correspond to the instruction HALT and P_1 be the first instruction. The input value x is expected to be in register 1 and the output value in register 3.

We construct a mutual mobile membrane $\Pi = (V, H, \mu, w_0, w_I, w_{op}, R, I)$:
$V = \{s\} \cup \{o_r \mid 1 \le r \le 3\} \cup \{P_k, P'_k \mid 1 \le k \le h\} \cup \{\beta, \overline{\beta}, \gamma, \overline{\gamma}\}$
$\quad \cup \{\beta_r \mid 1 \le r \le 3\}$
$H = \{0, I, op\} \quad \mu = \{(0, I); (0, op)\} \quad w_I = s\beta\overline{\gamma} \quad w_0 = \emptyset \quad w_{op} = \overline{\beta}\,\gamma$

(i) Generation of the initial contents x of register 1:

1. $[s\beta]_I[\overline{\beta}]_{op} \to [[s\beta]_I\overline{\beta}]_{op}$ (mendo)
   $[[s\beta]_I\overline{\beta}]_{op} \to [so_1\beta]_I[\overline{\beta}]_{op}$ (mexo)
2. $[[s\beta]_I\overline{\beta}]_{op} \to [P_1\beta]_I[\overline{\beta}]_{op}$ (mexo)

Rule 1 can be used any number of times, generating a number x (o_1^x) as the initial content of register 1. Rule 2 replaces s with the initial instruction P_1, and we are ready for the simulation of the register machine.

(ii) Simulation of an add rule $P_i = (INC(r), j)$, $1 \le r \le 3$, $1 \le i < h$, $1 \le j \le h$

3. $[P_i\beta]_I[\overline{\beta}]_{op} \to [[P_i\beta]_I\overline{\beta}]_{op}$ (mendo)
4. $[[P_i\beta]_I\overline{\beta}]_{op} \to [P_jo_r\beta]_I[\overline{\beta}]_{op}$ (mexo)

Membrane I enters membrane op using rule 3, and then exits it by replacing P_i with P_jo_r (rule 4), thus simulating an add instruction.

(iii) Simulation of a subtract rule $P_i = (DEC(r), j, k)$, $1 \le r \le 3$, $1 \le i < h$, $1 \le j, k \le h$

5. $[[P_i\beta]_I\overline{\beta}]_{op} \to [P'_j\beta_r\beta]_I[\overline{\beta}]_{op}$ (mexo)
6. $[o_r\beta_r\beta]_I[\overline{\beta}]_{op} \to [[\beta]_I\overline{\beta}]_{op}$ (mendo), otherwise
   $[P'_j\beta_r\beta]_I[\overline{\beta}]_{op} \to [[P'_k\beta]_I\overline{\beta}]_{op}$ (mendo)
7. $[[P'_j\beta]_I\overline{\beta}]_{op} \to [P_j\beta]_I[\overline{\beta}]_{op}$ (mexo)
   $[[P'_k\beta]_I\overline{\beta}]_{op} \to [P_k\beta]_I[\overline{\beta}]_{op}$ (mexo)

To simulate a subtract instruction, we start with rule 3, with membrane I entering membrane op. Then rule 5 is used, by which P_i is replaced with $P'_j\beta_r$, and membrane I exits membrane op. The newly created object β_r denotes the register which has to be decreased. If there is an o_r in membrane I, then by rule 6 the object o_r is erased together with β_r, and membrane I enters membrane op. This is followed by rule 7, where P'_j is replaced with P_j and membrane I is back inside the skin membrane. If there are no o_r's in membrane I, then by applying rule 6, P'_j together with β_r is replaced by P'_k. This is followed by rule 7, where P'_k is replaced with P_k and membrane I is inside the skin membrane, thus simulating a subtract instruction.

(iv) Halting:

8. $[\gamma]_{op}[P_h\overline{\gamma}]_I \rightarrow [[\gamma]_{op}\overline{\gamma}]_I$ (mendo)

To halt the computation, the halt instruction P_h must be obtained. Once we obtain P_h in membrane I, membrane op enters membrane I and the computation stops (rule 8). When the system halts, membrane I contains only o_3's, the content of register 3. □

This result reveals a different technique in proving the computational power of a system with three mutual mobile membranes.

There are many families of languages included in RE. Although Theorem 2.18 is valid for all of them, these families can have particular sets of rules simulating them. We exemplify this aspect by an effective construction of a system with three mutual membranes able to simulate an ET0L system in the normal form.

In order to get the power of an ET0L system by using the operations of mutual endocytosis and mutual exocytosis, we need only three membranes.

Proposition 2.1. *$PsET0L \subseteq PsMM_3(mendo, mexo)$.*

Proof. In what follows, we use the following normal form: each language $L \in ET0L$ can be generated by $G = (V, T, \omega, R_1, R_2)$. Moreover, from [141], any derivation starts with several steps of R_1, then R_2 is used exactly once, and the process is iterated; the derivation ends by using R_2.

Let $G = (V, T, \omega, R_1, R_2)$ be an ET0L system in the normal form. We construct the mutual mobile membrane system

$$\Pi = (V', H, \mu, w_0, w_1, w_2, R, 0)$$

as follows:

$V' = \{\dagger, \alpha, \overline{\alpha}, \beta, \overline{\beta}\} \cup \{\beta_i, \overline{\beta}_i \mid i = 1, 2\} \cup V \cup V_i$, where $V_i = \{a_i \mid a \in V\}, i = 1, 2$

$H = \{0, 1, 2\}$ $\quad \mu = \{(2,0); (2,1)\}$ $\quad w_0 = \omega\alpha\beta_1\overline{\beta}$ $\quad w_1 = \overline{\alpha}\beta\overline{\beta}_i$

<u>*Simulation of table R_i, $i = 1, 2$*</u>

1. $[\beta_i]_0[\overline{\beta_i}]_1 \rightarrow [[\beta_i]_0\overline{\beta}_i]_1$ (mendo)
2. $[[a\beta_i]_0\overline{\beta}_i]_1 \rightarrow [w_i\beta_i]_0[\overline{\beta}_i]_1$, if $a \rightarrow w \in R_i$ (mexo)
3. $[\beta]_1[a\overline{\beta}]_0 \rightarrow [[\beta]_1\dagger\overline{\beta}]_0$ (mendo)
4. $[[a_i\beta_i]_0\overline{\beta}_i]_1 \rightarrow [a\beta_i]_0[\overline{\beta}_i]_1$ (mexo)
5. $[\beta]_1[a_i\overline{\beta}]_0 \rightarrow [[\beta]_1\dagger\overline{\beta}]_0$ (mendo)

6. $[[\beta_1\alpha]_0\overline{\alpha}]_1 \to [\beta_i\alpha]_0[\overline{\alpha}]_1$ (mexo)
$[[\beta_2\alpha]_0\overline{\alpha}]_1 \to [\beta_1\alpha]_0[\overline{\alpha}]_1$ (mexo)
$[[\beta_2\alpha]_0\overline{\alpha}]_1 \to [\alpha]_0[\overline{\alpha}]_1$ (mexo)
7. $[[\beta]_1\dagger\overline{\beta}]_0 \to [\beta]_1[\dagger\overline{\beta}]_0$ (mexo)
$[\beta]_1[\dagger\overline{\beta}]_0 \to [[\beta]_1\dagger\overline{\beta}]_0$ (mendo)

In the initial configuration the string $\beta_1\omega$ is in membrane 0, where ω is the axiom, and β_1 indicates that table 1 should be simulated first. The simulation begins with rule 1: membrane 0 enters membrane 1. In membrane 1, the only applicable rule is 2, by which the symbols $a \in V$ are replaced by w_1 corresponding to the rule $a \to w \in R_1$. Rules 1 and 2 can be repeated until all the symbols $a \in V$ have been replaced according to a rule in R_1, thus obtaining only objects from the alphabet V_1. In order to keep track of which table R_i of rules is simulated, each rule of the form $a \to w \in R_i$ is rewritten as $a \to w_i$.

If any symbol $a \in V$ is still present in membrane 0, i.e., if some symbol $a \in V$ has been left out of the simulation, membrane 1 enters membrane 0, replacing it with the trap symbol † (rule 3), and this triggers a never ending computation (rule 7). Otherwise, rules 1 and 4 are applied as long as required until all the symbols of V_1 are replaced with the corresponding symbols of V. Next, if a symbol $a_1 \in V_1$ has not been replaced, membrane 1 enters membrane 0 and the computation stops, replacing it with the trap symbol † (rule 5), and this triggers a never ending computation (rule 7). Otherwise, we have three possible evolutions (rule 6):

(i) if β_1 is in membrane 0, then it is replaced by β_i, and the computation continues with the simulation of table i;
(ii) if β_2 is in membrane 0, then it is replaced by β_1, and the computation continues with the simulation of table 1;
(iii) if β_2 is in membrane 0, then it is deleted, and the computation stops.

It is clear that $Ps(\Pi)$ contains all the vectors in $Ps(L(G))$. □

Corollary 2.2. $PsE0L \subseteq PsMM_3(mendo, mexo)$.

We can interpret the multiset of objects present in the output membrane as a set of strings x such that the multiplicity of symbols in x is the same as the multiplicity of objects in the output membrane. In this way, the multiset of objects in the output membrane generates a language. For a system Π, let $L(\Pi)$ represent this language (all strings computed by Π), and let $LMM_n(\alpha)$ represent the family of languages $L(\Pi)$ generated by systems having $\leq n$ membranes, using a set of operations $\alpha \subseteq \{mendo, mexo\}$. We get the following result.

Lemma 2.1. $LMM_3(mendo, mexo) - ET0L \neq \emptyset$.

Proof. $L=\{x \in \{a,b\}^* \mid |x|_b = 2^{|x|_a}\} \notin ET0L$ [141]. We construct
$\Pi = (\{a, b, b', \dagger, \beta, \beta_1, \overline{\beta}, \overline{\beta}_1, \beta_2, \overline{\beta}_2\}, \beta_3, \overline{\beta}_3\}, \{0,\ldots,2\}, [[\eta\ b\ in]_1[\overline{in}]_2]_0, R, 1)$
with rules as given below to generate L.

1. $[\,\beta\,]_1\,[\,\overline{\beta}\,]_2 \rightarrow [\,[\,\beta\,]_1\,\overline{\beta}\,]_2$, (mendo),
   $[\,[\,b\,\beta\,]_1\,\overline{\beta}\,]_2 \rightarrow [\,b'\,b'\,\beta\,]_1[\,\overline{\beta}\,]_2$, (mexo),
   $[\,[\,\beta\,]_1\,\overline{\beta}\,]_2 \rightarrow [\,\beta\,]_1\,[\,\overline{\beta}\,]_2$, (mexo),
2. $[\,\beta\,]_1\,[\,\overline{\beta}\,]_2 \rightarrow [\,[\,a\,\beta_1\,]\,\overline{\beta}_1\,]_2$, (mendo),
3. $[\,[\,b\,\beta_1\,]_1\,\overline{\beta}_1\,]_2 \rightarrow [\,\dagger\,\beta_1\,]_1\,[\,\overline{\beta}_1\,]_2$, (mexo),
   $[\,\dagger\,\beta_1\,]_1\,[\,\overline{\beta}_1\,]_2 \rightarrow [\,[\,\dagger\,\beta_1\,]_1\,\overline{\beta}_1\,]_2$, (mendo),
   $[\,[\,\dagger\,\beta_1\,]_1\,\overline{\beta}_1\,]_2 \rightarrow [\,\dagger\,\beta_1\,]_1\,[\,\overline{\beta}_1\,]_2$, (mexo),
4. $[\,[\,\beta_1\,]\,\overline{\beta}_1\,]_2 \rightarrow [\,\beta_2\,]_1\,[\,\overline{\beta}_2\,]_2$, (mexo),
   $[\,b'\,\beta_2\,]_1\,[\,\overline{\beta}_2\,]_2 \rightarrow [\,[\,b\,\beta_2\,]_1\,\overline{\beta}_2\,]_2$, (mexo),
   $[\,[\,\beta_2\,]_1\,\overline{\beta}_2\,]_2 \rightarrow [\,\beta_2\,]_1\,[\,\overline{\beta}_2\,]_2$ (mexo)
5. $[\,\beta_2\,]_1\,[\,\overline{\beta}_2\,]_2 \rightarrow [\,[\,\beta_3\,]_1\,\overline{\beta}_3\,]_2$, (mendo),
6. $[\,[\,b'\,\beta_3\,]_1\,\overline{\beta}_3\,]_2 \rightarrow [\,\dagger\,\beta_3\,]_1\,[\,\overline{\beta}_3\,]_2$, (mexo),
   $[\,\dagger\,\beta_3\,]_1\,[\,\overline{\beta}_3\,]_2 \rightarrow [\,[\,\dagger\,\beta_3\,]_1\,\overline{\beta}_3\,]_2$, (mendo),
   $[\,[\,\dagger\,\beta_3\,]_1\,\overline{\beta}_3\,]_2 \rightarrow [\,\dagger\,\beta_3\,]_1\,[\,\overline{\beta}_3\,]_2$, (mexo),
7. $[\,[\,\beta_3\,]_1\,\overline{\beta}_3\,]_2 \rightarrow [\,]_1\,[\,]_2$, (mexo),
8. $[\,[\,\beta_3\,]_1\,\overline{\beta}_3\,]_2 \rightarrow [\,\beta\,]_1\,[\,\overline{\beta}\,]_2$, (mexo).

The system works as follows: Rule 1 is used to replace every b with $b'b'$. Rule 2 can be used at any moment to replace β and $\overline{\beta}$ with β_1 and $\overline{\beta}_1$ (guessing that all b's have been replaced) and also to create an object a. Rule 3 checks that every b has been replaced with $b'b'$, and if not an infinite computation is obtained. If there is no b then rule 4 replaces β_1 and $\overline{\beta}_1$ with β_2 and $\overline{\beta}_2$, and then is used to replace every b' with b. Rule 5 can be used at any moment to replace β_2 and $\overline{\beta}_2$ with β_3 and $\overline{\beta}_3$ (guessing that all b''s have been replaced). Rule 6 checks that every b' has been replaced with b, and if not an infinite computation is obtained. The computation can halt using rule 7, and can continue using rule 8. It is easy to see that membrane 1 contains strings of L at the end of a halting computation. □

Exercise 2.2. What is the minimum number of membranes used to get full computational power for the class of mutual mobile membranes, if instead of the mutual endocytosis rule

$$[uv]_h[\overline{u}v']_m \rightarrow [\,[w]_h w']_m$$

we consider the enhanced endocytosis rule

$$[v]_h[uv']_m \rightarrow [[w]_h w']_m?$$

Perform the proof using either matrix grammars or register machines.

Exercise 2.3. Consider other combinations of rules from the simple, enhanced and mutual mobile membranes and find the minimal set of ingredients in order to obtain the computational power of a Turing machine.

2.4.5 Mutual Mobile Membranes with Objects on Surface

We explore now the computational power of systems of mutual mobile membranes with objects on surface using $(pino, exo)$ and $(phago, exo)$ as applicable pairs of rules. The power of these classes was already investigated in [105]; we improve those results with respect to the number of membranes used in computation. A summary of the results (existing results, as well as new ones) is given in Table 2.2.

Table 2.2 Summary of Results

Operations	Number of membranes	Weights	RE	Ref.
Pino, exo	8	4,3	Yes	Theorem 6.1 [105]
Pino, exo	3	5,4	Yes	Theorem 2.19
Phago, exo	9	5,2	Yes	Theorem 6.2 [105]
Phago, exo	9	4,3	Yes	Theorem 6.2 [105]
Phago, exo	4	5,4	Yes	Theorem 2.20

In each of the combinations presented in Table 2.2, the rules *pino* and *phago* are used to increase the number of membranes, while rule *exo* is used to decrease the number of membranes. Thus we combine the rules *pino* and *phago* with *exo* just to balance the number of membranes.

The result of a computation is considered to be the multiplicity vector of the multiset representing the union of the multisets placed on all the membranes of a system. Thus, the result of a halting computation consists of all the vectors describing the multiplicity of objects from all the membranes (a non-halting computation provides no output). The number of objects on the right hand side of a rule is called its *weight*. The family of all sets $Ps(\Pi)$ generated by systems of mutual mobile membranes with objects on surface using at any moment during a computation at most n membranes, and any of the rules $r_1 \in \{pino, phago\}$ and $r_2 \in \{exo\}$ of weight at most r, s respectively, is denoted by $PsM^3OS_n(r_1(r), r_2(s))$.

In what follows, we study the computational power of the $(pino, exo)$ combination of operations, and prove their universality by using during the computation at most three membranes.

Theorem 2.19. *$PsRE = PsM^3OS_m(pino(r), exo(s))$, for all $m \geq 3$, $r \geq 5$, $s \geq 4$.*

Proof. The inclusion of $PsM^3OS_m(pino(r), exo(s))$ into $PsRE$ is assumed true by invoking the Church-Turing thesis. This implies that we have to prove only the inclusion $PsRE \subseteq PsM^3OS_3(pino(5), exo(4))$. For this, we construct a system Π of three mutual mobile membranes with objects on surface,

$$\Pi = (V, \{(3,1);(3,2)\}, A\overline{X}, \overline{A}X, \lambda, R).$$

The finite alphabet V of objects is defined as follows:

$$V = \{\beta, \overline{\beta}\} \cup \{X, \overline{X}, X'_l, \overline{X'_l}, X^{(j)}, \overline{X^{(j)}}, X_l^{(j)}, \overline{X_l^{(j)}}, X_l^{(j)\prime}, \overline{X_l^{(j)\prime}}$$
$$\mid X \in N_1, 1 \leq l \leq n_1 + n_2, 1 \leq j \leq 2\}$$

The set of types of rules R is constructed as follows:

(i) For each (nonterminal) matrix $m_l : (X \to Y, A \to x)$, $X, Y \in N_1$, $x \in (N_2 \cup T)^*$, $A \in N_2$, $|x| \leq 2$, with $1 \leq l \leq n_1$, we consider the rules:

1. $[\,[\,]_{Au}]_{\overline{A}v} \to_m [\,]_{A_l\overline{A_l}uv}$ (exo)
2. $[\,]_{X\overline{X}uv} \to_m [\,[\,]_{x'X'_lu}]_{\overline{X'_l}v}$ (pino)
   (If $m_l : (X \to Y, A \to \alpha_1\alpha_2)$ then $x' = \alpha'_1\alpha_2$ or $x' = \alpha_1\alpha'_2$, and if $m_l : (X \to Y, A \to \alpha_1)$ then $x' = \alpha'_1$)
3. $[\,[\,]_{A_lu}]_{\overline{A_l}v} \to_m [\,]_{\overline{\alpha'}uv}$ (exo)
4. $[\,]_{\alpha'\overline{\alpha'}uv} \to_m [\,[\,]_{\alpha u}]_v$ (pino)
5. $[\,[\,]_{X'_lu}]_{\overline{X'_l}v} \to_m [\,]_{\overline{X''_l}X''_luv}$ (exo)
6. $[\,]_{\overline{X''_l}X''_luv} \to_m [\,[\,]_u]_{Yv}$ (pino)
7. $[\,[\,]_{\overline{X}u}]_{Xv} \to_m [\,]_{\beta\overline{\beta}uv}$ (exo)
8. $[\,]_{\beta\overline{\beta}uv} \to_m [\,[\,]_{\beta u}]_{\overline{\beta}v}$ (pino)
9. $[\,[\,]_{\beta u}]_{\overline{\beta}v} \to_m [\,]_{\beta\overline{\beta}uv}$ (exo)

We start the simulation of a type (2) matrix by using rule 1; A and $\overline{A}$ are replaced by A_l and $\overline{A_l}$, marking the beginning of the simulation. This is followed by rule 2, where X and $\overline{X}$ are replaced by x', X'_l and $\overline{X'_l}$. Next, we apply rule 3 to replace A_l and $\overline{A_l}$ by $\overline{\alpha'}$, in order to prevent replacing more A's from now on. In rule 4 we replace α' and $\overline{\alpha'}$ by α, while rule 5 replaces X'_l and $\overline{X'_l}$ by X''_l and $\overline{X''_l}$, respectively. Rule 6 is used to replace X''_l and $\overline{X''_l}$ by Y, thus successfully simulating a type (2) matrix and returning to the initial membrane structure. In case the corresponding symbol $A \in N_2$ is not present (we cannot apply rule 1), rule 7 introduces two symbols β and $\overline{\beta}$ which lead to an infinite computation (by using rules 8 and 9). We now illustrate the evolution of the configurations during one simulation of a type (2) matrix.

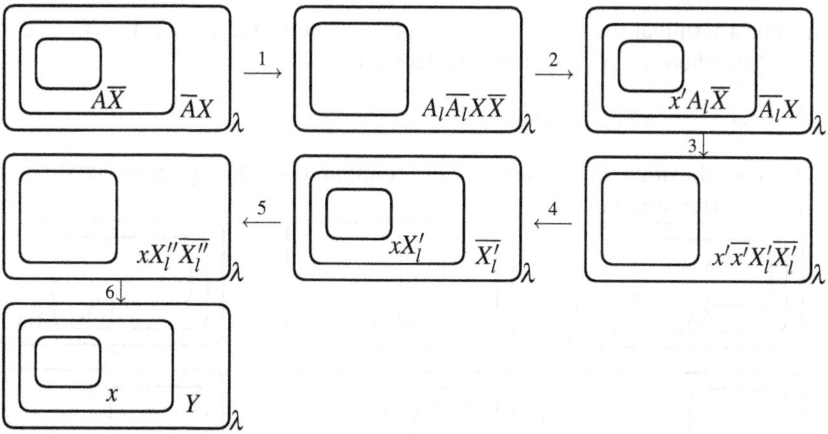

(ii) For each matrix $m'_l : (X \to Y, B^{(j)} \to \#)$, $X, Y \in N_1$, $A \in N_2$ and $B^{(j)} \to \# \in F$, where $n_1 + 1 \leq l \leq n_1 + n_2$, $l \in lab_j$, $j = 1, 2$, we consider the rules:

10. $[\,[\,]_{\overline{X}u}]_{Xv} \to_m [\,]_{X_l^{(j)}\overline{X_l^{(j)}}X^{(j)}\overline{X^{(j)}}uv}$ (exo)
11. $[\,]_{X^{(j)}\overline{X^{(j)}}uv} \to_m [\,[\,]_{\overline{X^{(j)}}u}]_{\overline{B^{(j)}}v}$ (pino)
12. $[\,[\,]_{B^{(j)}u}]_{\overline{B^{(j)}}v} \to_m [\,]_{\beta\overline{\beta}X^{(j)}uv}$ (exo)

13. $[\,[\,]_{X_l^{(j)}u}]_{\overline{X_l^{(j)}}v} \to_m [\,]_{B^{(j)}uv}$ (exo)
14. $[\,]_{B^{(j)}\overline{B^{(j)}}uv} \to_m [\,[\,]_u]_{Yv}$ (pino)

Rule 10 starts the simulation of a type (3) matrix by replacing $\overline{X}$ with $X_l^{(j)}X^{(j)}$ and X by $\overline{X_l^{(j)}}\overline{X^{(j)}}$, thereby remembering the index l of the matrix and the index j of the possibly present symbol $B^{(j)}$. This is followed by rule 11 in which $X^{(j)}$ and $\overline{X^{(j)}}$ are replaced by $\overline{X^{(j)}}$ and $\overline{B^{(j)}}$, respectively. At this step we need to check if the corresponding symbol $B^{(j)} \in N_2$ is present. If $B^{(j)}$ is present, rule 12 replaces it and its co-object $\overline{B^{(j)}}$ with $\beta\overline{\beta}$ together with $X^{(j)}$. In this case, by applying rule 11, we go to the configuration obtained before replacing $B^{(j)}$. Regardless of the presence of $B^{(j)}$, rule 13 is applied replacing $X_l^{(j)}$ and $\overline{X^{(j)}}$ by $B^{(j)}$. Rule 14 involves the creation of Y, thus successfully simulating a type (3) matrix and returning to the initial membrane structure. We now illustrate the evolution of the configurations during one simulation of a type (3) matrix.

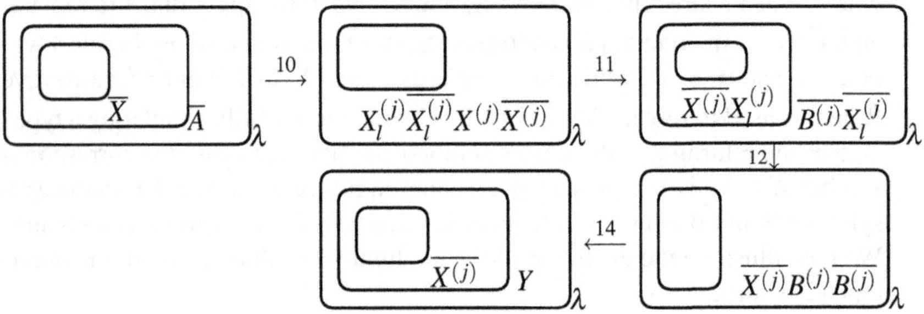

(iii) For a terminal matrix $m_l : (X \to a, A \to x)$, $X \in N_1$, $a \in T$, $A \in N_2$, $x \in T^*$, $|x| \leq 2$, where $1 \leq l \leq n_1$, we consider the rule

15. $[\,]_{\overline{X_l''}X_l''uv} \to_m [\,[\,]_u]_{av}$ (pino)

We now illustrate the evolution of the configurations during one simulation of a type (4) (terminal) matrix.

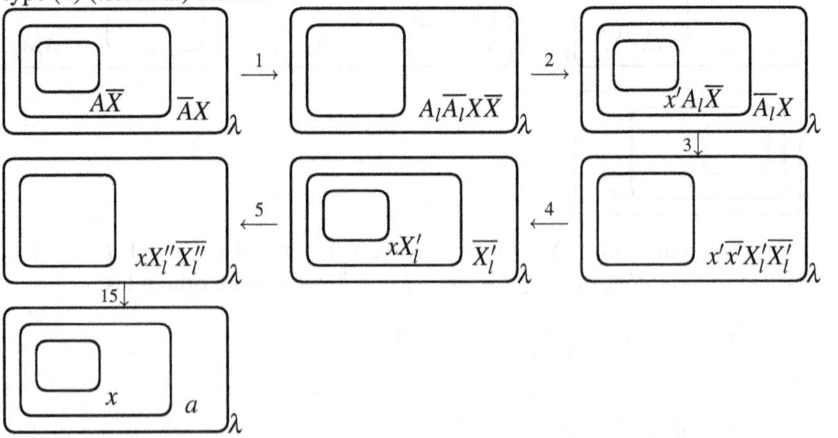

By replacing rule 6 with rule 15 in the sequence 1-9, we correctly simulate a terminal matrix. The result of a correct simulation is the multiset of all symbols present on the surfaces of all membranes. □

We also study the computational power of the $(phago, exo)$ combination of operations and prove their universality by using during the computation at most four membranes. We initially consider a system of three membranes. Compared with Theorem 2.19, the higher number of membranes is related to the (use of) *phago* operation.

Theorem 2.20. $PsRE = PsM^3OS_m(phago(r), exo(s))$, *for all* $m \geq 4$, $r \geq 5$, $s \geq 4$.

Proof. The inclusion of $PsM^3OS_m(phago(r), exo(s))$ in $PsRE$ is assumed true by invoking the Church-Turing thesis. This implies that we have to prove only the inclusion $PsRE \subseteq PsM^3OS_4(phago(5), exo(4))$. For this we construct a system Π of three mutual mobile membranes with objects on surface

$$\Pi = (V, \{(3,1);(3,2)\}, A\overline{X}, \overline{A}X, \lambda, R)$$

The finite alphabet V of objects is defined as

$$V = \{\beta, \overline{\beta}\} \cup \{X, \overline{X}, X_l, \overline{X_l}, X_l', \overline{X_l'}, X^{(j)}, \overline{X^{(j)}}, X_l^{(j)}, \overline{X_l^{(j)}}, X_l^{(j)'}, \overline{X_l^{(j)'}} \mid X \in N_1, 1 \leq l \leq n_1 + n_2, 1 \leq j \leq 2\}$$

The set of types of rules R is constructed as follows:

(i) For each (nonterminal) matrix $m_l : (X \to Y, A \to x)$, $X, Y \in N_1$, $x \in (N_2 \cup T)^*$, $A \in N_2$, $|x| \leq 2$, with $1 \leq i \leq n_1$, we consider the rules:

1. $[\,]_{Au}[\,]_{\overline{A}v} \to_m [[[\,]_{A_l u}]_{\overline{X}}]_v$ (phago)
2. $[[\,]_{\overline{X}u}]_{Xv} \to_m [\,]_{\overline{A_l} X_l uv}$ (exo)
3. $[\,]_{A_l u}[\,]_{\overline{A_l}v} \to_m [[[\,]_{x'u}]_{\overline{X_l}}]_v$ (phago)
   (If $m_l : (X \to Y, A \to \alpha_1\alpha_2)$ then $x' = \alpha_1'\alpha_2$ or $x' = \alpha_1\alpha_2'$, and if $m_l : (X \to Y, A \to \alpha_1)$ then $x' = \alpha_1'$)
4. $[[\,]_{\overline{X}_l u}]_{X_l v} \to_m [\,]_{\overline{\alpha}' X_l' uv}$ (exo)
5. $[\,]_{\alpha' u}[\,]_{\overline{\alpha}' v} \to_m [[[\,]_{\alpha u}]_{\overline{X}_l'}]_v$ (phago)
6. $[[\,]_{\overline{X}_l' u}]_{X_l' v} \to_m [\,]_{Yuv}$ (exo)
7. $[\,]_{\overline{X}u}[\,]_{Xv} \to_m [[[\,]_{\overline{\beta}u}]_{\beta}]_v$ (phago)
8. $[[\,]_{\beta u}]_{\overline{\beta}v} \to_m [\,]_{\beta\overline{\beta}uv}$ (exo)
9. $[\,]_{\overline{\beta}u}[\,]_{\beta v} \to_m [[[\,]_{\beta u}]_{\beta}]_v$ (phago)

We start the simulation of a type (2) matrix by using rule 1; A and $\overline{A}$ are replaced by A_l and $\overline{X}$, marking the beginning of the simulation. This is followed by rule 2 replacing X and $\overline{X}$ by $\overline{A_l}$ and X_l, respectively. In rule 3, A_l is replaced by x' in order to prevent replacing more A's from now on, while $\overline{A_l}$ is replaced by $\overline{X_l}$. This is followed by rule 4 in which X_l and $\overline{X_l}$ are replaced by $\overline{\alpha'}$ and X_l', respectively. Rule 5 replaces α' and $\overline{\alpha'}$ by α and X_l'. Rule 6 involves the replacing of X_l' and $\overline{X_l'}$ by Y, thus successfully simulating a type (2) matrix and returning to the initial membrane structure. If the corresponding symbol $A \in N_2$ is not present (i.e., we cannot apply rule 1), rule 7 introduces two symbols β and $\overline{\beta}$ which lead to an infinite computation (by using rules 8 and 9).

(ii) For each matrix $m'_l : (X \to Y, B^{(j)} \to \#)$, $X,Y \in N_1$, $A \in N_2$ and $B^{(j)} \to \# \in F$, where $n_1+1 \leq i \leq n_1+n_2$, $i \in lab_j$, $j=1,2$, we consider the rules:

10. $[\,]_{\overline{X}u}[\,]_{Xv} \to_m [[[\,]_{X_l^{(j)}X^{(j)}u}]_{\overline{X_l^{(j)}}\,\overline{X^{(j)}}}]_v$ (phago)
11. $[[\,]_{X^{(j)}u}]_{\overline{X^{(j)}}v} \to_m [\,]_{\overline{B^{(j)}}\,\overline{X^{(j)}}uv}$ (exo)
12. $[\,]_{B^{(j)}u}[\,]_{\overline{B^{(j)}}v} \to_m [[[\,]_{\overline{\beta}u}]_{\beta X^{(j)}}]_v$ (phago)
13. $[\,]_{X_l^{(j)}u}[\,]_{\overline{X_l^{(j)}}v} \to_m [[[\,]_u]_{B^{(j)}}]_v$ (phago)
14. $[[\,]_{B^{(j)}u}]_{\overline{B^{(j)}}v} \to_m [\,]_{Yuv}$ (exo)

Rule 10 starts the simulation of a type (3) matrix by first replacing $\overline{X}$ and X by $X_l^{(j)}$, $X^{(j)}$, $\overline{X_l^{(j)}}$ and $\overline{X^{(j)}}$, thereby remembering the index l of the matrix and the index j of the possibly present symbol $B^{(j)}$. This is followed by rule 11 in which $X^{(j)}$ and $\overline{X^{(j)}}$ are replaced by $\overline{B^{(j)}}$ and $\overline{X^{(j)}}$, respectively. At this step we need to verify if the corresponding symbol $B^{(j)} \in N_2$ is present. If $B^{(j)}$ is present, rule 12 replaces it and its co-object $\overline{B^{(j)}}$ with $\beta\overline{\beta}$ together with $X^{(j)}$. In this case, by applying rule 11, we go to the configuration obtained before replacing $B^{(j)}$. Regardless of the presence of $B^{(j)}$, rule 13 is applied; $X_l^{(j)}$ and $\overline{X^{(j)}}$ are replaced by $B^{(j)}$. Rule 14 involves the creation of Y, successfully simulating a type (3) matrix and returning to the initial membrane structure.

(iii) For a terminal matrix $m_l : (X \to a, A \to x)$, $X \in N_1$, $a \in T$, $A \in N_2$, $x \in T^*$, $|x| \leq 2$ where $1 \leq i \leq n_1$, we consider the rule

15. $[[\,]_{\overline{X'_l}u}]_{X'_l v} \to_m [\,]_{auv}$ (exo)

By replacing rule 6 with rule 15 in the sequence 1-9, we correctly simulate a terminal matrix. The result of a correct simulation is the multiset of all symbols present on the surfaces of all membranes. □

Exercise 2.4. Perform the proofs of this subsection using register machines instead of matrix grammars. Do the weights of the used operations differ?

2.5 Complexity of Mutual Mobile Membranes

Regarding the complexity aspects [126], polynomial time solutions to NP-complete problems in the framework of membrane computing are presented comprehensively in [131]. The authors of this survey use P systems with active membranes having associated electrical charges, membrane division and membrane creation. We present solutions to NP-complete problems by using systems of mutual mobile membranes that can perform only mobility and elementary division rules. In order to find such a solution, mutual mobile membranes are treated as deciding devices that respect the following conditions: (1) all computations halt, (2) two additional objects *yes* and *no* are used, (3) only one of the objects *yes* and *no* appears in the halting configuration. The computation is accepting if the *yes* object is present in the halting configuration, and rejecting if the *no* object is present in the halting configuration.

A family of mutual mobile membrane systems $\{\Pi\}$ solves a decision problem if there is a member of the family to recognize every instance of the problem. To ensure that the construction algorithm of each member of the family does not increase the set of problems decided by the family, we require that the algorithm, is computable within certain restricted resources (time/space). When the algorithm maps an instance size to a membrane system that decides all instances of that length, then the algorithm is called a *uniformity condition.* The notion of *uniformity* was first introduced by Borodin [30] for boolean circuits. When the algorithm maps a single instance to a membrane system that decides that instance, then the algorithm is called a *semi-uniformity* condition. The notions of uniformity and semi-uniformity were first applied to membrane systems in [132].

2.5.1 SAT Problem

The SAT problem checks the satisfiability of a propositional logic formula in conjunctive normal form (CNF). Let $\{x_1, x_2, \dots, x_n\}$ be a set of propositional variables. A formula in CNF is of the form $\varphi = C_1 \wedge C_2 \wedge \cdots \wedge C_m$ where each $C_i, 1 \le i \le m$ is a disjunction of the form $C_i = y_1 \vee y_2 \vee \cdots \vee y_r$ $(r \le n)$, where each y_j is either a variable x_k or its negation $\neg x_k$. In this section, we propose a uniform polynomial time solution to the SAT problem using the operations of *mendo*, *mexo* and elementary division (for any instance of SAT we construct a system of mutual mobile membranes which solves it). Consider the formula $\varphi = C_1 \wedge C_2 \wedge \dots C_m$, over the variables $\{x_1, \dots, x_n\}$. Consider a system of mutual mobile membranes having the initial configuration

$$[[c_0\, \beta]_J [\overline{\beta}]_K [c\, \overline{d}]_L [g^{n-1} g_0]_1 [a_1]_0]_2$$

and working over the alphabet:

$$V = \{c, \overline{c}, d, \overline{d}, g, g_0, \beta, \overline{\beta}, yes, no\} \cup \{a_i, t_i, f_i \mid 1 \le i \le n\}$$
$$\cup \{\beta_i, \overline{\beta}_i \mid 1 \le i \le m\} \cup \{c_i \mid 0 \le i \le n+2m+1\}$$

In addition to mutual endocytosis and mutual exocytosis rules, we use elementary division rules to generate all the possible assignments. An elementary division rule has the form:

$$[a]_h \to [u]_h [v]_h, \text{ for } h \in H,\, a \in V,\, u, v \in V^* \qquad \text{(div)}$$

where a copy of each object from membrane h is placed inside the newly created membranes, except for object a which is replaced by the multisets of objects u and v. If w is the multiset of objects from h except the a object, then the rule is illustrated as:

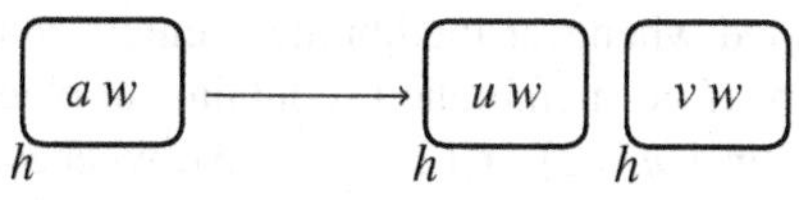

The system of mutual mobile membranes solving the SAT problem uses the rules:

(i) $[a_i]_0 \to [t_i\, a_{i+1}]_0[f_i\, a_{i+1}]_0$, for $1 \le i \le n-1$ (div)
$[a_n]_0 \to [tn\, \beta_1]_0[f_n\, \beta_1]_0$ (div)
$[g]_1 \to [\,]_1[\,]_1$ (div)
$[g_0]_1 \to [\overline{\beta_1}]_1[\overline{\beta_1}]_1$ (div)
The first two rules create 2^n membranes labelled by 0 containing all the possible assignments over variables $\{x_1,\dots,x_n\}$. In each membrane labelled by 0 is placed also a symbol β_1 . The next two rules create 2^n membranes labelled by 1 each containing an object $\overline{\beta_1}$. The symbols β_1 and $\overline{\beta_1}$ are used to determine in two steps which assignments are true for C_1.

(ii) $[t_j\, \beta_i]_0[\overline{\beta_i}]_1 \to [[t_j\, \beta_i]_0\overline{\beta_i}]_1$ (mendo)
$[[t_j\, \beta_i]_0\overline{\beta_i}]_1 \to [t_j\, \beta_{i+1}]_0[\overline{\beta_{i+1}}]_1$, $1 \le i \le m-1$, $1 \le j \le n$ (mexo)
(if clause C_i contains the literal x_j)
$[f_j\, \beta_i]_0[\overline{\beta_i}]_1 \to [[f_j\, \beta_i]_0\overline{\beta_i}]_1$ (mendo)
$[[f_j\, \beta_i]_0\overline{\beta_i}]_1 \to [f_j\, \beta_{i+1}]_0[\overline{\beta_{i+1}}]_1$, $1 \le i \le m-1$, $1 \le j \le n$ (mexo)
(if clause C_i contains the literal $\neg x_j$)
$[t_j\, \beta_m]_0[\overline{\beta_m}]_1 \to [[t_j\, \beta_m]_0\overline{\beta_m}]_1$ (mendo)
$[[t_j\, \beta_m]_0\overline{\beta_m}]_1 \to [t_j\, \overline{c}]_0[\overline{\beta_m}]_1$, $1 \le j \le n$ (mexo)
(if clause C_m contains the literal x_j)
$[f_j\, \beta_m]_0[\overline{\beta_m}]_1 \to [[f_j\, \beta_m]_0\overline{\beta_m}]_1$ (mendo)
$[[f_j\, \beta_m]_0\overline{\beta_m}]_1 \to [f_j\, \overline{c}]_0[\overline{\beta_m}]_1$, $1 \le j \le n$ (mexo)
(if clause C_m contains the literal $\neg x_j$)
If some assignments satisfy the clause C_i, $1 \le i < m$, then the objects β_i from the corresponding membranes 0 are replaced by β_{i+1}. The assignments from the membranes containing β_{i+1} satisfy the clauses $C_1,\dots,C_i$, the object β_{i+1} marking the fact that in the next step the clause C_{i+1} is checked. If there exist assignments which satisfy all the clauses, then the membranes containing these assignments contain an object $\overline{c}$ after $n+2m$ steps.

(iii) $[c_i\, \beta]_J[\overline{\beta}]_K \to [[c_{i+1}\, \beta]_J\overline{\beta}]_K$ (mendo)
$[[c_i\, \beta]_J\overline{\beta}]_K \to [c_{i+1}\, \beta]_J[\overline{\beta}]_K$, $0 \le i \le n+2m$ (mexo)
$[[c_{n+2m+1}\, \beta]_J\overline{\beta}]_K \to [d\, \beta]_J[\overline{\beta}]_K$ (mexo)
$[c_{n+2m+1}\, \beta]_J[\overline{\beta}]_K \to [[c_{n+2m+1}\, \beta]_J\overline{\beta}]_K$ (mendo)
These rules trace the number of steps performed. If this number is greater than $n+2m+1$, then an object d is created, which will subsequently create an object *no*; $n+2m+1$ is determined by: generating space (n steps), verifying assignments ($2m$ steps), creating a *yes* object (1 step). An extra step can be performed, such that membrane J containing the object c_{n+2m+1} becomes sibling with membrane K, thus increasing the number of steps needed to create d to $n+2m+2$.

(iv) $[\overline{c}]_0[c]_L \to [[yes]_L]_0$ (mendo)
$[d]_J[\overline{d}]_L \to [[no]_J]_L$ (mendo)
A *yes* object is created whenever membrane L enters some membrane 0 in the $(2m+n+1)$-th step. If no membrane 0 contains an object $\overline{c}$, then a *no* object is created, in step $(2m+n+2)$ or $(2m+n+3)$, whenever membrane J enters membrane L. By applying one of these two rules, the other one cannot by applied

anymore, so at the end of the computation the system contains either an object *yes* or an object *no*.

The number of membranes in the initial configuration is 6, and the number of objects is $n+6$. The size of the working alphabet is $4n+4m+13$. The number of rules in the above system: $n+2$ rules of type (i), $4nm$ rules of type (ii), $n+2m+3$ rules of type (iii), and 2 rules of type (iv). Hence, the size of the constructed system of mutual mobile membranes is $\mathcal{O}(mn)$.

Example 2.2. Consider the SAT problem with $\phi = C_1 \wedge C_2 \wedge C_3$ and $X = \{x_1, x_2, x_3\}$, $C_1 = x_1 \vee \neg x_3$, $C_2 = \neg x_1 \vee \neg x_2$ and $C_3 = x_2$. In this case, $n = 3$, $m = 3$ and

$$[[c_0\,\beta]_J[\overline{\beta}]_K[c\,\overline{d}]_L[g^2 g_0]_1[a_1]_0]_2$$

Graphically this is illustrated as:

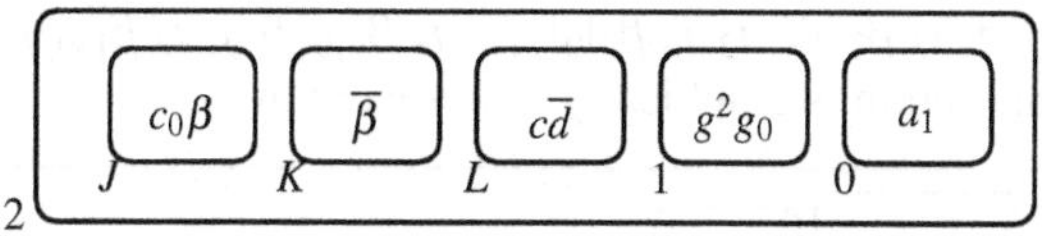

The evolution of the system is described by the following steps, where $[w]_i^n$ stands for n membranes $[w]_i$. The working space is created in $n = 3$ steps leading from the initial configuration 1 to configuration 4:

1. $[[c_0\,\beta]_J[\overline{\beta}]_K[c\,\overline{d}]_L[g^2 g_0]_1[a_1]_0]_2$
2. $[[[c_1\,\beta]_J\overline{\beta}]_K[c\,\overline{d}]_L[g^2\overline{\beta_1}]_{\overline{1}}^2[t_1\,a_2]_0[f_1\,a_2]_0]_2$
3. $[[c_2\,\beta]_J[\overline{\beta}]_K[c\,\overline{d}]_L[g\overline{\beta_1}]_{\overline{1}}^4[t_1\,t_2\,a_3]_0[t_1\,f_2\,a_3]_0[f_1\,t_2\,a_3]_0[f_1\,f_2\,a_3]_0]_2$
4. $[[[c_3\,\beta]_J\overline{\beta}]_K[c\,\overline{d}]_L[\overline{\beta_1}]_{\overline{1}}^8[t_1\,t_2\,t_3\,\beta_1]_0[t_1\,t_2\,f_3\,\beta_1]_0[t_1\,f_2\,t_3\,\beta_1]_0[t_1\,f_2\,f_3\,\beta_1]_0$
   $[f_1\,t_2\,t_3\,\beta_1]_0[f_1\,t_2\,f_3\,\beta_1]_0[f_1\,f_2\,t_3\,\beta_1]_0[f_1\,f_2\,f_3\,\beta_1]_0]_2$

   Graphically the working space is described by the following picture:

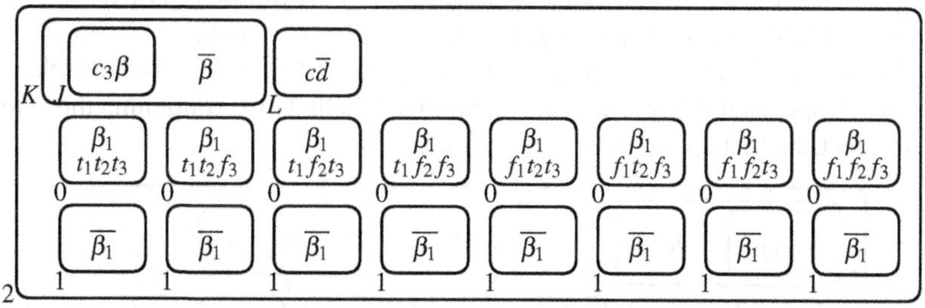

The next two steps mark the solutions of C_1 by replacing β_1 by β_2:

5. $[[c_4\,\beta]_J[\overline{\beta}]_K[c\,\overline{d}]_L[\overline{\beta_1}]_{\overline{1}}^2[\overline{\beta_1}[t_1\,t_2\,t_3\,\beta_1]_0]_1[\overline{\beta_1}[t_1\,t_2\,f_3\,\beta_1]_0]_1[\overline{\beta_1}[t_1\,f_2\,t_3\,\beta_1]_0]_1$
   $[\overline{\beta_1}[t_1\,f_2\,f_3\,\beta_1]_0]_1[\overline{\beta_1}[f_1\,t_2\,f_3\,\beta_1]_0]_1[\overline{\beta_1}[f_1\,f_2\,f_3\,\beta_1]_0]_1[f_1\,t_2\,t_3\,\beta_1]_0[f_1\,f_2\,t_3\,\beta_1]_0]_2$
6. $[[[c_5\,\beta]_J\overline{\beta}]_K[c\,\overline{d}]_L[\overline{\beta_1}]_{\overline{1}}^2[\overline{\beta_2}]_{\overline{1}}^6[t_1\,t_2\,t_3\,\beta_2]_0[t_1\,t_2\,f_3\,\beta_2]_0[t_1\,f_2\,t_3\,\beta_2]_0[t_1\,f_2\,f_3\,\beta_2]_0$
   $[f_1\,t_2\,f_3\,\beta_2]_0[f_1\,f_2\,f_3\,\beta_2]_0[f_1\,t_2\,t_3\,\beta_1]_0[f_1\,f_2\,t_3\,\beta_1]_0]_2$

   The new configuration is graphically represented by:

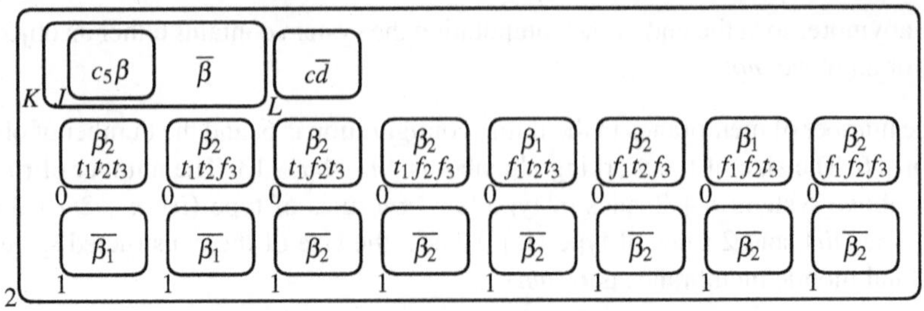

The next two steps mark the solutions of C_2 by replacing β_2 by β_3:

7. $[[c_6\ \beta]_J[\overline{\beta}]_K[c\ \overline{d}]_L[\overline{\beta_1}]_1^2[\overline{\beta_2}]_1^2[\overline{\beta_2}[t_1\ f_2\ t_3\ \beta_2]_0]_1[\overline{\beta_2}[t_1\ f_2\ f_3\ \beta_2]_0]_1[\overline{\beta_2}[f_1\ t_2\ f_3\ \beta_2]_0]_1$ $[\overline{\beta_2}[f_1\ f_2\ f_3\ \beta_2]_0]_1[t_1\ t_2\ t_3\ \beta_2]_0[t_1\ t_2\ f_3\ \beta_2]_0[f_1\ t_2\ t_3\ \beta_1]_0[f_1\ f_2\ t_3\ \beta_1]_0]_2$
8. $[[[c_7\ \beta]_J\overline{\beta}]_K[c\ \overline{d}]_L[\overline{\beta_1}]_1^2[\overline{\beta_2}]_1^2[\overline{\beta_3}]_1^4[t_1\ f_2\ t_3\ \beta_3]_0[t_1\ f_2\ f_3\ \beta_3]_0[f_1\ t_2\ f_3\ \beta_3]_0$ $[f_1\ f_2\ f_3\ \beta_3]_0[t_1\ t_2\ t_3\ \beta_2]_0[t_1\ t_2\ f_3\ \beta_2]_0[f_1\ t_2\ t_3\ \beta_1]_0[f_1\ f_2\ t_3\ \beta_1]_0]_2$

   The new configuration is graphically represented by:

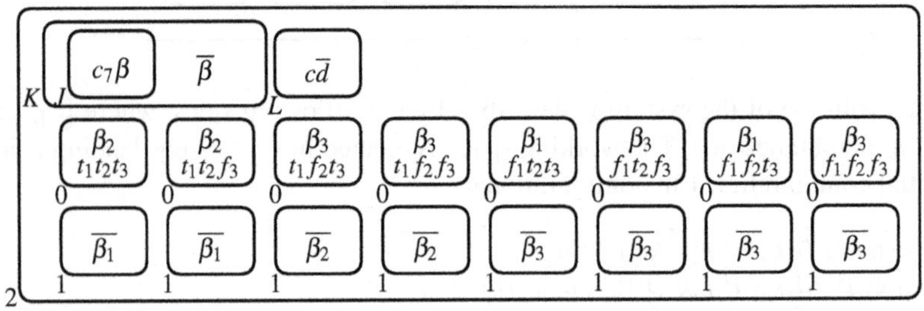

The next two steps mark the solutions of C_3 by replacing β_3 by $\overline{c}$:

9. $[[c_8\ \beta]_J[\overline{\beta}]_K[c\ \overline{d}]_L[\overline{\beta_1}]_1^2[\overline{\beta_2}]_1^2[\overline{\beta_3}]_1^3[\overline{\beta_3}[f_1\ t_2\ f_3\ \beta_3]_0]_1[t_1\ f_2\ t_3\ \beta_2]_0[t_1\ f_2\ f_3\ \beta_2]_0$ $[f_1\ f_2\ f_3\ \beta_2]_0[t_1\ t_2\ t_3\ \beta_2]_0[t_1\ t_2\ f_3\ \beta_2]_0[f_1\ t_2\ t_3\ \beta_1]_0[f_1\ f_2\ t_3\ \beta_1]_0]_2$
10. $[[[c_9\ \beta]_J\overline{\beta}]_K[c\ \overline{d}]_L[\overline{\beta_1}]_1^2[\overline{\beta_2}]_1^2[\overline{\beta_3}]_1^4[f_1\ t_2\ f_3\ \overline{c}]_0[t_1\ f_2\ t_3\ \beta_3]_0[t_1\ f_2\ f_3\ \beta_3]_0$ $[f_1\ f_2\ f_3\ \beta_3]_0[t_1\ t_2\ t_3\ \beta_2]_0[t_1\ t_2\ f_3\ \beta_2]_0[f_1\ t_2\ t_3\ \beta_1]_0[f_1\ f_2\ t_3\ \beta_1]_0]_2$

    The new configuration is graphically illustrated below, where we have placed the membrane labelled by L near the membrane labelled by 0 containing the symbol $\overline{c}$ to illustrate that an interaction is possible:

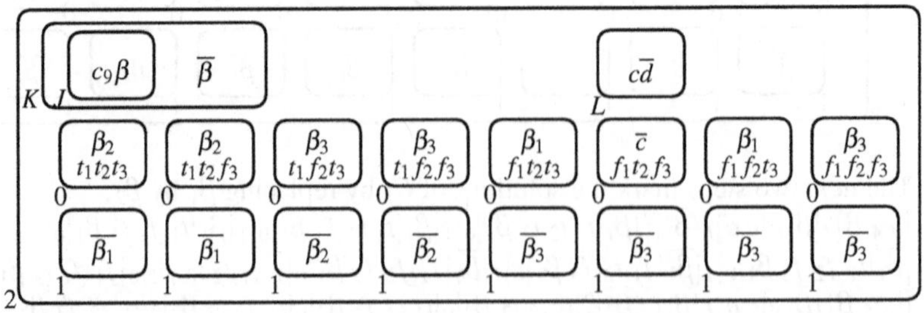

In the next step, an object *yes* is created and placed in membrane L, marking the fact that there exists an assignment such that the formula $(C_1 \wedge C_2 \wedge C_3)$ holds. The number of steps needed to create an object *yes* is $n+2m+1 = 3+6+1 = 10$.

11. $[[c_{10}\ \beta]_J[\overline{\beta}]_K[\overline{\beta_1}]_1^2[\overline{\beta_2}]_1^2[\overline{\beta_3}]_1^4[f_1\ t_2\ f_3\ [yes\ \overline{d}]_L]_0[t_1\ f_2\ t_3\ \beta_3]_0[t_1\ f_2\ f_3\ \beta_3]_0$
$[f_1\ f_2\ f_3\ \beta_3]_0[t_1\ t_2\ t_3\ \beta_2]_0[t_1\ t_2\ f_3\ \beta_2]_0[f_1\ t_2\ t_3\ \beta_1]_0[f_1\ f_2\ t_3\ \beta_1]_0]_2$
The new configuration is graphically illustrated as below:

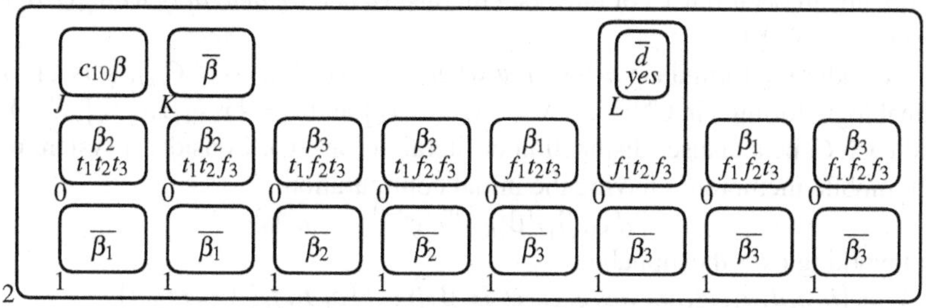

An object d used to create an object no is created after performing the steps:

12. $[[[c_{10}\ \beta]_J\overline{\beta}]_K[\overline{\beta_1}]_1^2[\overline{\beta_2}]_1^2[\overline{\beta_3}]_1^4[f_1\ t_2\ f_3\ [yes\ \overline{d}]_L]_0[t_1\ f_2\ t_3\ \beta_3]_0[t_1\ f_2\ f_3\ \beta_3]_0$
$[f_1\ f_2\ f_3\ \beta_3]_0[t_1\ t_2\ t_3\ \beta_2]_0[t_1\ t_2\ f_3\ \beta_2]_0[f_1\ t_2\ t_3\ \beta_1]_0[f_1\ f_2\ t_3\ \beta_1]_0]_2$
13. $[[d\ \beta]_J[\overline{\beta}]_K[\overline{\beta_1}]_1^2[\overline{\beta_2}]_1^2[\overline{\beta_3}]_1^4[f_1\ t_2\ f_3\ [yes\ \overline{d}]_L]_0[t_1\ f_2\ t_3\ \beta_3]_0[t_1\ f_2\ f_3\ \beta_3]_0$
$[f_1\ f_2\ f_3\ \beta_3]_0[t_1\ t_2\ t_3\ \beta_2]_0[t_1\ t_2\ f_3\ \beta_2]_0[f_1\ t_2\ t_3\ \beta_1]_0[f_1\ f_2\ t_3\ \beta_1]_0]_2$
The new configuration is graphically illustrated below, where we place the membrane labelled by J near the membrane 0 containing membrane L to illustrate that an interaction between membranes J and L is not possible, and so the computation stops after $n+2m+3=12$ steps.

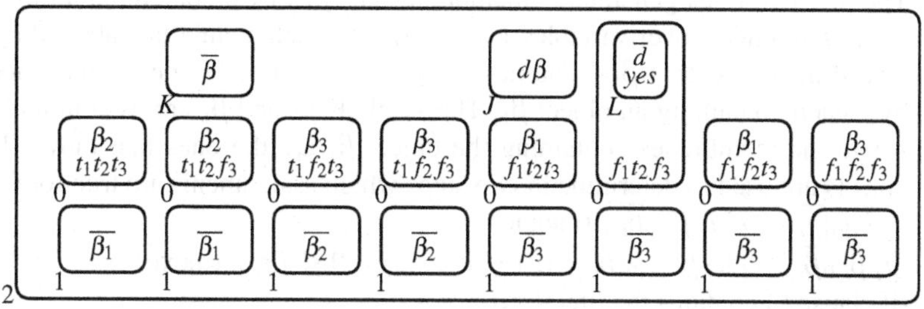

The fact that the computation ends in $n+2m+3$ steps is given by the fact that $n+2m$ is an odd number, and thus we had to perform an extra step before generating d from c_{n+2m+1}. If instead $n+2m$ is an even number, then d is created after $n+2m+2$ steps.

Exercise 2.5. Solve the SAT problem using other classes of mobile membranes.

2.5.2 2QBF Problem

In this section, we propose a polynomial time solution for solving satisfiability of 2QBF using mutual mobile membranes using the operations *mendo*, *mexo* and *div*. A quantified boolean formula is said to be in 2QBF if it is of the form $\varphi = \forall X \exists Y \psi$

or $\exists X \forall Y \psi$ where ψ is in CNF and X, Y partition the variables of ψ. By ψ is denoted the quantifier free part of φ. For 2QBF formulae of the form $\exists X \forall Y \psi$, satisfiability simplifies to the SAT problem $\exists X \psi'$ where ψ' is the CNF obtained from ψ by removing all occurrences of universal literals. *Hence we deal only with 2QBF of the form $\varphi = \forall X \exists Y \psi$.*

Consider the formula $\varphi = \forall X \exists Y \psi$ where $\psi = (C_1 \wedge C_2 \cdots \wedge C_m)$. ψ is a propositional logic formula in CNF. Let $X = \{x_1, \ldots, x_k\}$ and $Y = \{x_{k+1}, \ldots, x_n\}$, $X \cap Y = \emptyset$, and each C_i be a clause (disjunction of literals x_i or $\neg x_i$). Consider a system of mutual mobile membranes having the initial configuration

$$[[d\ c_0\ \beta]_J[\overline{\beta}]_K[\overline{d}]_L[g^{n-1}g_0]_1[a_1]_0]_2$$

and working over the alphabet:

$$V = \{b, c, d, \overline{d}, z, \overline{z}, no, yes, g, g_0, \alpha, \overline{\alpha}, \beta, \overline{\beta}\} \cup \{a_i, t_i, f_i \mid 1 \le i \le n\}$$
$$\cup \{\overline{t_i}, \overline{f_i} \mid 1 \le i \le k\} \cup \{\beta_i, \overline{\beta_i} \mid 1 \le i \le m\}$$
$$\cup \{c_i \mid 0 \le i \le n + 2m + 4k + 4\} \cup \{d_i, d_i', d_i'', d_i''', \overline{d_i}''' \mid 1 \le i \le k\} \cup \{\overline{d_k}''\}$$

In addition to mutual endocytosis and mutual exocytosis rules, we use elementary division rules to generate all the possible assignments. The system of mutual mobile membranes solving the 2QBF problem uses the rules:

(i) $[a_i]_0 \to [t_i\ a_{i+1}]_0[f_i\ a_{i+1}]_0$, for $1 \le i \le n-1$ (div)
$[a_n]_0 \to [tn\ \beta_1]_0[f_n\ \beta_1]_0$ (div)
$[g]_1 \to [\ ]_1[\ ]_1$ (div)
$[g_0]_1 \to [\overline{\beta_1}]_1[\overline{\beta_1}]_1$ (div)
The first two rules generate 2^n membranes labelled by 0 containing all the possible assignments over variables $\{x_1, \ldots, x_n\}$. In each membrane labelled by 0 is placed also a symbol β_1 . The next two rules generate 2^n membranes labelled by 1 each containing an object $\overline{\beta_1}$. The symbols β_1 and $\overline{\beta_1}$ are used in mobility, where the membranes containing the object β_1 are the ones that move. These objects are used to determine in two steps which assignments are true for C_1.

(ii) $[t_j\ \beta_i]_0[\overline{\beta_i}]_1 \to [[t_j\ \beta_i]_0\overline{\beta_i}]_1$ (mendo)
$[[t_j\ \beta_i]_0\overline{\beta_i}]_1 \to [t_j\ \beta_{i+1}]_0[\overline{\beta_{i+1}}]_1$, $1 \le i \le m-1, 0 \le j \le n$ (mexo)
(if clause C_i contains the literal x_j)
$[f_j\ \beta_i]_0[\overline{\beta_i}]_1 \to [[f_j\ \beta_i]_0\overline{\beta_i}]_1$ (mendo)
$[[f_j\ \beta_i]_0\overline{\beta_i}]_1 \to [f_j\ \beta_{i+1}]_0[\overline{\beta_{i+1}}]_1$, $1 \le i \le m-1, 0 \le j \le n$ (mexo)
(if clause C_i contains the literal $\neg x_j$)
$[t_j\ \beta_m]_0[\overline{\beta_m}]_1 \to [[t_j\ \beta_m]_0\overline{\beta_m}]_1$ (mendo)
$[[t_j\ \beta_m]_0\overline{\beta_m}]_1 \to [t_j\ c]_0[\overline{\beta_m}]_1$, $0 \le j \le n$ (mexo)
(if clause C_m contains the literal x_j)
$[f_j\ \beta_m]_0[\overline{\beta_m}]_1 \to [[f_j\ \beta_m]_0\overline{\beta_m}]_1$ (mendo)
$[[f_j\ \beta_m]_0\overline{\beta_m}]_1 \to [f_j\ c]_0[\overline{\beta_m}]_1$, $0 \le j \le n$ (mexo)
(if clause C_m contains the literal $\neg x_j$)
If some assignments satisfy the clause C_i, $1 \le i < n$, then the objects β_i from the corresponding membranes 0 are replaced by β_{i+1}. The object β_{i+1} marks the fact that the assignment satisfies clauses $C_1, \ldots, C_i$ and that in the next step the clause C_{i+1} is checked. If there exist assignments which verify all the clauses,

then the membranes containing these assignments contain an object $\overline{c}$ after $n+2m$ steps.

(iii) $[c_i\, \beta]_J[\overline{\beta}]_K \rightarrow [[c_{i+1}\, \beta]_J\overline{\beta}]_K$ (mendo)
$[[c_i\, \beta]_J\overline{\beta}]_K \rightarrow [c_{i+1}\, \beta]_J[\overline{\beta}]_K,\ 0 \leq i \leq n+2m-1$ (mexo)
$[[c_{n+2m}\, \beta]_J\overline{\beta}]_K \rightarrow [c_{n+2m}\, \beta]_J[\overline{\beta}]_K$ (mexo)
$[c_{n+2m}d]_J[\overline{d}]_L \rightarrow [[c_{n+2m}d]_J\overline{d}]_L$ (mendo)
$[[c_{n+2m}d]_J\overline{d}]_L \rightarrow [c_{n+2m+3}]_J[d_1]_L$ (mexo)
These rules trace the number of steps performed. If this number is greater than $n+2m$, then an object d_1 is created in membrane L that marks the end of checking ψ. If there are solutions for ψ, the corresponding membranes contain the object c. The number $n+2m+3$ is determined by: generating space (n steps), verifying assignments ($2m$ steps), creating a d_1 object (2 steps) and eventually one step to perform the third rule if necessary.

(iv) $[d_i]_L \rightarrow [d_i']_L[d_i'']_L$, for $1 \leq i \leq k$ (div)
$[d_i']_L \rightarrow [\overline{t_i}^{\,2^{n-i}}\overline{d_i'''}\overline{z}]_L[\overline{f_i}^{\,2^{n-i}}\overline{d_i'''}\overline{z}]_L$, for $1 \leq i \leq k-1$ (div)
$[d_k']_L \rightarrow [\overline{t_k}^{\,2^{n-k}}d_k''']_L[\overline{f_k}^{\,2^{n-k}}d_k''']_L$ (div)
$[d_i'']_L \rightarrow [d_i''']_L[d_i''']_L$, for $1 \leq i \leq k-1$ (div)
$[d_k'']_L \rightarrow [\overline{d_k''}]_L[\overline{d_k''}]_L$ (div)
$[\overline{d_k''}]_L \rightarrow [\overline{d_k''''}]_L[\overline{d_k''''}]_L$ (div)
$[d_i''']_L[\overline{d_i'''}]_L \rightarrow [[d_{i+1}]_L]_L$, for $1 \leq i \leq k-1$ (mendo)
$[t_ic]_0[\overline{t_i}]_L \rightarrow [[t_ic]_0\overline{t_i}]_L$, for $1 \leq i \leq k$ (mendo)
$[f_ic]_0[\overline{f_i}]_L \rightarrow [[f_ic]_0\overline{f_i}]_L$, for $1 \leq i \leq k$ (mendo)
Next, after finding the solutions of ψ, the $\forall$ part of the formula over variables $x_1,\ldots,x_k$ is checked. This amounts to checking if all the 2^k combinations of $t_i, f_i, 1 \leq i \leq k$ contain an object c. If so, then φ is true, and any last $n-k$ symbols will suffice for a solution. In order to check that all the 2^k combinations are in membranes containing a c object, membrane L is divided and a membrane structure is created in $3k$ steps. First, d_1 is replaced with d_1', d_1'' in two membranes. This is followed by the division of the membrane containing d_1' into two new membranes in which this object is replaced by a multiset containing $\overline{t_1}$, respectively $\overline{f_1}$. The membrane containing d_1'' is used to obtain two new membranes that are sent inside the membranes containing $\overline{t_1}$, respectively $\overline{f_1}$, in order to continue the construction until membranes containing $\overline{t_k}$, respectively $\overline{f_k}$, are obtained. In parallel, the membranes 0 containing an object c, enter the newly created structure.

(v) $[\overline{t_k}d_k''']_L[\overline{d_k''''}]_L \rightarrow [[b\, z]_L\overline{z}]_L$ (mendo)
$[\overline{f_k}d_k''']_L[\overline{d_k''''}]_L \rightarrow [[b\, z]_L\overline{z}]_L$ (mendo)
If at the end of the construction from step (iv) there exists an elementary membrane L containing an object $\overline{t_k}$ or $\overline{f_k}$, it means that not all possible assignments over the variables $x_1,\ldots,x_k$ are solutions, so the object b is created which will subsequently create an object *no*.

(vi) $[[z]_L\overline{z}]_L \rightarrow [z]_L[\overline{z}]_L$ (mexo)
$[[z]_L\overline{z}\,\overline{t_1}]_L \rightarrow [z\overline{\alpha}]_L[\overline{z}\,\overline{t_1}]_L$ (mexo)
$[[z]_L\overline{z}\overline{f_1}]_L \rightarrow [z\overline{\alpha}]_L[\overline{z}\overline{f_1}]_L$ (mexo)

If there exists a membrane L containing an object z together with the object *no*, then in k steps it reaches membrane 2, in order to deliver the answer inside membrane J.

(vii) $[c_i\,\beta]_J[\overline{\beta}]_K \to [[c_{i+1}\,\beta]_J\overline{\beta}]_K$ (mendo)
$[[c_i\,\beta]_J\overline{\beta}]_K \to [c_{i+1}\,\beta]_J[\overline{\beta}]_K$, $n+2m+3 \leq i \leq n+2m+4k+3$ (mexo)
In parallel, the counter in membrane J evolves until it reaches $n+2m+4k+4$. The extra $4k+1$ steps from $n+2m+3$ to $n+2m+4k+4$ is determined by: generating the structure starting from membrane L and movement of membranes 0 containing a c object inside membranes L containing $\overline{t_k}$ or $\overline{f_k}$ ($3k$ steps), creating a *no* object (1 step), and moving the membrane L containing an object z to membrane 2 (k steps).

(viii) $[[c_{n+2m+4k+4}\,\beta]_J\overline{\beta}]_K \to [c_{n+2m+4k+4}\,\beta\,\alpha]_J[\overline{\beta}]_K$ (mexo)
$[c_{n+2m+4k+4}\,\alpha]_J[\overline{\alpha}\,b]_L \to [[no]_J\overline{\alpha}]_L$ (mendo)
$[\overline{\beta}]_K \to [\overline{\alpha}]_K[\overline{\alpha}]_K$ (div)
$[c_{n+2m+4k+4}\,\alpha]_J[\overline{\alpha}]_K \to [[yes]_J\overline{\alpha}]_K$ (mendo)
In case all the 2^k assignments do not contain an object c, a membrane L containing an object b and an object z reaches membrane 2. Membrane J with the counter value $n+m+4k+4$ exits membrane K, and in the next step enters membrane L containing a b object and creates an object *no* inside J, deleting the one in K. In case all the 2^k assignments contain an object c, then there is no membrane L containing a b object that will ever reach membrane 2. In this case, a *yes* object is created in membrane J by allowing it to enter membrane K after a determined period of time. The maximum number of steps needed to obtain a *no* object is $n+2m+4k+6$, while to obtain a *yes* object is $n+2m+4k+7$.

The number of membranes in the initial configuration is 6, and the number of objects is $n+6$. The size of the working alphabet is $4n+4m+11k+18$. The number of rules in the above system is: $n+2$ rules of type (*i*), $2(n+1)(2m-1)$ rules of type (*ii*), $2(n+2m)+3$ rules of type (*iii*), $6k$ rules of type (*iv*), 2 rules of type (*v*), 3 of type (*vi*), $4k+1$ rules of type (*vii*) and 4 rules of type (*viii*). Hence, the size of the constructed system of mutual mobile membranes is $\mathcal{O}(mn)$.

Example 2.3. Consider the 2QBF problem with $\phi = \forall X \exists Y (C_1 \wedge C_2)$, $X = \{x_1, x_2\}$, $Y = \{x_3\}$, $C_1 = \neg x_1 \vee x_2$, $C_2 = x_1 \vee x_3$. In this case, $n=3$, $m=2$, $k=2$ and

$$[[c_0\,d\,\beta]_J[\overline{\beta}]_K[\overline{d}]_L[g^2 g_0]_1[a_1]_0]_2$$

Graphically this is illustrated as:

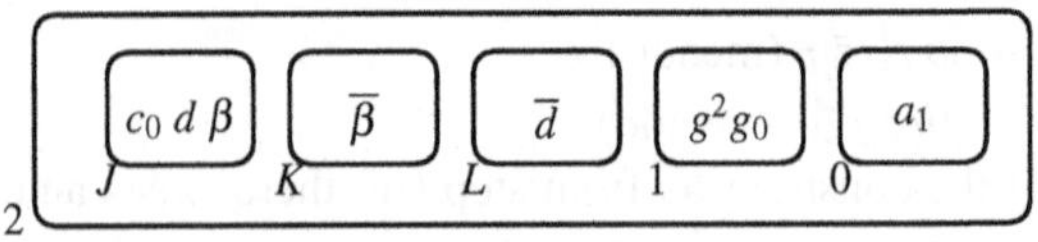

The evolution of the system is described by the following steps. The working space is generated in $n=3$ steps leading from the initial configuration 1 to configuration 4:

1. $[[c_0\,d\,\beta]_J[\overline{\beta}]_K[\overline{d}]_L[g^2 g_0]_1[a_1]_0]_2$
2. $[[[c_1\,d\,\beta]_J\overline{\beta}]_K[\overline{d}]_L[g^2\overline{\beta_1}]_1^2[t_1\,a_2]_0[f_1\,a_2]_0]_2$

3. $[[c_2\, d\, \beta]_J [\overline{\beta}]_K [\overline{d}]_L [g\overline{\beta_1}]_1^4 [t_1\, t_2\, a_3]_0 [t_1\, f_2\, a_3]_0 [f_1\, t_2\, a_3]_0 [f_1\, f_2\, a_3]_0]_2$
4. $[[[c_3\, d\, \beta]_J \overline{\beta}]_K [\overline{d}]_L [\overline{\beta_1}]_1^8 [t_1\, t_2\, t_3\, \beta_1]_0 [t_1\, t_2\, f_3\, \beta_1]_0 [t_1\, f_2\, t_3\, \beta_1]_0 [t_1\, f_2\, f_3\, \beta_1]_0$
$[f_1\, t_2\, t_3\, \beta_1]_0 [f_1\, t_2\, f_3\, \beta_1]_0 [f_1\, f_2\, t_3\, \beta_1]_0 [f_1\, f_2\, f_3\, \beta_1]_0]_2$
Graphically the working space is described by the following picture:

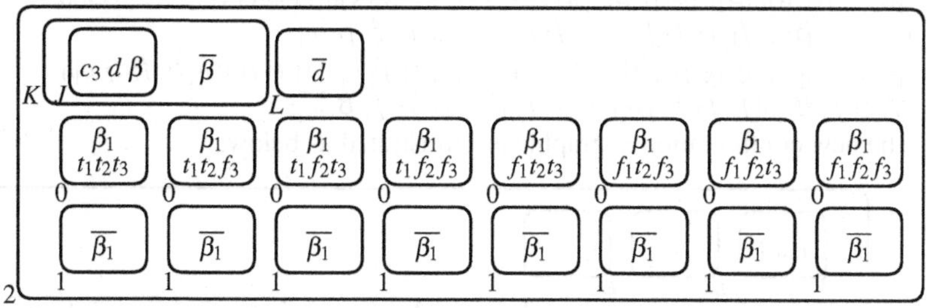

The next two steps mark the solutions of C_1 by replacing β_1 by β_2:

5. $[[c_4\, d\, \beta]_J [\overline{\beta}]_K [\overline{d}]_L [\overline{\beta_1}]_1^2 [\overline{\beta_1} [t_1\, t_2\, t_3\, \beta_1]_0]_1 [\overline{\beta_1} [t_1\, t_2\, f_3\, \beta_1]_0]_1 [\overline{\beta_1} [f_1\, t_2\, f_3\, \beta_1]_0]_1$
$[\overline{\beta_1} [f_1\, f_2\, f_3\, \beta_1]_0]_1 [\overline{\beta_1} [f_1\, t_2\, t_3\, \beta_1]_0]_1 [\overline{\beta_1} [f_1\, f_2\, t_3\, \beta_1]_0]_1 [t_1\, f_2\, f_3\, \beta_1]_0 [t_1\, f_2\, t_3\, \beta_1]_0]_2$
6. $[[[c_5\, d\, \beta]_J \overline{\beta}]_K [\overline{d}]_L [\overline{\beta_1}]_1^2 [\overline{\beta_2}]_1^6 [t_1\, t_2\, t_3\, \beta_2]_0 [t_1\, t_2\, f_3\, \beta_2]_0 [f_1\, t_2\, f_3\, \beta_2]_0 [f_1\, f_2\, f_3\, \beta_2]_0$
$[f_1\, t_2\, t_3\, \beta_2]_0 [f_1\, f_2\, t_3\, \beta_2]_0 [t_1\, f_2\, t_3\, \beta_1]_0 [t_1\, f_2\, f_3\, \beta_1]_0]_2$
The new configuration is graphically represented by:

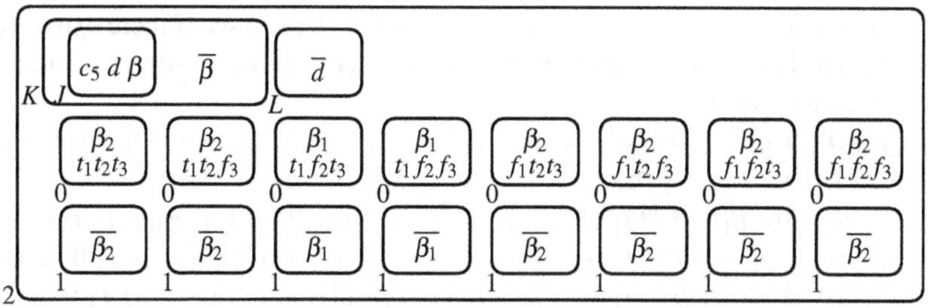

The next two steps mark the solutions of C_2 by replacing β_2 by c:

7. $[[c_6\, d\, \beta]_J [\overline{\beta}]_K [\overline{d}]_L [\overline{\beta_1}]_1^2 [\overline{\beta_2}]_1^2 [\overline{\beta_2} [t_1\, t_2\, t_3\, \beta_2]_0]_1 [\overline{\beta_2} [t_1\, t_2\, f_3\, \beta_2]_0]_1 [\overline{\beta_2} [f_1\, t_2\, t_3\, \beta_2]_0]_1$
$[\overline{\beta_2} [f_1\, f_2\, t_3\, \beta_2]_0]_1 [f_1\, t_2\, f_3\, \beta_2]_0 [f_1\, f_2\, f_3\, \beta_2]_0 [t_1\, f_2\, t_3\, \beta_1]_0 [t_1\, f_2\, f_3\, \beta_1]_0]_2$
8. $[[[c_7\, d\, \beta]_J \overline{\beta}]_K [\overline{d}]_L [\overline{\beta_1}]_1^2 [\overline{\beta_2}]_1^6 [t_1\, t_2\, t_3\, c]_0 [t_1\, t_2\, f_3\, c]_0 [f_1\, t_2\, t_3\, c]_0 [f_1\, f_2\, t_3\, c]_0$
$[f_1\, t_2\, f_3\, \beta_2]_0 [f_1\, f_2\, f_3\, \beta_2]_0 [t_1\, f_2\, t_3\, \beta_1]_0 [t_1\, f_2\, f_3\, \beta_1]_0]_2$
The new configuration is graphically represented by:

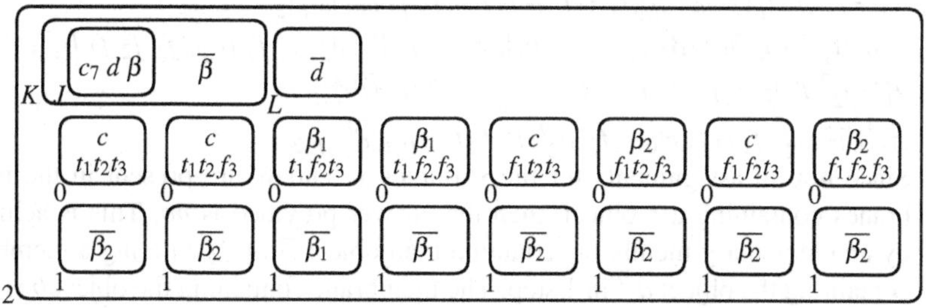

In the next step, an object d_1 is generated. The number of steps needed to create this object is $n+2m+3=3+4+3=10$.

9. $[[c_7\, d\, \beta]_J [\overline{\beta}]_K [\overline{d}]_L [\overline{\beta_1}]_1^2 [\overline{\beta_2}]_1^6 [t_1\, t_2\, t_3\, c]_0 [t_1\, t_2\, f_3\, c]_0 [f_1\, t_2\, t_3\, c]_0 [f_1\, f_2\, t_3\, c]_0$ $[f_1\, t_2\, f_3\, \beta_2]_0 [f_1\, f_2\, f_3\, \beta_2]_0 [t_1\, f_2\, t_3\, \beta_1]_0 [t_1\, f_2\, f_3\, \beta_1]_0]_2$
10. $[[[c_7\, d\, \beta]_J \overline{d}]_L [\overline{\beta}]_K [\overline{\beta_1}]_1^2 [\overline{\beta_2}]_1^6 [t_1\, t_2\, t_3\, c]_0 [t_1\, t_2\, f_3\, c]_0 [f_1\, t_2\, t_3\, c]_0 [f_1\, f_2\, t_3\, c]_0$ $[f_1\, t_2\, f_3\, \beta_2]_0 [f_1\, f_2\, f_3\, \beta_2]_0 [t_1\, f_2\, t_3\, \beta_1]_0 [t_1\, f_2\, f_3\, \beta_1]_0]_2$
11. $[[c_{10}\, \beta]_J [d_1]_L [\overline{\beta}]_K [\overline{\beta_1}]_1^2 [\overline{\beta_2}]_1^6 [t_1\, t_2\, t_3\, c]_0 [t_1\, t_2\, f_3\, c]_0 [f_1\, t_2\, t_3\, c]_0 [f_1\, f_2\, t_3\, c]_0$ $[f_1\, t_2\, f_3\, \beta_2]_0 [f_1\, f_2\, f_3\, \beta_2]_0 [t_1\, f_2\, t_3\, \beta_1]_0 [t_1\, f_2\, f_3\, \beta_1]_0]_2$

The new configuration is graphically illustrated as below:

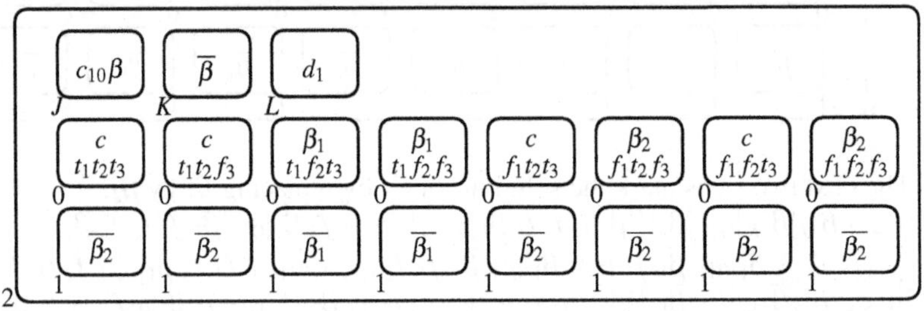

In what follows we check the ∀ part of the formula, namely whether all assignments over variables $x_1,\ldots,x_k$ appear in the existing solutions (the membranes 0 containing an object c). We need $3k=3*2=6$ steps to see if there exists a combination that is missing from the membranes containing an object c, in order to create an object *no*.

12. $[[[c_{11}\, \beta]_J \overline{\beta}]_K [\overline{\beta_1}]_1^2 [\overline{\beta_2}]_1^6 [d_1']_L [d_1'']_L [t_1\, t_2\, t_3\, c]_0 [t_1\, t_2\, f_3\, c]_0 [f_1\, t_2\, t_3\, c]_0 [f_1\, f_2\, t_3\, c]_0$ $[f_1\, t_2\, f_3\, \beta_2]_0 [f_1\, f_2\, f_3\, \beta_2]_0 [t_1\, f_2\, t_3\, \beta_1]_0 [t_1\, f_2\, f_3\, \beta_1]_0]_2$
13. $[[c_{12}\, \beta]_J [\overline{\beta}]_K [\overline{\beta_1}]_1^2 [\overline{\beta_2}]_1^6 [\overline{t_1}^4\, \overline{d_1''''}\, \overline{z}]_L [\overline{f_1}^4\, \overline{d_1''''}\, \overline{z}]_L [d_1''']_L [d_1''']_L [t_1\, t_2\, t_3\, c]_0 [t_1\, t_2\, f_3\, c]_0$ $[f_1\, t_2\, t_3\, c]_0 [f_1\, f_2\, t_3\, c]_0 [f_1\, t_2\, f_3\, \beta_2]_0 [f_1\, f_2\, f_3\, \beta_2]_0 [t_1\, f_2\, t_3\, \beta_1]_0 [t_1\, f_2\, f_3\, \beta_1]_0]_2$
14. $[[[c_{13}\, \beta]_J \overline{\beta}]_K [\overline{\beta_1}]_1^2 [\overline{\beta_2}]_1^6 [t_1\, f_2\, t_3\, \beta_1]_0 [t_1\, f_2\, f_3\, \beta_1]_0 [f_1\, t_2\, f_3\, \beta_2]_0 [f_1\, f_2\, f_3\, \beta_2]_0$ $[\overline{t_1}^4 \overline{z} [d_2]_L [t_1\, t_2\, t_3\, c]_0 [t_1\, t_2\, f_3\, c]_0]_L [\overline{f_1}^4 \overline{z} [d_2]_L [f_1\, t_2\, t_3\, c]_0 [f_1\, f_2\, t_3\, c]_0]_L]_2$
15. $[[c_{14}\, \beta]_J [\overline{\beta}]_K [\overline{\beta_1}]_1^2 [\overline{\beta_2}]_1^6 [t_1\, f_2\, t_3\, \beta_1]_0 [t_1\, f_2\, f_3\, \beta_1]_0 [f_1\, t_2\, f_3\, \beta_2]_0 [f_1\, f_2\, f_3\, \beta_2]_0$ $[\overline{t_1}^4 \overline{z} [d_2']_L [d_2'']_L [t_1\, t_2\, t_3\, c]_0 [t_1\, t_2\, f_3\, c]_0]_L$ $[\overline{f_1}^4 \overline{z} [d_2']_L [d_2'']_L [f_1\, t_2\, t_3\, c]_0 [f_1\, f_2\, t_3\, c]_0]_L]_2$
16. $[[[c_{15}\, \beta]_J \overline{\beta}]_K [\overline{\beta_1}]_1^2 [\overline{\beta_2}]_1^6 [t_1\, f_2\, t_3\, \beta_1]_0 [t_1\, f_2\, f_3\, \beta_1]_0 [f_1\, t_2\, f_3\, \beta_2]_0 [f_1\, f_2\, f_3\, \beta_2]_0$ $[\overline{t_1}^4 \overline{z} [\overline{t_2}^2 d_2'''']_L [\overline{f_2}^2 d_2'''']_L [\overline{d_2''}]_L^2 [t_1\, t_2\, t_3\, c]_0 [t_1\, t_2\, f_3\, c]_0]_L$ $[\overline{f_1}^4 \overline{z} [\overline{t_2}^2 d_2'''']_L [\overline{f_2}^2 d_2'''']_L [\overline{d_2''}]_L^2 [f_1\, t_2\, t_3\, c]_0 [f_1\, f_2\, t_3\, c]_0]_L]_2$
17. $[[c_{16}\, \beta]_J [\overline{\beta}]_K [\overline{\beta_1}]_1^2 [\overline{\beta_2}]_1^6 [t_1\, f_2\, t_3\, \beta_1]_0 [t_1\, f_2\, f_3\, \beta_1]_0 [f_1\, t_2\, f_3\, \beta_2]_0 [f_1\, f_2\, f_3\, \beta_2]_0$ $[\overline{t_1}^4 \overline{z} [\overline{t_2}^2 d_2'''' [t_1\, t_2\, t_3\, c]_0 [t_1\, t_2\, f_3\, c]_0]_L [\overline{f_2}^2 d_2'''']_L [\overline{d_2''''}]_L^4]_L$ $[\overline{f_1}^4 \overline{z} [\overline{t_2}^2 d_2'''' [f_1\, t_2\, t_3\, c]_0]_L [\overline{f_2}^2 d_2'''' [f_1\, f_2\, t_3\, c]_0]_L [\overline{d_2''''}]_L^4]_2$

Since not all assignments for the variables x_1 and x_2 are present in the membranes containing a c object, then the answer provided is *no*. This is achieved by an elementary membrane containing an object $\overline{t_2}$ or $\overline{f_2}$ entering a membrane containing the object $\overline{d_2''''}$ in 1 step. The membrane containing the object b is sent, in k steps to membrane 2 (skin membrane). The new configuration is graphically illustrated as below:

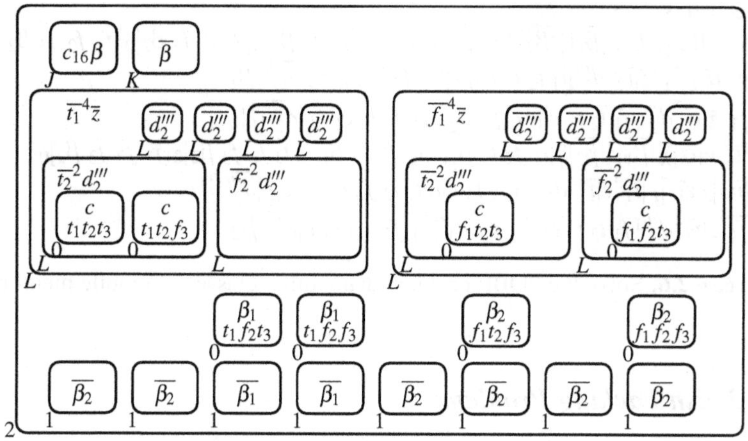

18. $[[[c_{17}\,\beta]_J\overline{\beta}]_K[\overline{\beta_1}]_1^2[\overline{\beta_2}]_1^6[t_1\,f_2\,t_3\,\beta_1]_0[t_1\,f_2\,f_3\,\beta_1]_0[f_1\,t_2\,f_3\,\beta_2]_0[f_1\,f_2\,f_3\,\beta_2]_0$
$[\overline{t_1}^4\overline{z}[\overline{t_2}^2 d_2'''[t_1\,t_2\,t_3\,c]_0[t_1\,t_2\,f_3\,c]_0]_L[\overline{z}[b\,z]_L]_L[\overline{d_2'''}]_L^3]_L$
$[\overline{f_1}^4\overline{z}[\overline{t_2}^2 d_2'''[f_1\,t_2\,t_3\,c]_0]_L[\overline{f_2}^2 d_2'''[f_1\,f_2\,t_3\,c]_0]_L[\overline{d_2'''}]_L^4]_2$
19. $[[c_{18}\,\beta]_J[\overline{\beta}]_K[\overline{\beta_1}]_1^2[\overline{\beta_2}]_1^6[t_1\,f_2\,t_3\,\beta_1]_0[t_1\,f_2\,f_3\,\beta_1]_0[f_1\,t_2\,f_3\,\beta_2]_0[f_1\,f_2\,f_3\,\beta_2]_0$
$[\overline{t_1}^4\overline{z}[\overline{t_2}^2 d_2'''[t_1\,t_2\,t_3\,c]_0[t_1\,t_2\,f_3\,c]_0]_L[\overline{z}]_L[b\,z]_L[\overline{d_2'''}]_L^3]_L$
$[\overline{f_1}^4\overline{z}[\overline{t_2}^2 d_2'''[f_1\,t_2\,t_3\,c]_0]_L[\overline{f_2}^2 d_2'''[f_1\,f_2\,t_3\,c]_0]_L[\overline{d_2'''}]_L^4]_2$
20. $[[[c_{19}\,\beta]_J\overline{\beta}]_K[\overline{\beta_1}]_1^2[\overline{\beta_2}]_1^6[t_1\,f_2\,t_3\,\beta_1]_0[t_1\,f_2\,f_3\,\beta_1]_0[f_1\,t_2\,f_3\,\beta_2]_0[f_1\,f_2\,f_3\,\beta_2]_0$
$[b\,z\overline{\alpha}]_L[\overline{t_1}^4\overline{z}[\overline{t_2}^2 d_2'''[t_1\,t_2\,t_3\,c]_0[t_1\,t_2\,f_3\,c]_0]_L[\overline{z}]_L[\overline{d_2'''}]_L^3]_L$
$[\overline{f_1}^4\overline{z}[\overline{t_2}^2 d_2'''[f_1\,t_2\,t_3\,c]_0]_L[\overline{f_2}^2 d_2'''[f_1\,f_2\,t_3\,c]_0]_L[\overline{d_2'''}]_L^4]_2$

The new configuration is graphically illustrated as below:

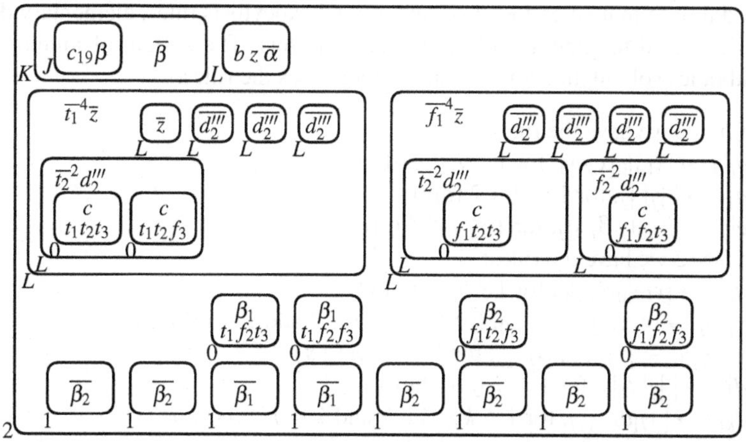

If a membrane L contains a b object and has reached membrane 2, then an object *no* is created inside membrane J. If not, an extra step is performed in order to have a *yes* object in membrane J. The maximum number of steps performed for an *no* object is $n+2m+4k+5=21$. For a positive answer the number of steps needed is $21+1=22$.

21. $[[c_{19}\,\beta\alpha]_J[\overline{\beta}]_K[\overline{\beta_1}]_1^2[\overline{\beta_2}]_1^6[t_1\,f_2\,t_3\,\beta_1]_0[t_1\,f_2\,f_3\,\beta_1]_0[f_1\,t_2\,f_3\,\beta_2]_0[f_1\,f_2\,f_3\,\beta_2]_0$
$[b\,z\overline{\alpha}]_L\overline{[t_1}^4\overline{z}\overline{[t_2}^2 d_2'''[t_1\,t_2\,t_3\,c]_0[t_1\,t_2\,f_3\,c]_0]_L[\overline{z}]_L[\overline{d_2''''}]_L^3]_L$
$\overline{[f_1}^4\overline{z}\overline{[t_2}^2 d_2'''[f_1\,t_2\,t_3\,c]_0]_L\overline{[f_2}^2 d_2'''[f_1\,f_2\,t_3\,c]_0]_L[\overline{d_2''''}]_L^4]_2$
22. $[[\overline{\alpha}]_K[\overline{\alpha}]_K[\overline{\beta_1}]_1^2[\overline{\beta_2}]_1^6[t_1\,f_2\,t_3\,\beta_1]_0[t_1\,f_2\,f_3\,\beta_1]_0[f_1\,t_2\,f_3\,\beta_2]_0[f_1\,f_2\,f_3\,\beta_2]_0$
$[[no]_J\,z\overline{\alpha}]_L\overline{[t_1}^4\overline{z}\overline{[t_2}^2 d_2'''[t_1\,t_2\,t_3\,c]_0[t_1\,t_2\,f_3\,c]_0]_L[\overline{z}]_L[\overline{d_2''''}]_L^3]_L$
$\overline{[f_1}^4\overline{z}\overline{[t_2}^2 d_2'''[f_1\,t_2\,t_3\,c]_0]_L\overline{[f_2}^2 d_2'''[f_1\,f_2\,t_3\,c]_0]_L[\overline{d_2''''}]_L^4]_2$

Exercise 2.6. Solve the 2QBF problem using other classes of mobile membranes.

2.5.3 Bin Packing Problem

The Bin Packing problem can be stated as follows: given a finite set A, a weight function $g : A \to \mathbb{N}$, and two constants $b \in \mathbb{N}$ and $c \in \mathbb{N}$, decide whether or not there exists a partition of A into b subsets such that the weight of each subset does not exceed c.

Consider $A = \{a_1, \ldots, a_n\}$, and a system of mutual mobile membranes having the initial configuration

$$[[\overline{\alpha}]_M[\alpha e_0]_J[\overline{\alpha}]_K[d^{n-1}e]_1\ldots[d^{n-1}e]_b[a_1\ldots a_n c_{1,0}\ldots c_{b,0}\overline{\beta_{1,0}}\ldots\overline{\beta_{b,0}}\beta]_0]_L$$

and working over the alphabet

$$V = \{\alpha,\overline{\alpha},\alpha_1,\beta,\overline{\beta},no,yes,e,d\}\cup\{\gamma_i \mid 1\le i\le n-1\}\cup\{d_i \mid 1\le i\le b-2\}$$
$$\cup\{w_i \mid 1\le i\le b\}\cup\{a_i \mid 1\le i\le n\}\cup\{e_i \mid 1\le i\le 2bn+2c+3\}$$
$$\cup\{\psi_{i,j},\beta_{i,j},x_{i,j},y_{i,j},z_{i,j} \mid 1\le i\le b, 1\le j\le n\}$$
$$\cup\{c_{i,j} \mid 1\le i\le b, 0\le j\le 2bn+2c+1\}$$

In addition to mutual endocytosis and mutual exocytosis rules, elementary division rules are used to generate all the possible subsets. The system of mutual mobile membranes solving the bin packing problem uses the rules:

(i) If $b=2$ we have the rules:
$[a_i]_0 \to [x_{1,i}]_0[x_{2,i}]_0$, for $1\le i\le n$ (div)
$[d]_j \to [\,]_j[\,]_j$, for $1\le j\le 2$ (div)
$[e]_j \to [\beta_{j,0}]_j[\beta_{j,0}]_j$, for $1\le j\le 2$ (div)
If $b>2$ we have the rules:
$[a_i]_0 \to [x_{1,i}]_0[y_{1,i}]_0$, for $1\le i\le n$ (div)
$[y_{j,i}]_0 \to [x_{j+1,i}]_0[y_{j+1,i}]_0$, for $1\le j\le b-3$ and $1\le i\le n$ (div)
$[y_{b-2,i}]_0 \to [x_{b-1,i}]_0[x_{b,i}]_0$, for $1\le i\le n$ (div)
$[d]_j \to [\,]_j[d_1]_j$, for $1\le j\le b$ (div)
$[d_k]_j \to [\,]_j[d_{k+1}]_j$, for $1\le k\le b-3$ and $1\le j\le b$ (div)
$[d_{b-2}]_j \to [\,]_j[\,]_j$, for $1\le j\le b$ (div)
$[e]_j \to [\beta_{j,0}]_j[e_1]_j$ (div)
$[e_k]_j \to [\beta_{j,0}]_j[e_{k+1}]_j$, for $1\le k\le b-3$ and $1\le j\le b$ (div)

$[e_{b-2}]_j \to [\beta_{j,0}]_j[\beta_{j,0}]_j$, for $1 \le j \le b$ (div)

Different sets of rules are needed, depending on the number of bins (2 or more). The rules containing the objects a, x or y are used to create b^n membranes labelled by 0 containing all the possible subsets over $\{a_1,\dots,a_n\}$. In $x_{j,i}$, the bin is denoted by j, while i denotes the variable placed in bin j. The objects $x_{j,i}$ are used to introduce objects that represent the weight of the object a_i. In each membrane are placed also the objects $\beta_{j,0}$, $1 \le j \le b$, that are used to count the weight of each bin. The next rules create, for each $1 \le j \le b$, b^n membranes labelled by j each containing an object $\beta_{j,0}$.

(ii) $[\beta_{j,i}]_j[x_{j,i}\overline{\beta_{j,i}}]_0 \to [[\beta_{j,i+1}]_j x_{j,i}\overline{\beta_{j,i+1}}]_0$, for $0 \le i \le n-1$ and $1 \le j \le b$ (mendo)
$[[\beta_{j,i}]_j x_{j,i}\overline{\beta_{j,i}}]_0 \to [\beta_{j,i}]_j[z_{j,i}\, w_j^{g(a_i)}\overline{\beta_{j,i}}]_0$, for $1 \le i \le n$ and $1 \le j \le b$ (mexo)
$[[\beta_{j,i}]_j\overline{\beta_{j,i}}]_0 \to [\beta_{j,i}]_j[\overline{\beta_{j,i}}]_0$, for $1 \le i \le n$ and $1 \le j \le b$ (mexo)
These rules are used to replace $x_{j,i}$ by $z_{j,i}\, w_j^{g(a_i)}$. If $x_{j,i}$ is not present in a membrane 0, then that membrane just increments the second index associated to β. After applying these rules, each membrane 0 contains a number of objects w_j equal to the weight contained in the bin j, and also objects $z_{j,i}$ used to remember which objects are contained in each bin.

(iii) $[\beta_{j,n}]_j[w_j\,\overline{\beta_{j,n}}]_0 \to [[\beta_{j,n}]_j\overline{\beta_{j,n}}]_0$ (mendo)
$[[\beta_{j,n}]_j c_{j,i}\,\overline{\beta_{j,n}}]_0 \to [\beta_{j,n}]_j[c_{j,i+1}\,\overline{\beta_{j,n}}]_0$, for $0 \le i$ and $1 \le j \le b$ (mexo)
These rules are used to calculate the weights of each bin j, by using the objects $c_{j,0}$ that appear in all 0 membranes.

(iv) $[\alpha\, e_i]_J[\overline{\alpha}]_K \to [[\alpha\, e_{i+1}]_J\overline{\alpha}]_K$ for $0 \le i < (b+1)n+2c-1$ (mendo)
$[[\alpha\, e_i]_J\overline{\alpha}]_K \to [\alpha\, e_{i+1}]_J[\overline{\alpha}]_K$, for $0 \le i < (b+1)n+2c-1$ (mexo)
$[[\alpha\, e_i]_J\overline{\alpha}]_K \to [\alpha\, e_{i+1}]_J[\alpha e_{i+1}]_K$, for $i = (b+1)n+2c-1$ (mexo)
The above rules are used in parallel to calculate the number of steps performed. The number $(b+1)n+2c$ is determined by: generating space ($(b-1)n$ steps), replacing $x_{j,i}$ by corresponding weight ($2n$ steps) and calculating weights ($2c$ steps).

(v) $[[\alpha\, e_{(b+1)n+2c}]_J\overline{\alpha}]_K \to [\alpha\, e_{(b+1)n+2c}]_J[\alpha e_{(b+1)n+2c}]_K$ (mexo)
If $b = 2$ we have the rules
$[\alpha\, e_{(b+1)n+2c}]_J \to [\gamma_1]_J[\psi_1]_J$ (div)
$[\gamma_i]_J \to [\gamma_{i+1}]_J[\gamma_{i+1}]_J$, for $1 \le i \le n-2$ (div)
$[\gamma_{n-1}]_J \to [\overline{\beta}]_J[\overline{\beta}]_J$ (div)
If $b > 2$ we have the rules
$[\alpha\, e_{(b+1)n+2c}]_J \to [\gamma_1]_J[\psi_{1,1}]_J$ (div)
$[\psi_{i,j}]_J \to [\gamma_i]_J[\psi_{i,j+1}]_J$, for $1 \le j \le b-3$ and $1 \le i \le n-1$ (div)
$[\psi_{i,b-2}]_J \to [\gamma_i]_J[\gamma_i]_J$, for $1 \le i \le n-1$ (div)
$[\gamma_i]_J \to [\gamma_{i+1}]_J[\psi_{i+1,1}]_J$, for $1 \le i \le n-2$ (div)
$[\psi_{n,j}]_J \to [\overline{\beta}]_J[\psi_{n,j+1}]_J$, for $1 \le j \le b-3$ (div)
$[\psi_{n,b-2}]_J \to [\overline{\beta}]_J[\overline{\beta}]_J$ (div)
For $b \ge 2$ we have the rules
$[\overline{\beta}]_J[c_{k,j_k}\beta]_0 \to [[\ ]_0\ ]_J$, if $j_k > c$ (mendo)
After an extra step needed to prepare the membrane J for division, b^n membranes J are created ($(b-1)n$ steps) to check which membranes respect the

weight condition. All membranes 0 that do not respect the condition are blocked inside the J membranes. In this way by choosing any membrane that is not placed inside a J membrane, we obtain a solution to the problem.

(vi) $[\alpha\, e_i]_K[\overline{\alpha}]_M \to [[\alpha\, e_{i+1}]_K\overline{\alpha}]_M$ for $(b+1)n+2c \le i \le 2bn+2c+2$ (mendo)
$[[\alpha\, e_i]_K\overline{\alpha}]_M \to [\alpha\, e_{i+1}]_K[\overline{\alpha}]_M$, for $(b+1)n+2c \le i < 2bn+2c+2$ (mexo)
$[[\alpha\, e_i]_K\overline{\alpha}]_M \to [\overline{\beta}\beta]_K[\alpha_1]_M$, for $i = 2bn+2c+2$ or $i = 2bn+2c+3$ (mexo)
$[\overline{\beta}]_K[\beta]_0 \to [yes[\ ]_0]_K$ (mendo)
$[\alpha_1]_M \to [\overline{\beta}]_M[\overline{\beta}]_M$ (mendo)
$[\beta]_K[\overline{\beta}]_M \to [[no]_K]_M$ (mendo)
The above rules are used in parallel to calculate the number of steps performed. The number $2bn+2c+2$ is determined by: generating space ($(b-1)n$ steps), replacing $x_{j,i}$ by corresponding weight ($2n$ steps), calculating weights ($2c$ steps), generating J membranes ($(b-1)n$ steps), preparing division of J (1 step), blocking all membranes 0 that do not satisfy conditions (1 step). If there still exists a membrane 0 that is not inside a membrane J, then the object *yes* is created inside membrane K. Otherwise, after one more step, the *no* object is created inside membrane K. The computation stops after $2bn+2c+5$ steps, with the answer placed inside membrane K.

The number of membranes in the initial configuration is $b+5$, and the number of objects is $nb+n+2b+3$. The size of the working alphabet is $7bn+3b+2n+2c+2b^2n+2bc+9$. The number of rules in the above system is: $n+4$ rules of type (i) if $b=2$ or $3bn-7n+4b$ if $b>2$, $3bn-b$ rules of type (ii), $2cb$ rules of type (iii), $2(b+1)n+4c$ rules of type (iv), and $n+1$ rules of type (v) if $b=2$ or $bn+bc-n+b$ if $b>2$. Hence, the size of the constructed system of mutual mobile membranes is $max\{\mathcal{O}(bn), \mathcal{O}(bc)\}$.

Example 2.4. Consider the bin packing problem with $A = \{a_1, a_2, a_3\}$, $g(a_1) = 1$, $g(a_2) = 3$, $g(a_3) = 2$, $b = 3$ and $c = 3$. In this case, $n = b = c = 3$ and
$$[[\overline{\alpha}]_M[\alpha e_0]_J[\overline{\alpha}]_K[d^2e]_1[d^2e]_2[d^2e]_3[a_1a_2a_3c_{1,0}c_{2,0}c_{3,0}\overline{\beta_{1,0}}\ \overline{\beta_{2,0}}\ \overline{\beta_{3,0}}\beta]_0]_L$$
Graphically this is illustrated as:

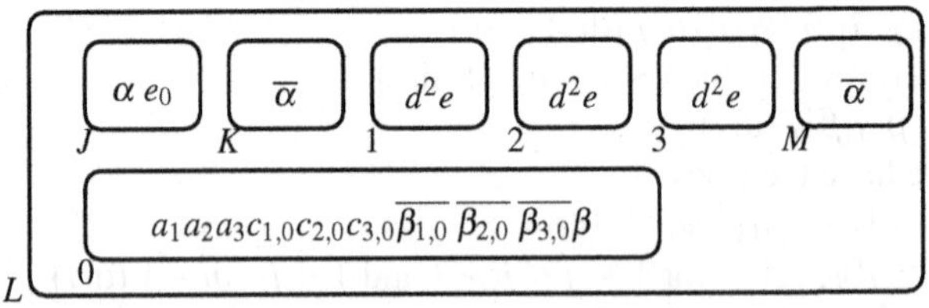

The evolution of the system is described by the following steps. The working space is created in $2n = 2*3 = 6$ steps leading from the initial configuration 1 to configuration 7.

1. $[[\overline{\alpha}]_M[\alpha e_0]_J[\overline{\alpha}]_K[d^2e]_1[d^2e]_2[d^2e]_3[a_1a_2a_3c_{1,0}c_{2,0}c_{3,0}\overline{\beta_{1,0}}\ \overline{\beta_{2,0}}\ \overline{\beta_{3,0}}\beta]_0]_L$
2. $[[\overline{\alpha}]_M[[\alpha e_1]_J\overline{\alpha}]_K[d\ e]_1[d_1d\ e]_1[d\ e]_2[d_1d\ e]_2[d\ e]_3[d_1d\ e]_3$
$[x_{1,1}a_2a_3c_{1,0}c_{2,0}c_{3,0}\overline{\beta_{1,0}}\ \overline{\beta_{2,0}}\ \overline{\beta_{3,0}}\beta]_0[y_{1,1}a_2a_3c_{1,0}c_{2,0}c_{3,0}\overline{\beta_{1,0}}\ \overline{\beta_{2,0}}\ \overline{\beta_{3,0}}\beta]_0]_L$

3. $[[\overline{\alpha}]_M[\alpha e_2]_J[\overline{\alpha}]_K[e]_1[d_1\ e]_1[d\ e]^2_1[e]_2[d_1\ e]_2[d\ e]^2_2[e]_3[d_1\ e]_3[d\ e]^2_3$
$[x_{1,1}x_{1,2}a_3c_{1,0}c_{2,0}c_{3,0}\overline{\beta_{1,0}}\ \overline{\beta_{2,0}}\ \overline{\beta_{3,0}}\beta]_0[x_{1,1}y_{1,2}a_3c_{1,0}c_{2,0}c_{3,0}\overline{\beta_{1,0}}\ \overline{\beta_{2,0}}\ \overline{\beta_{3,0}}\beta]_0$
$[x_{2,1}a_2a_3c_{1,0}c_{2,0}c_{3,0}\overline{\beta_{1,0}}\ \overline{\beta_{2,0}}\ \overline{\beta_{3,0}}\beta]_0[x_{3,1}a_2a_3c_{1,0}c_{2,0}c_{3,0}\overline{\beta_{1,0}}\ \overline{\beta_{2,0}}\ \overline{\beta_{3,0}}\beta]_0]_L$

4. $[[\overline{\alpha}]_M[[\alpha e_3]_J\overline{\alpha}]_K[\beta_{1,0}]_1[e_1]_1[e]^4_1[d_1\ e]^2_1[\beta_{2,0}]_2[e_1]_2[e]^4_2[d_1\ e]^2_2[\beta_{3,0}]_3[e_1]_3[e]^4_3[d_1\ e]^2_3$
$[x_{1,1}x_{1,2}x_{1,3}c_{1,0}c_{2,0}c_{3,0}\overline{\beta_{1,0}}\ \overline{\beta_{2,0}}\ \overline{\beta_{3,0}}\beta]_0[x_{1,1}x_{1,2}y_{1,3}c_{1,0}c_{2,0}c_{3,0}\overline{\beta_{1,0}}\ \overline{\beta_{2,0}}\ \overline{\beta_{3,0}}\beta]_0$
$[x_{1,1}x_{2,2}a_3c_{1,0}c_{2,0}c_{3,0}\overline{\beta_{1,0}}\ \overline{\beta_{2,0}}\ \overline{\beta_{3,0}}\beta]_0[x_{1,1}x_{3,2}a_3c_{1,0}c_{2,0}c_{3,0}\overline{\beta_{1,0}}\ \overline{\beta_{2,0}}\ \overline{\beta_{3,0}}\beta]_0$
$[x_{2,1}x_{1,2}a_3c_{1,0}c_{2,0}c_{3,0}\overline{\beta_{1,0}}\ \overline{\beta_{2,0}}\ \overline{\beta_{3,0}}\beta]_0[x_{2,1}y_{1,2}a_3c_{1,0}c_{2,0}c_{3,0}\overline{\beta_{1,0}}\ \overline{\beta_{2,0}}\ \overline{\beta_{3,0}}\beta]_0$
$[x_{3,1}x_{1,2}a_3c_{1,0}c_{2,0}c_{3,0}\overline{\beta_{1,0}}\ \overline{\beta_{2,0}}\ \overline{\beta_{3,0}}\beta]_0[x_{3,1}y_{1,2}a_3c_{1,0}c_{2,0}c_{3,0}\overline{\beta_{1,0}}\ \overline{\beta_{2,0}}\ \overline{\beta_{3,0}}\beta]_0]_L$

5. $[[\overline{\alpha}]_M[\alpha e_4]_J[\overline{\alpha}]_K[\beta_{1,0}]^7_1[e_1]^4_1[e]^4_1[\beta_{2,0}]^7_2[e_1]^4_2[e]^4_2[\beta_{3,0}]^7_3[e_1]^4_3[e]^4_3$
$[x_{1,1}x_{1,2}x_{1,3}c_{1,0}c_{2,0}c_{3,0}\overline{\beta_{1,0}}\ \overline{\beta_{2,0}}\ \overline{\beta_{3,0}}\beta]_0[x_{1,1}x_{1,2}x_{2,3}c_{1,0}c_{2,0}c_{3,0}\overline{\beta_{1,0}}\ \overline{\beta_{2,0}}\ \overline{\beta_{3,0}}\beta]_0$
$[x_{1,1}x_{1,2}x_{3,3}c_{1,0}c_{2,0}c_{3,0}\overline{\beta_{1,0}}\ \overline{\beta_{2,0}}\ \overline{\beta_{3,0}}\beta]_0[x_{1,1}x_{2,2}x_{1,3}c_{1,0}c_{2,0}c_{3,0}\overline{\beta_{1,0}}\ \overline{\beta_{2,0}}\ \overline{\beta_{3,0}}\beta]_0$
$[x_{1,1}x_{2,2}y_{1,3}c_{1,0}c_{2,0}c_{3,0}\overline{\beta_{1,0}}\ \overline{\beta_{2,0}}\ \overline{\beta_{3,0}}\beta]_0[x_{1,1}x_{3,2}x_{1,3}c_{1,0}c_{2,0}c_{3,0}\overline{\beta_{1,0}}\ \overline{\beta_{2,0}}\ \overline{\beta_{3,0}}\beta]_0$
$[x_{1,1}x_{3,2}y_{1,3}c_{1,0}c_{2,0}c_{3,0}\overline{\beta_{1,0}}\ \overline{\beta_{2,0}}\ \overline{\beta_{3,0}}\beta]_0[x_{2,1}x_{1,2}x_{1,3}c_{1,0}c_{2,0}c_{3,0}\overline{\beta_{1,0}}\ \overline{\beta_{2,0}}\ \overline{\beta_{3,0}}\beta]_0$
$[x_{2,1}x_{1,2}y_{1,3}c_{1,0}c_{2,0}c_{3,0}\overline{\beta_{1,0}}\ \overline{\beta_{2,0}}\ \overline{\beta_{3,0}}\beta]_0[x_{2,1}x_{2,2}a_3c_{1,0}c_{2,0}c_{3,0}\overline{\beta_{1,0}}\ \overline{\beta_{2,0}}\ \overline{\beta_{3,0}}\beta]_0$
$[x_{2,1}x_{3,2}a_3c_{1,0}c_{2,0}c_{3,0}\overline{\beta_{1,0}}\ \overline{\beta_{2,0}}\ \overline{\beta_{3,0}}\beta]_0[x_{3,1}x_{1,2}x_{1,3}c_{1,0}c_{2,0}c_{3,0}\overline{\beta_{1,0}}\ \overline{\beta_{2,0}}\ \overline{\beta_{3,0}}\beta]_0$
$[x_{3,1}x_{1,2}y_{1,3}c_{1,0}c_{2,0}c_{3,0}\overline{\beta_{1,0}}\ \overline{\beta_{2,0}}\ \overline{\beta_{3,0}}\beta]_0[x_{3,1}x_{2,2}a_3c_{1,0}c_{2,0}c_{3,0}\overline{\beta_{1,0}}\ \overline{\beta_{2,0}}\ \overline{\beta_{3,0}}\beta]_0$
$[x_{3,1}x_{3,2}a_3c_{1,0}c_{2,0}c_{3,0}\overline{\beta_{1,0}}\ \overline{\beta_{2,0}}\ \overline{\beta_{3,0}}\beta]_0]_L$

6. $[[\overline{\alpha}]_M[[\alpha e_5]_J\overline{\alpha}]_K[\beta_{1,0}]^{19}_1[e_1]^4_1[\beta_{2,0}]^{19}_2[e_1]^4_2[\beta_{3,0}]^{19}_3[e_1]^4_3$
$[x_{1,1}x_{1,2}x_{1,3}c_{1,0}c_{2,0}c_{3,0}\overline{\beta_{1,0}}\ \overline{\beta_{2,0}}\ \overline{\beta_{3,0}}\beta]_0[x_{1,1}x_{1,2}x_{2,3}c_{1,0}c_{2,0}c_{3,0}\overline{\beta_{1,0}}\ \overline{\beta_{2,0}}\ \overline{\beta_{3,0}}\beta]_0$
$[x_{1,1}x_{1,2}x_{3,3}c_{1,0}c_{2,0}c_{3,0}\overline{\beta_{1,0}}\ \overline{\beta_{2,0}}\ \overline{\beta_{3,0}}\beta]_0[x_{1,1}x_{2,2}x_{1,3}c_{1,0}c_{2,0}c_{3,0}\overline{\beta_{1,0}}\ \overline{\beta_{2,0}}\ \overline{\beta_{3,0}}\beta]_0$
$[x_{1,1}x_{2,2}x_{2,3}c_{1,0}c_{2,0}c_{3,0}\overline{\beta_{1,0}}\ \overline{\beta_{2,0}}\ \overline{\beta_{3,0}}\beta]_0[x_{1,1}x_{2,2}x_{3,3}c_{1,0}c_{2,0}c_{3,0}\overline{\beta_{1,0}}\ \overline{\beta_{2,0}}\ \overline{\beta_{3,0}}\beta]_0$
$[x_{1,1}x_{3,2}x_{1,3}c_{1,0}c_{2,0}c_{3,0}\overline{\beta_{1,0}}\ \overline{\beta_{2,0}}\ \overline{\beta_{3,0}}\beta]_0[x_{1,1}x_{3,2}x_{2,3}c_{1,0}c_{2,0}c_{3,0}\overline{\beta_{1,0}}\ \overline{\beta_{2,0}}\ \overline{\beta_{3,0}}\beta]_0$
$[x_{1,1}x_{3,2}x_{3,3}c_{1,0}c_{2,0}c_{3,0}\overline{\beta_{1,0}}\ \overline{\beta_{2,0}}\ \overline{\beta_{3,0}}\beta]_0[x_{2,1}x_{1,2}x_{1,3}c_{1,0}c_{2,0}c_{3,0}\overline{\beta_{1,0}}\ \overline{\beta_{2,0}}\ \overline{\beta_{3,0}}\beta]_0$
$[x_{2,1}x_{1,2}x_{2,3}c_{1,0}c_{2,0}c_{3,0}\overline{\beta_{1,0}}\ \overline{\beta_{2,0}}\ \overline{\beta_{3,0}}\beta]_0[x_{2,1}x_{1,2}x_{3,3}c_{1,0}c_{2,0}c_{3,0}\overline{\beta_{1,0}}\ \overline{\beta_{2,0}}\ \overline{\beta_{3,0}}\beta]_0$
$[x_{2,1}x_{2,2}x_{1,3}c_{1,0}c_{2,0}c_{3,0}\overline{\beta_{1,0}}\ \overline{\beta_{2,0}}\ \overline{\beta_{3,0}}\beta]_0[x_{2,1}x_{2,2}y_{1,3}c_{1,0}c_{2,0}c_{3,0}\overline{\beta_{1,0}}\ \overline{\beta_{2,0}}\ \overline{\beta_{3,0}}\beta]_0$
$[x_{2,1}x_{3,2}x_{1,3}c_{1,0}c_{2,0}c_{3,0}\overline{\beta_{1,0}}\ \overline{\beta_{2,0}}\ \overline{\beta_{3,0}}\beta]_0[x_{2,1}x_{3,2}y_{1,3}c_{1,0}c_{2,0}c_{3,0}\overline{\beta_{1,0}}\ \overline{\beta_{2,0}}\ \overline{\beta_{3,0}}\beta]_0$
$[x_{3,1}x_{1,2}x_{1,3}c_{1,0}c_{2,0}c_{3,0}\overline{\beta_{1,0}}\ \overline{\beta_{2,0}}\ \overline{\beta_{3,0}}\beta]_0[x_{3,1}x_{1,2}x_{2,3}c_{1,0}c_{2,0}c_{3,0}\overline{\beta_{1,0}}\ \overline{\beta_{2,0}}\ \overline{\beta_{3,0}}\beta]_0$
$[x_{3,1}x_{1,2}x_{3,3}c_{1,0}c_{2,0}c_{3,0}\overline{\beta_{1,0}}\ \overline{\beta_{2,0}}\ \overline{\beta_{3,0}}\beta]_0[x_{3,1}x_{2,2}x_{1,3}c_{1,0}c_{2,0}c_{3,0}\overline{\beta_{1,0}}\ \overline{\beta_{2,0}}\ \overline{\beta_{3,0}}\beta]_0$
$[x_{3,1}x_{2,2}y_{1,3}c_{1,0}c_{2,0}c_{3,0}\overline{\beta_{1,0}}\ \overline{\beta_{2,0}}\ \overline{\beta_{3,0}}\beta]_0[x_{3,1}x_{3,2}x_{1,3}c_{1,0}c_{2,0}c_{3,0}\overline{\beta_{1,0}}\ \overline{\beta_{2,0}}\ \overline{\beta_{3,0}}\beta]_0$
$[x_{3,1}x_{3,2}y_{1,3}c_{1,0}c_{2,0}c_{3,0}\overline{\beta_{1,0}}\ \overline{\beta_{2,0}}\ \overline{\beta_{3,0}}\beta]_0]_L$

7. $[[\overline{\alpha}]_M[\alpha e_6]_J[\overline{\alpha}]_K[\beta_{1,0}]^{27}_1[\beta_{2,0}]^{27}_2[\beta_{3,0}]^{27}_3$
$[x_{1,1}x_{1,2}x_{1,3}c_{1,0}c_{2,0}c_{3,0}\overline{\beta_{1,0}}\ \overline{\beta_{2,0}}\ \overline{\beta_{3,0}}\beta]_0[x_{1,1}x_{1,2}x_{2,3}c_{1,0}c_{2,0}c_{3,0}\overline{\beta_{1,0}}\ \overline{\beta_{2,0}}\ \overline{\beta_{3,0}}\beta]_0$
$[x_{1,1}x_{1,2}x_{3,3}c_{1,0}c_{2,0}c_{3,0}\overline{\beta_{1,0}}\ \overline{\beta_{2,0}}\ \overline{\beta_{3,0}}\beta]_0[x_{1,1}x_{2,2}x_{1,3}c_{1,0}c_{2,0}c_{3,0}\overline{\beta_{1,0}}\ \overline{\beta_{2,0}}\ \overline{\beta_{3,0}}\beta]_0$
$[x_{1,1}x_{2,2}x_{2,3}c_{1,0}c_{2,0}c_{3,0}\overline{\beta_{1,0}}\ \overline{\beta_{2,0}}\ \overline{\beta_{3,0}}\beta]_0[x_{1,1}x_{2,2}x_{3,3}c_{1,0}c_{2,0}c_{3,0}\overline{\beta_{1,0}}\ \overline{\beta_{2,0}}\ \overline{\beta_{3,0}}\beta]_0$
$[x_{1,1}x_{3,2}x_{1,3}c_{1,0}c_{2,0}c_{3,0}\overline{\beta_{1,0}}\ \overline{\beta_{2,0}}\ \overline{\beta_{3,0}}\beta]_0[x_{1,1}x_{3,2}x_{2,3}c_{1,0}c_{2,0}c_{3,0}\overline{\beta_{1,0}}\ \overline{\beta_{2,0}}\ \overline{\beta_{3,0}}\beta]_0$
$[x_{1,1}x_{3,2}x_{3,3}c_{1,0}c_{2,0}c_{3,0}\overline{\beta_{1,0}}\ \overline{\beta_{2,0}}\ \overline{\beta_{3,0}}\beta]_0[x_{2,1}x_{1,2}x_{1,3}c_{1,0}c_{2,0}c_{3,0}\overline{\beta_{1,0}}\ \overline{\beta_{2,0}}\ \overline{\beta_{3,0}}\beta]_0$
$[x_{2,1}x_{1,2}x_{2,3}c_{1,0}c_{2,0}c_{3,0}\overline{\beta_{1,0}}\ \overline{\beta_{2,0}}\ \overline{\beta_{3,0}}\beta]_0[x_{2,1}x_{1,2}x_{3,3}c_{1,0}c_{2,0}c_{3,0}\overline{\beta_{1,0}}\ \overline{\beta_{2,0}}\ \overline{\beta_{3,0}}\beta]_0$
$[x_{2,1}x_{2,2}x_{1,3}c_{1,0}c_{2,0}c_{3,0}\overline{\beta_{1,0}}\ \overline{\beta_{2,0}}\ \overline{\beta_{3,0}}\beta]_0[x_{2,1}x_{2,2}x_{2,3}c_{1,0}c_{2,0}c_{3,0}\overline{\beta_{1,0}}\ \overline{\beta_{2,0}}\ \overline{\beta_{3,0}}\beta]_0$
$[x_{2,1}x_{2,2}x_{3,3}c_{1,0}c_{2,0}c_{3,0}\overline{\beta_{1,0}}\ \overline{\beta_{2,0}}\ \overline{\beta_{3,0}}\beta]_0[x_{2,1}x_{3,2}x_{1,3}c_{1,0}c_{2,0}c_{3,0}\overline{\beta_{1,0}}\ \overline{\beta_{2,0}}\ \overline{\beta_{3,0}}\beta]_0$
$[x_{2,1}x_{3,2}x_{2,3}c_{1,0}c_{2,0}c_{3,0}\overline{\beta_{1,0}}\ \overline{\beta_{2,0}}\ \overline{\beta_{3,0}}\beta]_0[x_{2,1}x_{3,2}x_{3,3}c_{1,0}c_{2,0}c_{3,0}\overline{\beta_{1,0}}\ \overline{\beta_{2,0}}\ \overline{\beta_{3,0}}\beta]_0$
$[x_{3,1}x_{1,2}x_{1,3}c_{1,0}c_{2,0}c_{3,0}\overline{\beta_{1,0}}\ \overline{\beta_{2,0}}\ \overline{\beta_{3,0}}\beta]_0[x_{3,1}x_{1,2}x_{2,3}c_{1,0}c_{2,0}c_{3,0}\overline{\beta_{1,0}}\ \overline{\beta_{2,0}}\ \overline{\beta_{3,0}}\beta]_0$

$[x_{3,1}x_{1,2}x_{3,3}c_{1,0}c_{2,0}c_{3,0}\overline{\beta_{1,0}}\ \overline{\beta_{2,0}}\ \overline{\beta_{3,0}}\beta]_0[x_{3,1}x_{2,2}x_{1,3}c_{1,0}c_{2,0}c_{3,0}\overline{\beta_{1,0}}\ \overline{\beta_{2,0}}\ \overline{\beta_{3,0}}\beta]_0$
$[x_{3,1}x_{2,2}x_{2,3}c_{1,0}c_{2,0}c_{3,0}\overline{\beta_{1,0}}\ \overline{\beta_{2,0}}\ \overline{\beta_{3,0}}\beta]_0[x_{3,1}x_{2,2}x_{3,3}c_{1,0}c_{2,0}c_{3,0}\overline{\beta_{1,0}}\ \overline{\beta_{2,0}}\ \overline{\beta_{3,0}}\beta]_0$
$[x_{3,1}x_{3,2}x_{1,3}c_{1,0}c_{2,0}c_{3,0}\overline{\beta_{1,0}}\ \overline{\beta_{2,0}}\ \overline{\beta_{3,0}}\beta]_0[x_{3,1}x_{3,2}x_{2,3}c_{1,0}c_{2,0}c_{3,0}\overline{\beta_{1,0}}\ \overline{\beta_{2,0}}\ \overline{\beta_{3,0}}\beta]_0$
$[x_{3,1}x_{3,2}x_{3,3}c_{1,0}c_{2,0}c_{3,0}\overline{\beta_{1,0}}\ \overline{\beta_{2,0}}\ \overline{\beta_{3,0}}\beta]_0]_L$

Graphically the working space is described by the following picture, where for membranes labelled by 1, 2 and 3 we draw only two representatives:

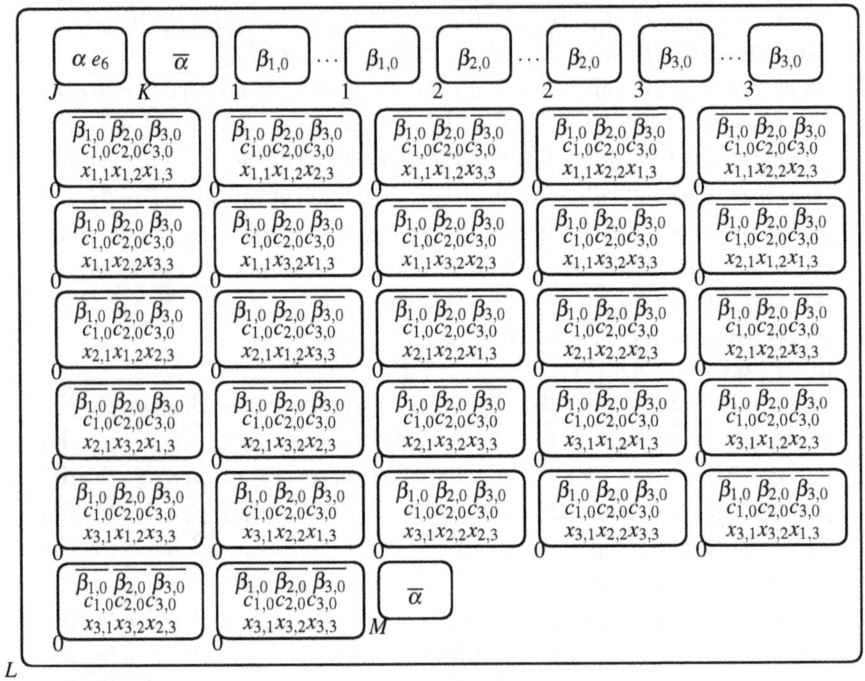

In what follows we replace $x_{j,i}$ by $z_{j,i}w_j^{g(a_i)}$, such that in each membrane we obtain a number of objects w_j equal to the weight of the objects contained in the bin j. The objects $z_{j,i}$ are used to remember which objects are contained in each membrane 0.

8. $[[\overline{\alpha}]_M[[\alpha e_7]_J\overline{\alpha}]_K$
$[[\beta_{1,1}]_1[\beta_{2,1}]_2[\beta_{3,1}]_3x_{1,1}x_{1,2}x_{1,3}c_{1,0}c_{2,0}c_{3,0}\overline{\beta_{1,1}}\ \overline{\beta_{2,1}}\ \overline{\beta_{3,1}}\beta]_0$
$[[\beta_{1,1}]_1[\beta_{2,1}]_2[\beta_{3,1}]_3x_{1,1}x_{1,2}x_{2,3}c_{1,0}c_{2,0}c_{3,0}\overline{\beta_{1,1}}\ \overline{\beta_{2,1}}\ \overline{\beta_{3,1}}\beta]_0$
$[[\beta_{1,1}]_1[\beta_{2,1}]_2[\beta_{3,1}]_3x_{1,1}x_{1,2}x_{3,3}c_{1,0}c_{2,0}c_{3,0}\overline{\beta_{1,1}}\ \overline{\beta_{2,1}}\ \overline{\beta_{3,1}}\beta]_0$
$[[\beta_{1,1}]_1[\beta_{2,1}]_2[\beta_{3,1}]_3x_{1,1}x_{2,2}x_{1,3}c_{1,0}c_{2,0}c_{3,0}\overline{\beta_{1,1}}\ \overline{\beta_{2,1}}\ \overline{\beta_{3,1}}\beta]_0$
$[[\beta_{1,1}]_1[\beta_{2,1}]_2[\beta_{3,1}]_3x_{1,1}x_{2,2}x_{2,3}c_{1,0}c_{2,0}c_{3,0}\overline{\beta_{1,1}}\ \overline{\beta_{2,1}}\ \overline{\beta_{3,1}}\beta]_0$
$[[\beta_{1,1}]_1[\beta_{2,1}]_2[\beta_{3,1}]_3x_{1,1}x_{2,2}x_{3,3}c_{1,0}c_{2,0}c_{3,0}\overline{\beta_{1,1}}\ \overline{\beta_{2,1}}\ \overline{\beta_{3,1}}\beta]_0$
$[[\beta_{1,1}]_1[\beta_{2,1}]_2[\beta_{3,1}]_3x_{1,1}x_{3,2}x_{1,3}c_{1,0}c_{2,0}c_{3,0}\overline{\beta_{1,1}}\ \overline{\beta_{2,1}}\ \overline{\beta_{3,1}}\beta]_0$
$[[\beta_{1,1}]_1[\beta_{2,1}]_2[\beta_{3,1}]_3x_{1,1}x_{3,2}x_{2,3}c_{1,0}c_{2,0}c_{3,0}\overline{\beta_{1,1}}\ \overline{\beta_{2,1}}\ \overline{\beta_{3,1}}\beta]_0$
$[[\beta_{1,1}]_1[\beta_{2,1}]_2[\beta_{3,1}]_3x_{1,1}x_{3,2}x_{3,3}c_{1,0}c_{2,0}c_{3,0}\overline{\beta_{1,1}}\ \overline{\beta_{2,1}}\ \overline{\beta_{3,1}}\beta]_0$

$[[\beta_{1,1}]_1[\beta_{2,1}]_2[\beta_{3,1}]_3 x_{2,1}x_{1,2}x_{1,3}c_{1,0}c_{2,0}c_{3,0}\overline{\beta_{1,1}}\ \overline{\beta_{2,1}}\ \overline{\beta_{3,1}}\beta]_0$
$[[\beta_{1,1}]_1[\beta_{2,1}]_2[\beta_{3,1}]_3 x_{2,1}x_{1,2}x_{2,3}c_{1,0}c_{2,0}c_{3,0}\overline{\beta_{1,1}}\ \overline{\beta_{2,1}}\ \overline{\beta_{3,1}}\beta]_0$
$[[\beta_{1,1}]_1[\beta_{2,1}]_2[\beta_{3,1}]_3 x_{2,1}x_{1,2}x_{3,3}c_{1,0}c_{2,0}c_{3,0}\overline{\beta_{1,1}}\ \overline{\beta_{2,1}}\ \overline{\beta_{3,1}}\beta]_0$
$[[\beta_{1,1}]_1[\beta_{2,1}]_2[\beta_{3,1}]_3 x_{2,1}x_{2,2}x_{1,3}c_{1,0}c_{2,0}c_{3,0}\overline{\beta_{1,1}}\ \overline{\beta_{2,1}}\ \overline{\beta_{3,1}}\beta]_0$
$[[\beta_{1,1}]_1[\beta_{2,1}]_2[\beta_{3,1}]_3 x_{2,1}x_{2,2}x_{2,3}c_{1,0}c_{2,0}c_{3,0}\overline{\beta_{1,1}}\ \overline{\beta_{2,1}}\ \overline{\beta_{3,1}}\beta]_0$
$[[\beta_{1,1}]_1[\beta_{2,1}]_2[\beta_{3,1}]_3 x_{2,1}x_{2,2}x_{3,3}c_{1,0}c_{2,0}c_{3,0}\overline{\beta_{1,1}}\ \overline{\beta_{2,1}}\ \overline{\beta_{3,1}}\beta]_0$
$[[\beta_{1,1}]_1[\beta_{2,1}]_2[\beta_{3,1}]_3 x_{2,1}x_{3,2}x_{1,3}c_{1,0}c_{2,0}c_{3,0}\overline{\beta_{1,1}}\ \overline{\beta_{2,1}}\ \overline{\beta_{3,1}}\beta]_0$
$[[\beta_{1,1}]_1[\beta_{2,1}]_2[\beta_{3,1}]_3 x_{2,1}x_{3,2}x_{2,3}c_{1,0}c_{2,0}c_{3,0}\overline{\beta_{1,1}}\ \overline{\beta_{2,1}}\ \overline{\beta_{3,1}}\beta]_0$
$[[\beta_{1,1}]_1[\beta_{2,1}]_2[\beta_{3,1}]_3 x_{2,1}x_{3,2}x_{3,3}c_{1,0}c_{2,0}c_{3,0}\overline{\beta_{1,1}}\ \overline{\beta_{2,1}}\ \overline{\beta_{3,1}}\beta]_0$
$[[\beta_{1,1}]_1[\beta_{2,1}]_2[\beta_{3,1}]_3 x_{3,1}x_{1,2}x_{1,3}c_{1,0}c_{2,0}c_{3,0}\overline{\beta_{1,1}}\ \overline{\beta_{2,1}}\ \overline{\beta_{3,1}}\beta]_0$
$[[\beta_{1,1}]_1[\beta_{2,1}]_2[\beta_{3,1}]_3 x_{3,1}x_{1,2}x_{2,3}c_{1,0}c_{2,0}c_{3,0}\overline{\beta_{1,1}}\ \overline{\beta_{2,1}}\ \overline{\beta_{3,1}}\beta]_0$
$[[\beta_{1,1}]_1[\beta_{2,1}]_2[\beta_{3,1}]_3 x_{3,1}x_{1,2}x_{3,3}c_{1,0}c_{2,0}c_{3,0}\overline{\beta_{1,1}}\ \overline{\beta_{2,1}}\ \overline{\beta_{3,1}}\beta]_0$
$[[\beta_{1,1}]_1[\beta_{2,1}]_2[\beta_{3,1}]_3 x_{3,1}x_{2,2}x_{1,3}c_{1,0}c_{2,0}c_{3,0}\overline{\beta_{1,1}}\ \overline{\beta_{2,1}}\ \overline{\beta_{3,1}}\beta]_0$
$[[\beta_{1,1}]_1[\beta_{2,1}]_2[\beta_{3,1}]_3 x_{3,1}x_{2,2}x_{2,3}c_{1,0}c_{2,0}c_{3,0}\overline{\beta_{1,1}}\ \overline{\beta_{2,1}}\ \overline{\beta_{3,1}}\beta]_0$
$[[\beta_{1,1}]_1[\beta_{2,1}]_2[\beta_{3,1}]_3 x_{3,1}x_{2,2}x_{3,3}c_{1,0}c_{2,0}c_{3,0}\overline{\beta_{1,1}}\ \overline{\beta_{2,1}}\ \overline{\beta_{3,1}}\beta]_0$
$[[\beta_{1,1}]_1[\beta_{2,1}]_2[\beta_{3,1}]_3 x_{3,1}x_{3,2}x_{1,3}c_{1,0}c_{2,0}c_{3,0}\overline{\beta_{1,1}}\ \overline{\beta_{2,1}}\ \overline{\beta_{3,1}}\beta]_0$
$[[\beta_{1,1}]_1[\beta_{2,1}]_2[\beta_{3,1}]_3 x_{3,1}x_{3,2}x_{2,3}c_{1,0}c_{2,0}c_{3,0}\overline{\beta_{1,1}}\ \overline{\beta_{2,1}}\ \overline{\beta_{3,1}}\beta]_0$
$[[\beta_{1,1}]_1[\beta_{2,1}]_2[\beta_{3,1}]_3 x_{3,1}x_{3,2}x_{3,3}c_{1,0}c_{2,0}c_{3,0}\overline{\beta_{1,1}}\ \overline{\beta_{2,1}}\ \overline{\beta_{3,1}}\beta]_0]_L$

9. $[[\overline{\alpha}]_M[\alpha e_8]_J[\overline{\alpha}]_K[\beta_{1,1}]_1^{27}[\beta_{2,1}]_2^{27}[\beta_{3,1}]_3^{27}$
$[w_1z_{1,1}x_{1,2}x_{1,3}c_{1,0}c_{2,0}c_{3,0}\overline{\beta_{1,1}}\ \overline{\beta_{2,1}}\ \overline{\beta_{3,1}}\beta]_0[w_1z_{1,1}x_{1,2}x_{2,3}c_{1,0}c_{2,0}c_{3,0}\overline{\beta_{1,1}}\ \overline{\beta_{2,1}}\ \overline{\beta_{3,1}}\beta]_0$
$[w_1z_{1,1}x_{1,2}x_{3,3}c_{1,0}c_{2,0}c_{3,0}\overline{\beta_{1,1}}\ \overline{\beta_{2,1}}\ \overline{\beta_{3,1}}\beta]_0[w_1z_{1,1}x_{2,2}x_{1,3}c_{1,0}c_{2,0}c_{3,0}\overline{\beta_{1,1}}\ \overline{\beta_{2,1}}\ \overline{\beta_{3,1}}\beta]_0$
$[w_1z_{1,1}x_{2,2}x_{2,3}c_{1,0}c_{2,0}c_{3,0}\overline{\beta_{1,1}}\ \overline{\beta_{2,1}}\ \overline{\beta_{3,1}}\beta]_0[w_1z_{1,1}x_{2,2}x_{3,3}c_{1,0}c_{2,0}c_{3,0}\overline{\beta_{1,1}}\ \overline{\beta_{2,1}}\ \overline{\beta_{3,1}}\beta]_0$
$[w_1z_{1,1}x_{3,2}x_{1,3}c_{1,0}c_{2,0}c_{3,0}\overline{\beta_{1,1}}\ \overline{\beta_{2,1}}\ \overline{\beta_{3,1}}\beta]_0[w_1z_{1,1}x_{3,2}x_{2,3}c_{1,0}c_{2,0}c_{3,0}\overline{\beta_{1,1}}\ \overline{\beta_{2,1}}\ \overline{\beta_{3,1}}\beta]_0$
$[w_1z_{1,1}x_{3,2}x_{3,3}c_{1,0}c_{2,0}c_{3,0}\overline{\beta_{1,1}}\ \overline{\beta_{2,1}}\ \overline{\beta_{3,1}}\beta]_0[w_2z_{2,1}x_{1,2}x_{1,3}c_{1,0}c_{2,0}c_{3,0}\overline{\beta_{1,1}}\ \overline{\beta_{2,1}}\ \overline{\beta_{3,1}}\beta]_0$
$[w_2z_{2,1}x_{1,2}x_{2,3}c_{1,0}c_{2,0}c_{3,0}\overline{\beta_{1,1}}\ \overline{\beta_{2,1}}\ \overline{\beta_{3,1}}\beta]_0[w_2z_{2,1}x_{1,2}x_{3,3}c_{1,0}c_{2,0}c_{3,0}\overline{\beta_{1,1}}\ \overline{\beta_{2,1}}\ \overline{\beta_{3,1}}\beta]_0$
$[w_2z_{2,1}x_{2,2}x_{1,3}c_{1,0}c_{2,0}c_{3,0}\overline{\beta_{1,1}}\ \overline{\beta_{2,1}}\ \overline{\beta_{3,1}}\beta]_0[w_2z_{2,1}x_{2,2}x_{2,3}c_{1,0}c_{2,0}c_{3,0}\overline{\beta_{1,1}}\ \overline{\beta_{2,1}}\ \overline{\beta_{3,1}}\beta]_0$
$[w_2z_{2,1}x_{2,2}x_{3,3}c_{1,0}c_{2,0}c_{3,0}\overline{\beta_{1,1}}\ \overline{\beta_{2,1}}\ \overline{\beta_{3,1}}\beta]_0[w_2z_{2,1}x_{3,2}x_{1,3}c_{1,0}c_{2,0}c_{3,0}\overline{\beta_{1,1}}\ \overline{\beta_{2,1}}\ \overline{\beta_{3,1}}\beta]_0$
$[w_2z_{2,1}x_{3,2}x_{2,3}c_{1,0}c_{2,0}c_{3,0}\overline{\beta_{1,1}}\ \overline{\beta_{2,1}}\ \overline{\beta_{3,1}}\beta]_0[w_2z_{2,1}x_{3,2}x_{3,3}c_{1,0}c_{2,0}c_{3,0}\overline{\beta_{1,1}}\ \overline{\beta_{2,1}}\ \overline{\beta_{3,1}}\beta]_0$
$[w_3z_{3,1}x_{1,2}x_{1,3}c_{1,0}c_{2,0}c_{3,0}\overline{\beta_{1,1}}\ \overline{\beta_{2,1}}\ \overline{\beta_{3,1}}\beta]_0[w_3z_{3,1}x_{1,2}x_{2,3}c_{1,0}c_{2,0}c_{3,0}\overline{\beta_{1,1}}\ \overline{\beta_{2,1}}\ \overline{\beta_{3,1}}\beta]_0$
$[w_3z_{3,1}x_{1,2}x_{3,3}c_{1,0}c_{2,0}c_{3,0}\overline{\beta_{1,1}}\ \overline{\beta_{2,1}}\ \overline{\beta_{3,1}}\beta]_0[w_3z_{3,1}x_{2,2}x_{1,3}c_{1,0}c_{2,0}c_{3,0}\overline{\beta_{1,1}}\ \overline{\beta_{2,1}}\ \overline{\beta_{3,1}}\beta]_0$
$[w_3z_{3,1}x_{2,2}x_{2,3}c_{1,0}c_{2,0}c_{3,0}\overline{\beta_{1,1}}\ \overline{\beta_{2,1}}\ \overline{\beta_{3,1}}\beta]_0[w_3z_{3,1}x_{2,2}x_{3,3}c_{1,0}c_{2,0}c_{3,0}\overline{\beta_{1,1}}\ \overline{\beta_{2,1}}\ \overline{\beta_{3,1}}\beta]_0$
$[w_3z_{3,1}x_{3,2}x_{1,3}c_{1,0}c_{2,0}c_{3,0}\overline{\beta_{1,1}}\ \overline{\beta_{2,1}}\ \overline{\beta_{3,1}}\beta]_0[w_3z_{3,1}x_{3,2}x_{2,3}c_{1,0}c_{2,0}c_{3,0}\overline{\beta_{1,1}}\ \overline{\beta_{2,1}}\ \overline{\beta_{3,1}}\beta]_0$
$[w_3z_{3,1}x_{3,2}x_{3,3}c_{1,0}c_{2,0}c_{3,0}\overline{\beta_{1,1}}\ \overline{\beta_{2,1}}\ \overline{\beta_{3,1}}\beta]_0]_L$

By applying in a similar way the set of rules (ii) for $2 \le i \le 3$, after 4 steps we obtain the configuration.

13. $[[\overline{\alpha}]_M[\alpha e_{12}]_J[\overline{\alpha}]_K[\beta_{1,3}]_1^{27}[\beta_{2,3}]_2^{27}[\beta_{3,3}^{27}]_3$
$[w_1^6z_{1,1}z_{1,2}z_{1,3}c_{1,0}c_{2,0}c_{3,0}\overline{\beta_{1,3}}\ \overline{\beta_{2,3}}\ \overline{\beta_{3,3}}\beta]_0$
$[w_1^4w_2^2z_{1,1}z_{1,2}z_{2,3}c_{1,0}c_{2,0}c_{3,0}\overline{\beta_{1,3}}\ \overline{\beta_{2,3}}\ \overline{\beta_{3,3}}\beta]_0$
$[w_1^4w_3^2z_{1,1}z_{1,2}z_{3,3}c_{1,0}c_{2,0}c_{3,0}\overline{\beta_{1,3}}\ \overline{\beta_{2,3}}\ \overline{\beta_{3,3}}\beta]_0$
$[w_1^3w_2^3z_{1,1}z_{2,2}z_{1,3}c_{1,0}c_{2,0}c_{3,0}\overline{\beta_{1,3}}\ \overline{\beta_{2,3}}\ \overline{\beta_{3,3}}\beta]_0$
$[w_1w_2^5z_{1,1}z_{2,2}z_{2,3}c_{1,0}c_{2,0}c_{3,0}\overline{\beta_{1,3}}\ \overline{\beta_{2,3}}\ \overline{\beta_{3,3}}\beta]_0$

$[w_1 w_2^3 w_3^2 z_{1,1} z_{2,2} z_{3,3} c_{1,0} c_{2,0} c_{3,0} \overline{\beta_{1,3}}\ \overline{\beta_{2,3}}\ \overline{\beta_{3,3}} \beta]_0$
$[w_1^3 w_3^3 z_{1,1} z_{3,2} z_{1,3} c_{1,0} c_{2,0} c_{3,0} \overline{\beta_{1,3}}\ \overline{\beta_{2,3}}\ \overline{\beta_{3,3}} \beta]_0$
$[w_1 w_3^3 w_2^2 z_{1,1} z_{3,2} z_{2,3} c_{1,0} c_{2,0} c_{3,0} \overline{\beta_{1,3}}\ \overline{\beta_{2,3}}\ \overline{\beta_{3,3}} \beta]_0$
$[w_1 w_3^5 z_{1,1} z_{3,2} z_{3,3} c_{1,0} c_{2,0} c_{3,0} \overline{\beta_{1,3}}\ \overline{\beta_{2,3}}\ \overline{\beta_{3,3}} \beta]_0$
$[w_2 w_1^5 z_{2,1} z_{1,2} z_{1,3} c_{1,0} c_{2,0} c_{3,0} \overline{\beta_{1,3}}\ \overline{\beta_{2,3}}\ \overline{\beta_{3,3}} \beta]_0$
$[w_2^3 w_1^3 z_{2,1} z_{1,2} z_{2,3} c_{1,0} c_{2,0} c_{3,0} \overline{\beta_{1,3}}\ \overline{\beta_{2,3}}\ \overline{\beta_{3,3}} \beta]_0$
$[w_2 w_1^3 w_3^2 z_{2,1} z_{1,2} z_{3,3} c_{1,0} c_{2,0} c_{3,0} \overline{\beta_{1,3}}\ \overline{\beta_{2,3}}\ \overline{\beta_{3,3}} \beta]_0$
$[w_2^4 w_1^2 z_{2,1} z_{2,2} z_{1,3} c_{1,0} c_{2,0} c_{3,0} \overline{\beta_{1,3}}\ \overline{\beta_{2,3}}\ \overline{\beta_{3,3}} \beta]_0$
$[w_2^6 z_{2,1} z_{2,2} z_{2,3} c_{1,0} c_{2,0} c_{3,0} \overline{\beta_{1,3}}\ \overline{\beta_{2,3}}\ \overline{\beta_{3,3}} \beta]_0$
$[w_2^4 w_3^2 z_{2,1} z_{2,2} z_{3,3} c_{1,0} c_{2,0} c_{3,0} \overline{\beta_{1,3}}\ \overline{\beta_{2,3}}\ \overline{\beta_{3,3}} \beta]_0$
$[w_2 w_3^3 w_1^2 z_{2,1} z_{3,2} z_{1,3} c_{1,0} c_{2,0} c_{3,0} \overline{\beta_{1,3}}\ \overline{\beta_{2,3}}\ \overline{\beta_{3,3}} \beta]_0$
$[w_2 w_3^3 w_1^2 z_{2,1} z_{3,2} z_{2,3} c_{1,0} c_{2,0} c_{3,0} \overline{\beta_{1,3}}\ \overline{\beta_{2,3}}\ \overline{\beta_{3,3}} \beta]_0$
$[w_2 w_3^5 z_{2,1} z_{3,2} z_{3,3} c_{1,0} c_{2,0} c_{3,0} \overline{\beta_{1,3}}\ \overline{\beta_{2,3}}\ \overline{\beta_{3,3}} \beta]_0$
$[w_3 w_1^5 z_{3,1} z_{1,2} z_{1,3} c_{1,0} c_{2,0} c_{3,0} \overline{\beta_{1,3}}\ \overline{\beta_{2,3}}\ \overline{\beta_{3,3}} \beta]_0$
$[w_3 w_1^3 w_2^2 z_{3,1} z_{1,2} z_{2,3} c_{1,0} c_{2,0} c_{3,0} \overline{\beta_{1,3}}\ \overline{\beta_{2,3}}\ \overline{\beta_{3,3}} \beta]_0$
$[w_3^3 w_1^3 z_{3,1} z_{1,2} z_{3,3} c_{1,0} c_{2,0} c_{3,0} \overline{\beta_{1,3}}\ \overline{\beta_{2,3}}\ \overline{\beta_{3,3}} \beta]_0$
$[w_3 w_3^2 w_1^2 z_{3,1} z_{2,2} z_{1,3} c_{1,0} c_{2,0} c_{3,0} \overline{\beta_{1,3}}\ \overline{\beta_{2,3}}\ \overline{\beta_{3,3}} \beta]_0$
$[w_3 w_2^5 z_{3,1} z_{2,2} z_{2,3} c_{1,0} c_{2,0} c_{3,0} \overline{\beta_{1,3}}\ \overline{\beta_{2,3}}\ \overline{\beta_{3,3}} \beta]_0$
$[w_3^3 w_2^3 z_{3,1} z_{2,2} z_{3,3} c_{1,0} c_{2,0} c_{3,0} \overline{\beta_{1,3}}\ \overline{\beta_{2,3}}\ \overline{\beta_{3,3}} \beta]_0$
$[w_3^4 w_1^2 z_{3,1} z_{3,2} z_{1,3} c_{1,0} c_{2,0} c_{3,0} \overline{\beta_{1,3}}\ \overline{\beta_{2,3}}\ \overline{\beta_{3,3}} \beta]_0$
$[w_3^4 w_2^2 z_{3,1} z_{3,2} z_{2,3} c_{1,0} c_{2,0} c_{3,0} \overline{\beta_{1,3}}\ \overline{\beta_{2,3}}\ \overline{\beta_{3,3}} \beta]_0$
$[w_3^6 z_{3,1} z_{3,2} z_{3,3} c_{1,0} c_{2,0} c_{3,0} \overline{\beta_{1,3}}\ \overline{\beta_{2,3}}\ \overline{\beta_{3,3}} \beta]_0]_L$

By applying in a similar way the set of rules (*iii*) for $1 \leq i \leq 3$, after 6 steps we obtain the next configuration, in which the second index of $c_{l,m}$ objects equals the number of w_j objects.

19. $[[\overline{\alpha}]_M [\alpha e_{18}]_J [\overline{\alpha}]_K [\beta_{1,3}]_1^{27} [\beta_{2,3}]_2^{27} [\beta_{3,3}^{27}]_3$
$[z_{1,1} z_{1,2} z_{1,3} c_{1,6} c_{2,0} c_{3,0} \overline{\beta_{1,3}}\ \overline{\beta_{2,3}}\ \overline{\beta_{3,3}} \beta]_0 [z_{1,1} z_{1,2} z_{2,3} c_{1,4} c_{2,2} c_{3,0} \overline{\beta_{1,3}}\ \overline{\beta_{2,3}}\ \overline{\beta_{3,3}} \beta]_0$
$[z_{1,1} z_{1,2} z_{3,3} c_{1,4} c_{2,0} c_{3,2} \overline{\beta_{1,3}}\ \overline{\beta_{2,3}}\ \overline{\beta_{3,3}} \beta]_0 [z_{1,1} z_{2,2} z_{1,3} c_{1,3} c_{2,3} c_{3,0} \overline{\beta_{1,3}}\ \overline{\beta_{2,3}}\ \overline{\beta_{3,3}} \beta]_0$
$[z_{1,1} z_{2,2} z_{2,3} c_{1,1} c_{2,5} c_{3,0} \overline{\beta_{1,3}}\ \overline{\beta_{2,3}}\ \overline{\beta_{3,3}} \beta]_0 [z_{1,1} z_{2,2} z_{3,3} c_{1,1} c_{2,3} c_{3,2} \overline{\beta_{1,3}}\ \overline{\beta_{2,3}}\ \overline{\beta_{3,3}} \beta]_0$
$[z_{1,1} z_{3,2} z_{1,3} c_{1,3} c_{2,0} c_{3,3} \overline{\beta_{1,3}}\ \overline{\beta_{2,3}}\ \overline{\beta_{3,3}} \beta]_0 [z_{1,1} z_{3,2} z_{2,3} c_{1,1} c_{2,2} c_{3,3} \overline{\beta_{1,3}}\ \overline{\beta_{2,3}}\ \overline{\beta_{3,3}} \beta]_0$
$[z_{1,1} z_{3,2} z_{3,3} c_{1,1} c_{2,0} c_{3,5} \overline{\beta_{1,3}}\ \overline{\beta_{2,3}}\ \overline{\beta_{3,3}} \beta]_0 [z_{2,1} z_{1,2} z_{1,3} c_{1,5} c_{2,1} c_{3,0} \overline{\beta_{1,3}}\ \overline{\beta_{2,3}}\ \overline{\beta_{3,3}} \beta]_0$
$[z_{2,1} z_{1,2} z_{2,3} c_{1,3} c_{2,3} c_{3,0} \overline{\beta_{1,3}}\ \overline{\beta_{2,3}}\ \overline{\beta_{3,3}} \beta]_0 [z_{2,1} z_{1,2} z_{3,3} c_{1,3} c_{2,1} c_{3,2} \overline{\beta_{1,3}}\ \overline{\beta_{2,3}}\ \overline{\beta_{3,3}} \beta]_0$
$[z_{2,1} z_{2,2} z_{1,3} c_{1,2} c_{2,4} c_{3,0} \overline{\beta_{1,3}}\ \overline{\beta_{2,3}}\ \overline{\beta_{3,3}} \beta]_0 [z_{2,1} z_{2,2} z_{2,3} c_{1,0} c_{2,6} c_{3,0} \overline{\beta_{1,3}}\ \overline{\beta_{2,3}}\ \overline{\beta_{3,3}} \beta]_0$
$[z_{2,1} z_{2,2} z_{3,3} c_{1,0} c_{2,4} c_{3,2} \overline{\beta_{1,3}}\ \overline{\beta_{2,3}}\ \overline{\beta_{3,3}} \beta]_0 [z_{2,1} z_{3,2} z_{1,3} c_{1,2} c_{2,1} c_{3,3} \overline{\beta_{1,3}}\ \overline{\beta_{2,3}}\ \overline{\beta_{3,3}} \beta]_0$
$[z_{2,1} z_{3,2} z_{2,3} c_{1,0} c_{2,3} c_{3,3} \overline{\beta_{1,3}}\ \overline{\beta_{2,3}}\ \overline{\beta_{3,3}} \beta]_0 [z_{2,1} z_{3,2} z_{3,3} c_{1,0} c_{2,1} c_{3,5} \overline{\beta_{1,3}}\ \overline{\beta_{2,3}}\ \overline{\beta_{3,3}} \beta]_0$
$[z_{3,1} z_{1,2} z_{1,3} c_{1,5} c_{2,0} c_{3,1} \overline{\beta_{1,3}}\ \overline{\beta_{2,3}}\ \overline{\beta_{3,3}} \beta]_0 [z_{3,1} z_{1,2} z_{2,3} c_{1,3} c_{2,2} c_{3,1} \overline{\beta_{1,3}}\ \overline{\beta_{2,3}}\ \overline{\beta_{3,3}} \beta]_0$
$[z_{3,1} z_{1,2} z_{3,3} c_{1,3} c_{2,0} c_{3,3} \overline{\beta_{1,3}}\ \overline{\beta_{2,3}}\ \overline{\beta_{3,3}} \beta]_0 [z_{3,1} z_{2,2} z_{1,3} c_{1,2} c_{2,3} c_{3,1} \overline{\beta_{1,3}}\ \overline{\beta_{2,3}}\ \overline{\beta_{3,3}} \beta]_0$
$[z_{3,1} z_{2,2} z_{2,3} c_{1,0} c_{2,5} c_{3,1} \overline{\beta_{1,3}}\ \overline{\beta_{2,3}}\ \overline{\beta_{3,3}} \beta]_0 [z_{3,1} z_{2,2} z_{3,3} c_{1,0} c_{2,3} c_{3,3} \overline{\beta_{1,3}}\ \overline{\beta_{2,3}}\ \overline{\beta_{3,3}} \beta]_0$
$[z_{3,1} z_{3,2} z_{1,3} c_{1,2} c_{2,0} c_{3,4} \overline{\beta_{1,3}}\ \overline{\beta_{2,3}}\ \overline{\beta_{3,3}} \beta]_0 [z_{3,1} z_{3,2} z_{2,3} c_{1,0} c_{2,2} c_{3,4} \overline{\beta_{1,3}}\ \overline{\beta_{2,3}}\ \overline{\beta_{3,3}} \beta]_0$
$[z_{3,1} z_{3,2} z_{3,3} c_{1,0} c_{2,0} c_{3,6} \overline{\beta_{1,3}}\ \overline{\beta_{2,3}}\ \overline{\beta_{3,3}} \beta]_0]_L$

Graphically the working space is described by the following picture:

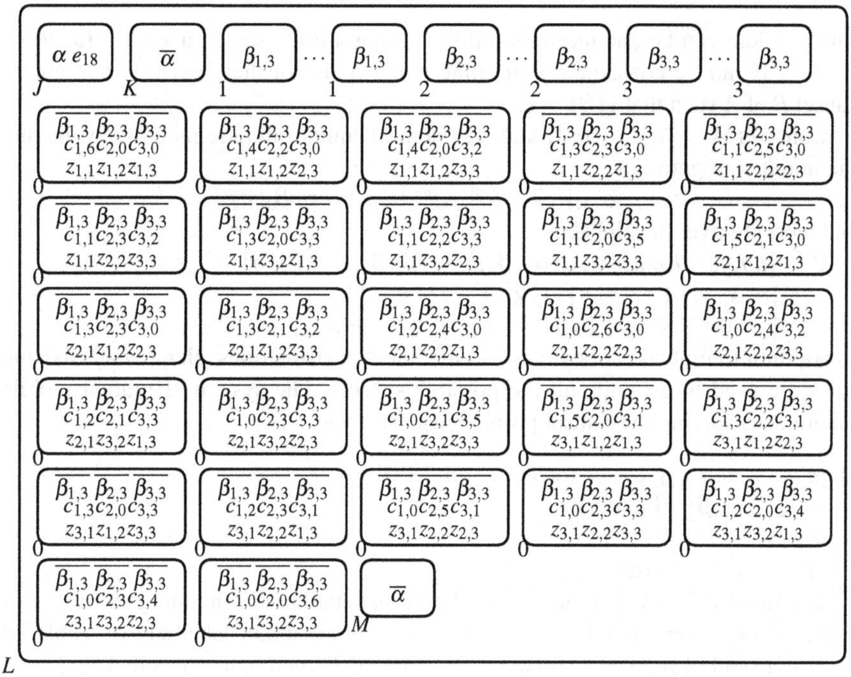

The next steps are used to create a *yes* object inside membrane K. We present only the final configuration:

29. $[[\overline{\beta}]^2_M [\overline{\beta}]^{12}_J [\beta_{1,3}]^{27}_1 [\beta_{2,3}]^{27}_2 [\beta^{27}_{3,3}]_3$

$[[z_{1,1}z_{1,2}z_{1,3}c_{1,6}c_{2,0}c_{3,0}\overline{\beta_{1,3}}\ \overline{\beta_{2,3}}\ \overline{\beta_{3,3}}]_0]_J [[z_{1,1}z_{1,2}z_{2,3}c_{1,4}c_{2,2}c_{3,0}\overline{\beta_{1,3}}\ \overline{\beta_{2,3}}\ \overline{\beta_{3,3}}]_0]_J$
$[[z_{1,1}z_{1,2}z_{3,3}c_{1,4}c_{2,0}c_{3,2}\overline{\beta_{1,3}}\ \overline{\beta_{2,3}}\ \overline{\beta_{3,3}}]_0]_J [[z_{1,1}z_{2,2}z_{2,3}c_{1,1}c_{2,5}c_{3,0}\overline{\beta_{1,3}}\ \overline{\beta_{2,3}}\ \overline{\beta_{3,3}}]_0]_J$
$[z_{1,1}z_{2,2}z_{3,3}c_{1,1}c_{2,3}c_{3,2}\overline{\beta_{1,3}}\ \overline{\beta_{2,3}}\ \overline{\beta_{3,3}}\beta]_0 [z_{1,1}z_{3,2}z_{1,3}c_{1,3}c_{2,0}c_{3,3}\overline{\beta_{1,3}}\ \overline{\beta_{2,3}}\ \overline{\beta_{3,3}}\beta]_0$
$[z_{1,1}z_{3,2}z_{2,3}c_{1,1}c_{2,2}c_{3,3}\overline{\beta_{1,3}}\ \overline{\beta_{2,3}}\ \overline{\beta_{3,3}}\beta]_0 [[z_{1,1}z_{3,2}z_{3,3}c_{1,1}c_{2,0}c_{3,5}\overline{\beta_{1,3}}\ \overline{\beta_{2,3}}\ \overline{\beta_{3,3}}]_0]_J$
$[[z_{2,1}z_{1,2}z_{1,3}c_{1,5}c_{2,1}c_{3,0}\overline{\beta_{1,3}}\ \overline{\beta_{2,3}}\ \overline{\beta_{3,3}}]_0]_J [z_{2,1}z_{1,2}z_{2,3}c_{1,3}c_{2,3}c_{3,0}\overline{\beta_{1,3}}\ \overline{\beta_{2,3}}\ \overline{\beta_{3,3}}\beta]_0$
$[z_{2,1}z_{1,2}z_{3,3}c_{1,3}c_{2,1}c_{3,2}\overline{\beta_{1,3}}\ \overline{\beta_{2,3}}\ \overline{\beta_{3,3}}\beta]_0 [[z_{2,1}z_{2,2}z_{1,3}c_{1,2}c_{2,4}c_{3,0}\overline{\beta_{1,3}}\ \overline{\beta_{2,3}}\ \overline{\beta_{3,3}}]_0]_J$
$[[z_{2,1}z_{2,2}z_{2,3}c_{1,0}c_{2,6}c_{3,0}\overline{\beta_{1,3}}\ \overline{\beta_{2,3}}\ \overline{\beta_{3,3}}]_0]_J [[z_{2,1}z_{2,2}z_{3,3}c_{1,0}c_{2,4}c_{3,2}\overline{\beta_{1,3}}\ \overline{\beta_{2,3}}\ \overline{\beta_{3,3}}]_0]_J$
$[z_{2,1}z_{3,2}z_{1,3}c_{1,2}c_{2,1}c_{3,3}\overline{\beta_{1,3}}\ \overline{\beta_{2,3}}\ \overline{\beta_{3,3}}\beta]_0 [z_{2,1}z_{3,2}z_{2,3}c_{1,0}c_{2,3}c_{3,3}\overline{\beta_{1,3}}\ \overline{\beta_{2,3}}\ \overline{\beta_{3,3}}\beta]_0$
$[[z_{2,1}z_{3,2}z_{3,3}c_{1,0}c_{2,1}c_{3,5}\overline{\beta_{1,3}}\ \overline{\beta_{2,3}}\ \overline{\beta_{3,3}}]_0]_J [[z_{3,1}z_{1,2}z_{1,3}c_{1,5}c_{2,0}c_{3,1}\overline{\beta_{1,3}}\ \overline{\beta_{2,3}}\ \overline{\beta_{3,3}}]_0]_J$
$[z_{3,1}z_{1,2}z_{2,3}c_{1,3}c_{2,2}c_{3,1}\overline{\beta_{1,3}}\ \overline{\beta_{2,3}}\ \overline{\beta_{3,3}}\beta]_0 [z_{3,1}z_{1,2}z_{3,3}c_{1,3}c_{2,0}c_{3,3}\overline{\beta_{1,3}}\ \overline{\beta_{2,3}}\ \overline{\beta_{3,3}}\beta]_0$
$[z_{3,1}z_{2,2}z_{1,3}c_{1,2}c_{2,3}c_{3,1}\overline{\beta_{1,3}}\ \overline{\beta_{2,3}}\ \overline{\beta_{3,3}}\beta]_0 [[z_{3,1}z_{2,2}z_{2,3}c_{1,0}c_{2,5}c_{3,1}\overline{\beta_{1,3}}\ \overline{\beta_{2,3}}\ \overline{\beta_{3,3}}]_0]_J$
$[z_{3,1}z_{2,2}z_{3,3}c_{1,0}c_{2,3}c_{3,3}\overline{\beta_{1,3}}\ \overline{\beta_{2,3}}\ \overline{\beta_{3,3}}\beta]_0 [[z_{3,1}z_{3,2}z_{1,3}c_{1,2}c_{2,0}c_{3,4}\overline{\beta_{1,3}}\ \overline{\beta_{2,3}}\ \overline{\beta_{3,3}}]_0]_J$
$[[z_{3,1}z_{3,2}z_{2,3}c_{1,0}c_{2,2}c_{3,4}\overline{\beta_{1,3}}\ \overline{\beta_{2,3}}\ \overline{\beta_{3,3}}]_0]_J [[z_{3,1}z_{3,2}z_{3,3}c_{1,0}c_{2,0}c_{3,6}\overline{\beta_{1,3}}\ \overline{\beta_{2,3}}\ \overline{\beta_{3,3}}]_0]_J$
$[yes[z_{1,1}z_{2,2}z_{1,3}c_{1,3}c_{2,3}c_{3,0}\overline{\beta_{1,3}}\ \overline{\beta_{2,3}}\ \overline{\beta_{3,3}}]_0\beta]_K]_L$

Exercise 2.7. Solve the Bin Packing problem using other classes of mobile membranes.

2.5.4 Subset Sum Problem

The problem can be enounced as follows: given a finite set, A, a weight function, $g : A \to \mathbb{N}$, and a constant s, determine whether or not there exists a non-empty subset B of A such that $g(B) = s$.

Consider $A = \{a_1, \ldots, a_n\}$, and a system of mutual mobile membranes having the initial configuration

$$[[\overline{\alpha}]_M[\alpha\, e_0]_J[\overline{\alpha}]_K[d^{n-1}d_0]_1[a_1\, c_0\, \beta_0]_0]_2$$

and working over the alphabet

$$V = \{no, yes, d, d_0, b, \alpha, \overline{\alpha}, \alpha_1, \beta, \overline{\beta}\} \cup \{\beta_i, \overline{\beta_i} \mid 0 \le i \le n\}$$
$$\cup \{a_i, x_i, y_i \mid 1 \le i \le n\} \cup \{\gamma_i \mid 1 \le i \le n-1\}$$
$$\cup \{e_i \mid 0 \le i \le 4n+2s+1\} \cup \{c_i \mid 0 \le i \le g(A)\}$$

In addition to mutual endocytosis and mutual exocytosis rules, elementary division rules are used to generate all the possible subsets. The system of mutual mobile membranes solving the Subset problem uses the rules:

(i) $[a_i]_0 \to [x_i a_{i+1}]_0[a_{i+1}]_0$, for $1 \le i \le n-1$ (div)
$[a_n]_0 \to [x_n\beta]_0[\beta]_0$ (div)
$[d]_1 \to [\,]_1[\,]_1$ (div)
$[d_0]_1 \to [\overline{\beta}]_1[\overline{\beta}]_1$ (div)
The first two rules generate 2^n membranes labelled by 0 containing all the possible subsets over variables $\{x_1, \ldots, x_n\}$. In each membrane labelled by 0 is placed also a symbol β . The next two rules generate 2^n membranes labelled by 1 each containing an object $\overline{\beta}$. The symbols β and $\overline{\beta}$ are used in mobility, where the membranes containing the object β are the ones that move.

(ii) $[\beta]_0[\overline{\beta}]_1 \to [[\beta_1]_0\overline{\beta_1}]_1$ (mendo)
$[\beta_i]_0[\overline{\beta_i}]_1 \to [[\beta_{i+1}]_0\overline{\beta_{i+1}}]_1$, for $1 \le i \le n-1$ (mendo)
$[[x_i\beta_i]_0\overline{\beta_i}]_1 \to [y_i\, b^{g(a_i)}\beta_i]_0[\overline{\beta_i}]_1$, for $1 \le i \le n$ (mexo)
$[[\beta_i]_0\overline{\beta_i}]_1 \to [\beta_i]_0[\overline{\beta_i}]_1$, for $1 \le i \le n$ (mexo)
These rules are used to replace x_i by $y_i\, b^{g(a_i)}$, for $1 \le i \le n$. If x_i is not present in a membrane 0, then the indexes of β and $\overline{\beta}$ are incremented. After applying these rules, each membrane 0 contains a number of objects b equal to the weight of the contained subset of A, and also objects y_i used to remember which objects are contained in this membrane.

(iii) $[b\, \beta_n]_0[\overline{\beta_n}]_1 \to [[\beta_n]_0\overline{\beta_n}]_1$ (mendo)
$[[c_i\, \beta_n]_0\overline{\beta_n}]_1 \to [c_{i+1}\, \beta_n]_0[\overline{\beta_n}]_1$, for $0 \le i$ (mexo)
These rules are used to calculate the weights of the subsets B of A, by using the objects c_0 that appear in all 0 membranes. For each b present in a membrane 0, the subscript of c, present in the same membrane, is incremented.

(iv) $[\alpha\, e_i]_J[\overline{\alpha}]_K \to [[\alpha\, e_{i+1}]_J\overline{\alpha}]_K$, for $0 \le i < 3n+2s-1$ (mendo)
$[[\alpha\, e_i]_J\overline{\alpha}]_K \to [\alpha\, e_{i+1}]_J[\overline{\alpha}]_K$, for $0 \le i < 3n+2s-1$ (mexo)
$[[\alpha\, e_i]_J\overline{\alpha}]_K \to [\alpha\, e_{i+1}]_J[\alpha\, e_{i+1}]_K$, for $i = 3n+2s-1$ (mexo)
These rules are used in parallel to calculate the number of steps performed. The counting stops after $3n+2s$ steps, a number determined by: generating space (n

steps), replacing x_i by corresponding weight ($2n$ steps) and calculating weights ($2s$ steps).

(v) $[[\alpha\ e_{3n+2s}]_J\overline{\alpha}]_K \to [\alpha\ e_{3n+2s}]_J[\alpha e_{3n+2s}]_K$ (mexo)
$[\alpha\ e_{3n+2s}]_J \to [\gamma_1]_J[\gamma_1]_J$ (div)
$[\gamma_i]_J \to [\gamma_{i+1}]_J[\gamma_{i+1}]_J$, for $1 \leq i \leq n-2$ (div)
$[\gamma_{n-1}]_J \to [\beta_0]_J[\beta_0]_J$ (div)
$[c_i\beta_0]_0[\overline{\beta_0}]_J \to [[\ ]_0]_J$, for $0 \leq i,\ i \neq s$ (mendo)
An extra step, needed to prepare the membrane J for division, is performed if needed. If membrane J contains an object e_{3n+2s}, and is placed near membrane K, then 2^n membranes are generated (n steps) in order to check which membranes contain the object c_s. All membranes 0 that do not respect the condition are blocked inside a J membrane. In this way by choosing any membrane that it is not placed inside a J membrane, we obtain a solution to the problem.

(vi) $[\alpha\ e_i]_K[\overline{\alpha}]_M \to [[\alpha\ e_{i+1}]_K\overline{\alpha}]_M$ for $3n+2s \leq i \leq 4n+2s$ (mendo)
$[[\alpha\ e_i]_K\overline{\alpha}]_M \to [\alpha\ e_{i+1}]_K[\overline{\alpha}]_M$, for $3n+2s \leq i < 4n+2s$ (mexo)
$[[\alpha\ e_i]_K\overline{\alpha}]_M \to [\beta_0\beta_0]_K[\alpha_1]_M$, for $i = 4n+2s$ or $i = 4n+2s+1$ (mexo)
$[\beta_0]_K[\overline{\beta_0}]_0 \to [yes[\ ]_0]_K$ (mendo)
$[\alpha_1]_M \to [\overline{\beta_0}]_M[\overline{\beta_0}]_M$ (div)
$[\beta_0]_K[\overline{\beta_0}]_M \to [[no]_K]_M$ (mendo)
The above rules are used in parallel to calculate the number of steps performed. The number $4n+2s+1$ is determined by: generating space (n steps), replacing x_i by corresponding weight ($2n$ steps),calculating weights ($2s$ steps), generating the J membranes ($n-1$ steps), preparing division of J (1 step), blocking all membranes 0 that do not satisfy conditions (1 step). If there still exists a membrane 0 that is not inside a membrane J, then the object *yes* is created inside membrane K. Otherwise, after one more step, the *no* object is created inside membrane K. The computation stops after $4n+2s+3$ steps, with the answer placed inside membrane K.

The number of membranes in the initial configuration is 6, and the number of objects is $n+7$. The size of the working alphabet is $10n+4s+12$. The number of rules in the above system is: $n+2$ rules of type (i), $3n$ rules of type (ii), $2(s+n-1)$ rules of type (iii), $3n+2s$ rules of type (iv), $n+2$ rules of type (v) and $2n+5$ rules of type (vi). Hence, the size of the constructed system is $\mathcal{O}(n)$.

Example 2.5. Consider the Subset problem with $A = \{a_1, a_2, a_3\}$, $s = 2$, $g(a_1) = 1$, $g(a_2) = 2$ and $g(a_3) = 1$. In this case, $n = 3$, $s = 2$, and the initial configuration of the system of mutual mobile membranes

$$[[\overline{\alpha}]_M[\alpha\ e_0]_J[\overline{\alpha}]_K[d^2d_0]_1[a_1\ c_0\ \beta_0]_0]_2$$

Graphically this is illustrated as:

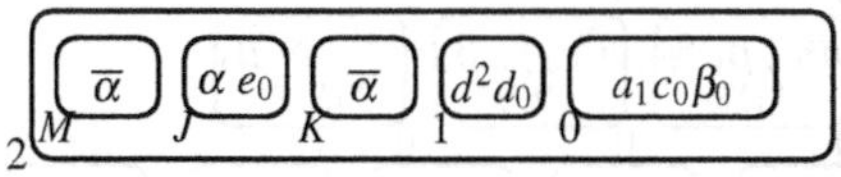

The evolution of the system is described by the following steps. The working space is generated in $n = 3$ steps leading from the initial configuration 1 to configuration 4:

1. $[[\overline{\alpha}]_M[\alpha\, e_0]_J[\overline{\alpha}]_K[d^2\, d_0]_1[a_1\, c_0\, \beta_0]_0]_2$
2. $[[\overline{\alpha}]_M[[\alpha\, e_1]_J\overline{\alpha}]_K[d\, d_0]_1^2[x_1\, a_2\, c_0\, \beta_0]_0[a_2\, c_0\, \beta_0]_0]_2$
3. $[[\overline{\alpha}]_M[\alpha\, e_2]_J[\overline{\alpha}]_K[d_0]_1^4[x_1\, x_2\, a_3\, c_0\, \beta_0]_0[x_1\, a_3\, c_0\, \beta_0]_0[x_2\, a_3\, c_0\, \beta_0]_0[a_3\, c_0\, \beta_0]_0]_2$
4. $[[\overline{\alpha}]_M[[\alpha\, e_3]_J\overline{\alpha}]_K[\overline{\beta}]_1^8[x_1\, x_2\, x_3\, c_0\, \beta\beta_0]_0[x_1\, x_2\, c_0\, \beta\beta_0]_0[x_1\, x_3\, c_0\, \beta\beta_0]_0[x_1\, c_0\, \beta\beta_0]_0$
$[x_2\, x_3\, c_0\, \beta\beta_0]_0[x_2\, c_0\, \beta\beta_0]_0[x_3\, c_0\, \beta\beta_0]_0[c_0\, \beta\beta_0]_0]_2$
Graphically the working space is described by the following picture:

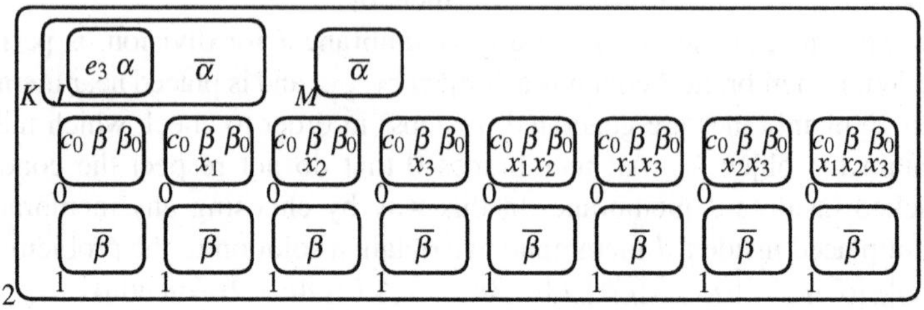

In the next steps we replace x_i by y_i and $b^{g(a_i)}$. We use y_i to mark the fact that in the membrane 0 containing it there is a subset containing a_i. The multiset of objects $b^{g(a_i)}$ is used to denote the weight of object a_i.

5. $[[\overline{\alpha}]_M[\alpha\, e_4]_J[\overline{\alpha}]_K[[x_1\, x_2\, x_3\, c_0\, \beta_1\, \beta_0]_0\overline{\beta_1}]_1[[x_1\, x_2\, c_0\, \beta_1\, \beta_0]_0\overline{\beta_1}]_1[[x_1\, x_3\, c_0\, \beta_1\, \beta_0]_0\overline{\beta_1}]_1$
$[[\overline{\alpha}]_M[x_1\, c_0\, \beta_1]_0\overline{\beta_1}]_1[[x_2\, x_3\, c_0\, \beta_1\, \beta_0]_0\overline{\beta_1}]_1[[x_2\, c_0\, \beta_1\, \beta_0]_0\overline{\beta_1}]_1$
$[[x_3\, c_0\, \beta_1\, \beta_0]_0\overline{\beta_1}]_1[[c_0\, \beta_1\, \beta_0]_0\overline{\beta_1}]_1]_2$
6. $[[\overline{\alpha}]_M[[\alpha\, e_5]_J\overline{\alpha}]_K[\overline{\beta_1}]_1^8[b\, y_1\, x_2\, x_3\, c_0\, \beta_1\, \beta_0]_0[b\, y_1\, x_2\, c_0\, \beta_1\, \beta_0]_0[b\, y_1\, x_3\, c_0\, \beta_1\, \beta_0]_0$
$[b\, y_1\, c_0\, \beta_1\, \beta_0]_0[x_2\, x_3\, c_0\, \beta_1\, \beta_0]_0[x_2\, c_0\, \beta_1\, \beta_0]_0[x_3\, c_0\, \beta_1]_0[c_0\, \beta_1\, \beta_0]_0]_2$
7. $[[\overline{\alpha}]_M[\alpha\, e_6]_J[\overline{\alpha}]_K[[b\, y_1\, x_2\, x_3\, c_0\, \beta_2\, \beta_0]_0\overline{\beta_2}]_1[[b\, y_1\, x_2\, c_0\, \beta_2\, \beta_0]_0\overline{\beta_2}]_1$
$[[b\, y_1\, x_3\, c_0\, \beta_2\, \beta_0]_0\overline{\beta_2}]_1[[b\, y_1\, c_0\, \beta_2\, \beta_0]_0\overline{\beta_2}]_1[[x_2\, x_3\, c_0\, \beta_2\, \beta_0]_0\overline{\beta_2}]_1$
$[[x_2\, c_0\, \beta_2\, \beta_0]_0\overline{\beta_2}]_1[[x_3\, c_0\, \beta_2\, \beta_0]_0\overline{\beta_2}]_1[[c_0\, \beta_2\, \beta_0]_0\overline{\beta_2}]_1]_2$
8. $[[\overline{\alpha}]_M[[\alpha\, e_7]_J\overline{\alpha}]_K[\overline{\beta_2}]_1^8[b^3\, y_1\, y_2\, x_3\, c_0\, \beta_2\, \beta_0]_0[b^3\, y_1\, y_2\, c_0\, \beta_2\, \beta_0]_0[b\, y_1\, x_3\, c_0\, \beta_2\, \beta_0]_0$
$[b\, y_1\, c_0\, \beta_2\, \beta_0]_0[b^2\, y_2\, x_3\, c_0\, \beta_2\, \beta_0]_0[b^2\, y_2\, c_0\, \beta_2\, \beta_0]_0[x_3\, c_0\, \beta_2\, \beta_0]_0[c_0\, \beta_2\, \beta_0]_0]_2$
9. $[[\overline{\alpha}]_M[\alpha\, e_8]_J[\overline{\alpha}]_K[[b\, y_1\, x_2\, x_3\, c_0\, \beta_3\, \beta_0]_0\overline{\beta_3}]_1[[b\, y_1\, x_2\, c_0\, \beta_3\, \beta_0]_0\overline{\beta_3}]_1$
$[[b\, y_1\, x_3\, c_0\, \beta_3\, \beta_0]_0\overline{\beta_3}]_1[[b\, y_1\, c_0\, \beta_3\, \beta_0]_0\overline{\beta_3}]_1[[x_2\, x_3\, c_0\, \beta_3\, \beta_0]_0\overline{\beta_3}]_1$
$[[x_2\, c_0\, \beta_3\, \beta_0]_0\overline{\beta_3}]_1[[x_3\, c_0\, \beta_3\, \beta_0]_0\overline{\beta_3}]_1[[c_0\, \beta_3\, \beta_0]_0\overline{\beta_3}]_1]_2$
10. $[[\overline{\alpha}]_M[[\alpha\, e_9]_J\overline{\alpha}]_K[\overline{\beta_3}]_1^8[b^4\, y_1\, y_2\, y_3\, c_0\, \beta_3\, \beta_0]_0[b^3\, y_1\, y_2\, c_0\, \beta_3\, \beta_0]_0[b^2\, y_1\, y_3\, c_0\, \beta_3\, \beta_0]_0$
$[b\, y_1\, c_0\, \beta_3\, \beta_0]_0[b^3\, y_2\, y_3\, c_0\, \beta_3\, \beta_0]_0[b^2\, y_2\, c_0\, \beta_3\, \beta_0]_0[b\, y_3\, c_0\, \beta_3\, \beta_0]_0[c_0\, \beta_3\, \beta_0]_0]_2$
Graphically the working space is described by the following picture:

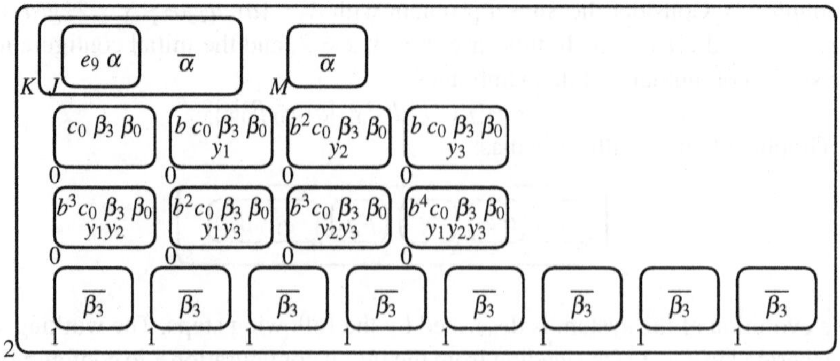

In what follows we calculate the weight of each membrane 0 by using the object c_0, until we reach c_2 since $s = 2$.

11. $[[\overline{\alpha}]_M [\alpha\, e_{10}]_J [\overline{\alpha}]_K [\overline{\beta_3}]_1 [[b^3\, y_1\, y_2\, y_3\, c_0\, \beta_3\, \beta_0]_0 \overline{\beta_3}]_1 [[b^2\, y_1\, y_2\, c_0\, \beta_3\, \beta_0]_0 \overline{\beta_3}]_1$
$[[b\, y_1\, y_3\, c_0\, \beta_3\, \beta_0]_0 \overline{\beta_3}]_1 [[y_1\, c_0\, \beta_3\, \beta_0]_0 \overline{\beta_3}]_1 [[b^2\, y_2\, y_3\, c_0\, \beta_3\, \beta_0]_0 \overline{\beta_3}]_1$
$[[b\, y_2\, c_0\, \beta_3\, \beta_0]_0 \overline{\beta_3}]_1 [[y_3\, c_0\, \beta_3\, \beta_0]_0 \overline{\beta_3}]_1 [c_0\, \beta_3]_0]_2$
12. $[[\overline{\alpha}]_M [[\alpha\, e_{11}]_J \overline{\alpha}]_K [\overline{\beta_3}]_1^8 [b^3\, y_1\, y_2\, y_3\, c_1\, \beta_3\, \beta_0]_0 [b^2\, y_1\, y_2\, c_1\, \beta_3\, \beta_0]_0 [b\, y_1\, y_3\, c_1\, \beta_3\, \beta_0]_0$
$[y_1\, c_1\, \beta_3\, \beta_0]_0 [b^2\, y_2\, y_3\, c_1\, \beta_3\, \beta_0]_0 [b\, y_2\, c_1\, \beta_3\, \beta_0]_0 [y_3\, c_1\, \beta_3\, \beta_0]_0 [c_0\, \beta_3\, \beta_0]_0]_2$
13. $[[\overline{\alpha}]_M [\alpha\, e_{12}]_J [\overline{\alpha}]_K [\overline{\beta_3}]_1^3 [[b^2\, y_1\, y_2\, y_3\, c_0\, \beta_3\, \beta_0]_0 \overline{\beta_3}]_1 [[b\, y_1\, y_2\, c_0\, \beta_3\, \beta_0]_0 \overline{\beta_3}]_1$
$[[y_1\, y_3\, c_0\, \beta_3\, \beta_0]_0 \overline{\beta_3}]_1 [y_1\, c_0\, \beta_3\, \beta_0]_0 [[b\, y_2\, y_3\, c_0\, \beta_3\, \beta_0]_0 \overline{\beta_3}]_1$
$[[y_2\, c_0\, \beta_3\, \beta_0]_0 \overline{\beta_3}]_1 [y_3\, c_0\, \beta_3\, \beta_0]_0 [c_0\, \beta_3\, \beta_0]_0]_2$
14. $[[\overline{\alpha}]_M [[\alpha\, e_{13}]_J \overline{\alpha}]_K [\overline{\beta_3}]_1^8 [b^2\, y_1\, y_2\, y_3\, c_2\, \beta_3\, \beta_0]_0 [b\, y_1\, y_2\, c_2\, \beta_3\, \beta_0]_0 [y_1\, y_3\, c_2\, \beta_3\, \beta_0]_0$
$[y_1\, c_1\, \beta_3\, \beta_0]_0 [b\, y_2\, y_3\, c_2\, \beta_3\, \beta_0]_0 [y_2\, c_2\, \beta_3\, \beta_0]_0 [y_3\, c_1\, \beta_3\, \beta_0]_0 [c_0\, \beta_3\, \beta_0]_0]_2$

Graphically the working space is described by the following picture:

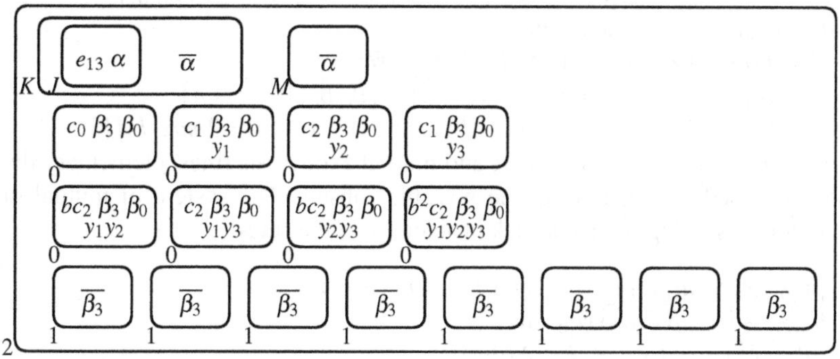

Now we divide membrane J, after performing an extra step that moves membrane J out of membrane K, in order to check if the index of c_i from the 0 membranes equals s. In parallel, if there still exist membranes 0 containing b objects, then the subscript of c_i is increased.

15. $[[\overline{\alpha}]_M [\alpha\, e_{13}]_J [\alpha\, e_{13}]_K [\overline{\beta_3}]_1^5 [[b\, y_1\, y_2\, y_3\, c_2\, \beta_3\, \beta_0]_0 \overline{\beta_3}]_1 [[y_1\, y_2\, c_2\, \beta_3\, \beta_0]_0 \overline{\beta_3}]_1$
$[y_1\, y_3\, c_2\, \beta_3\, \beta_0]_0 [y_1\, c_1\, \beta_3\, \beta_0]_0 [[y_2\, y_3\, c_2\, \beta_3\, \beta_0]_0 \overline{\beta_3}]_1 [y_2\, c_2\, \beta_3\, \beta_0]_0 [y_3\, c_1\, \beta_3\, \beta_0]_0$
$[c_0\, \beta_3\, \beta_0]_0]_2$
16. $[[\overline{\alpha}]_M [\overline{\alpha} [\alpha\, e_{14}]_K]_M [\gamma_1]_J^2 [\overline{\beta_3}]_1^8 [b\, y_1\, y_2\, y_3\, c_3\, \beta_3\, \beta_0]_0 [y_1\, y_2\, c_3\, \beta_3\, \beta_0]_0 [y_1\, y_3\, c_2\, \beta_3\, \beta_0]_0$
$[y_1\, c_1\, \beta_3\, \beta_0]_0 [y_2\, y_3\, c_3\, \beta_3\, \beta_0]_0 [y_2\, c_2\, \beta_3\, \beta_0]_0 [y_3\, c_1\, \beta_3\, \beta_0]_0 [c_0\, \beta_3\, \beta_0]_0]_2$
17. $[[\overline{\alpha}]_M [\gamma_2]_J^4 [\alpha\, e_{15}]_K [\overline{\beta_3}]_1^7 [[y_1\, y_2\, y_3\, c_3\, \beta_3\, \beta_0]_0 \overline{\beta_3}]_1 [y_1\, y_2\, c_3\, \beta_3\, \beta_0]_0 [y_1\, y_3\, c_2\, \beta_3\, \beta_0]_0$
$[y_1\, c_1\, \beta_3\, \beta_0]_0 [y_2\, y_3\, c_3\, \beta_3\, \beta_0]_0 [y_2\, c_2\, \beta_3\, \beta_0]_0 [y_3\, c_1\, \beta_3\, \beta_0]_0 [c_0\, \beta_3\, \beta_0]_0]_2$
18. $[[\overline{\alpha} [\alpha\, e_{16}]_K]_M [\overline{\beta_0}]_J^8 [\overline{\beta_3}]_1^8 [y_1\, y_2\, y_3\, c_4\, \beta_3\, \beta_0]_0 [y_1\, y_2\, c_3\, \beta_3\, \beta_0]_0 [y_1\, y_3\, c_2\, \beta_3\, \beta_0]_0$
$[y_1\, c_1\, \beta_3\, \beta_0]_0 [y_2\, y_3\, c_3\, \beta_3\, \beta_0]_0 [y_2\, c_2\, \beta_3\, \beta_0]_0 [y_3\, c_1\, \beta_3\, \beta_0]_0 [c_0\, \beta_3\, \beta_0]_0]_2$
19. $[[\alpha_1]_M [\beta_0 \overline{\beta_0}]_K [\overline{\beta_0}]_J^8 [\overline{\beta_3}]_1^8 [[y_1\, y_2\, y_3\, \beta_3]_0]_J [[y_1\, y_2\, \beta_3]_0]_J [y_1\, y_3\, c_2\, \beta_3\, \beta_0]_0$
$[[y_1\, \beta_3]_0]_J [[y_2\, y_3\, \beta_3]_0]_J [y_2\, c_2\, \beta_3\, \beta_0]_0 [[y_3\, \beta_3]_0]_J [[\beta_3]_0]_J]_2$

The next steps are used to generate a *yes* object inside membrane K. We present only the final configuration obtained after $4n + 2s + 3 = 12 + 4 + 3 = 19$ steps:

20. $[[\overline{\beta}_1]_M^2 [\beta_0\, yes [y_1\, y_3\, c_2\, \beta_3]_0]_K [\overline{\beta_0}]_J^8 [\overline{\beta_3}]_1^8 [[y_1\, y_2\, y_3\, \beta_3]_0]_J [[y_1\, y_2\, \beta_3]_0]_J$
$[[y_1\, \beta_3]_0]_J [[y_2\, y_3\, \beta_3]_0]_J [y_2\, c_2\, \beta_3\, \beta_0]_0 [[y_3\, \beta_3]_0]_J [[\beta_3]_0]_J]_2$

Exercise 2.8. Solve the Subset Sum problem using other classes of mobile membranes.

2.5.5 Knapsack Problem (0/1)

The decision Knapsack problem can be stated as follows: Given a knapsack of capacity $k \in \mathbb{N}$, a set A of n elements, a weight function $g : A \to \mathbb{N}$, a value function $r : A \to \mathbb{N}$, and a constant $l \in \mathbb{N}$, decide whether or not there exists a subset of A such that its weight does not exceed k and its value is greater than or equal than l.

Consider $A = \{a_1, \ldots, a_n\}$, and a system of mutual mobile membranes having the initial configuration

$$[[\overline{\psi_n}]_M [\alpha\ e_0]_J [\overline{\alpha}]_K [d^{n-1} d_0]_1 [d^{n-1} d_0]_2 [a_1\ b_0\ c_0\ \beta_0]_0]_3$$

and working over the alphabet

$$V = \{yes, no, d, d_0, b, c, \varphi, \alpha, \overline{\alpha}, \beta, \overline{\beta}, \psi, \overline{\psi}\}$$
$$\cup \{a_i, x_i, y_i, \psi_i, \overline{\psi_i}, \varphi_i, \gamma_i \mid 1 \le i \le n\} \cup \{\beta_i, \overline{\beta_i} \mid 0 \le i \le n\}$$
$$\cup \{b_i, c_i \mid 0 \le i \le n + max\{k, l\}\} \cup \{e_i \mid 0 \le i \le 3n + 2max\{k, l\}\}$$

In addition to mutual endocytosis and mutual exocytosis rules, elementary division rules are used to generate all the possible subsets. The system of mutual mobile membranes solving the Knapsack problem uses the rules:

(i) $[a_i]_0 \to [x_i z_i a_{i+1}]_0 [a_{i+1}]_0$, for $1 \le i \le n-1$ (div)
$[a_n]_0 \to [x_n z_n \overline{\beta}\,\overline{\psi}]_0 [\overline{\beta}\,\overline{\psi}]_0$ (div)
$[d]_j \to [\,]_j [\,]_j$, for $1 \le j \le 2$ (div)
$[d_0]_1 \to [\beta]_1 [\beta]_1$ (div)
$[d_0]_2 \to [\psi]_2 [\psi]_2$ (div)
The first two rules generate 2^n membranes labelled by 0 containing all the possible subsets over variables $\{x_1, \ldots, x_n\}$. When an object x_i appears in a membrane 0, there appears also an object z_i. The objects x_i and z_i are used to introduce objects that represent the weight and value, respectively, of each object a_i from a subset of A. In each membrane are placed also two symbols $\overline{\beta}$ and $\overline{\psi}$. The next two rules generate 2^n membranes labelled by 1 each containing an object β and 2^n membranes labelled by 2 each containing an object ψ. The symbols β, $\overline{\beta}$, ψ, and $\overline{\psi}$ are used in the mobility of membranes 1 and 2.

(ii) $[\beta]_1 [\overline{\beta}]_0 \to [[\beta_1]_1 \overline{\beta_1}]_0$ (mendo)
$[\beta_i]_1 [\overline{\beta_i}]_0 \to [[\beta_{i+1}]_1 \overline{\beta_{i+1}}]_0$, for $1 \le i \le n-1$ (mendo)
$[[\beta_i]_1 x_i \overline{\beta_i}]_0 \to [\beta_i]_1 [y_i\ b^{g(a_i)} \overline{\beta_i}]_0$, for $1 \le i \le n$ (mexo)
$[[\beta_i]_1 \overline{\beta_i}]_0 \to [\beta_i]_1 [\overline{\beta_i}]_0$, for $1 \le i \le n$ (mexo)
These rules are used to replace x_i by $y_i\ b^{g(a_i)}$, for $1 \le i \le n$. If x_i is not present in a membrane 0, then the indexes of β and $\overline{\beta}$ are incremented. After applying these rules, each membrane 0 contains a number of objects b equal to the weight of the contained subset of A, and also objects y_i used to remember which objects are contained in this membrane.

(iii) $[\psi]_2 [\overline{\psi}]_0 \to [[\psi_1]_2 \overline{\psi_1}]_0$ (mendo)

$[\psi_i]_2[\overline{\psi_i}]_0 \rightarrow [[\psi_{i+1}]_2\overline{\psi_{i+1}}]_0$, for $1 \leq i \leq n-1$ (mendo)
$[[\psi_i]_2 z_i \overline{\psi_i}]_0 \rightarrow [\psi_i]_2[c^{r(a_i)}\overline{\psi_i}]_0$, for $1 \leq i \leq n$ (mexo)
$[[\psi_i]_2\overline{\psi_i}]_0 \rightarrow [\psi_i]_2[\overline{\psi_i}]_0$, for $1 \leq i \leq n$ (mexo)
In parallel with the rules (ii), these rules are used to replace z_i by $c^{r(a_i)}$, for $1 \leq i \leq n$. If z_i is not present in a membrane 0, then the indexes of ψ and $\overline{\psi}$ are incremented. After applying these rules, each membrane 0 contains a number of objects c equal to the value of the contained subset of A.

(iv) $[\beta_n]_1[b\,\overline{\beta_n}]_0 \rightarrow [[\beta_n]_1\overline{\beta_n}]_0$ (mendo)
$[[\beta_n]_1 b_i\,\overline{\beta_n}]_0 \rightarrow [\beta_n]_1[b_{i+1}\,\overline{\beta_n}]_0$, for $0 \leq i$ (mexo)
These rules are used to calculate the weights of the subsets B of A, by using the objects b_0 that appear in all 0 membranes. If an object b is present, then the index of b_i, placed in the same membrane as b, is incremented.

(v) $[\psi_n]_2[c\,\overline{\psi_n}]_0 \rightarrow [[\psi_n]_2\overline{\psi_n}]_0$ (mendo)
$[[\psi_n]_2 c_i\,\overline{\psi_n}]_0 \rightarrow [\psi_n]_2[c_{i+1}\,\overline{\psi_n}]_0$, for $0 \leq i$ (mexo)
In parallel with the rules (iv), these rules are used to calculate the value of the subsets B of A, by using the objects c_0 that appear in all 0 membranes. If an object c is present, then the index of c_i, placed in the same membrane as c, is incremented.

(vi) $[\alpha\, e_i]_J[\overline{\alpha}]_K \rightarrow [[\alpha\, e_{i+1}]_J\overline{\alpha}]_K$ (mendo)
$[[\alpha\, e_i]_J\overline{\alpha}]_K \rightarrow [\alpha\, e_{i+1}]_J[\overline{\alpha}]_K$, for $0 \leq i \leq 3n+2max\{k,l\}-1$ (mexo)
We use these rules in parallel to calculate the number of steps performed. The counting stops after $3n+2max\{k,l\}$ steps, a number determined by: generating space (n steps), replacing x_i by the corresponding weight ($2n$ steps) and calculating weights and values ($2max\{k,l\}$ steps).

(vii) $[\alpha\, e_{3n+2max\{k,l\}}]_J[\overline{\alpha}]_K \rightarrow [[\alpha\, e_{3n+2max\{k,l\}}]_J\overline{\alpha}]_K$ (mendo)
$[[\alpha\, e_{3n+2max\{k,l\}}]_J\overline{\alpha}]_K \rightarrow [\alpha\, e_{3n+2max\{k,l\}}]_J[\varphi]_K$ (mexo)
$[\alpha\, e_{3n+2max\{k,l\}}]_J \rightarrow [\gamma_1]_J[\gamma_1]_J$ (div)
$[\gamma_i]_J \rightarrow [\gamma_{i+1}]_J[\gamma_{i+1}]_J$, for $1 \leq i \leq n-2$ (div)
$[\gamma_{n-1}]_J \rightarrow [\beta_n]_J[\beta_n]_J$ (div)
$[\varphi]_K \rightarrow [\varphi_1]_K[\varphi_1]_K$ (div)
$[\varphi_i]_K \rightarrow [\varphi_{i+1}]_K[\varphi_{i+1}]_K$, for $1 \leq i \leq n-2$ (div)
$[\varphi_{n-1}]_K \rightarrow [\psi_n]_K[\psi_n]_K$ (div)
$[\beta_n]_J[b_i\overline{\beta_n}]_0 \rightarrow [[\ ]_J]_0$, for $k < i$ (mendo)
$[\psi_n]_K[c_i\overline{\psi_n}]_0 \rightarrow [[\ ]_K]_0$, for $0 \leq i < l$ (mendo)
One or two extra steps, needed to prepare the membranes J and K for division, are performed if needed. If membrane J contains two objects α and $3n+2max\{k,l\}$, and is placed near membrane K containing an object ϕ, then 2^n membranes J are generated to check which membranes respect the weight condition, and 2^n membranes K are generated to check which membranes respect the value condition. All membranes 0 that do not respect the conditions are blocked inside a J or K membrane. In this way by choosing any membrane that is not placed inside a J or K membrane, we obtain a solution to the problem.

(viii) $[\overline{\psi_n}]_M[\psi_n]_K \rightarrow [\overline{\psi_n}[\psi_n]_K]_M$ (mendo)
$[\overline{\psi_n}[\psi_n]_K]_M \rightarrow [\beta_0\beta_0]_M[\beta_0]_K$ (mexo)
$[\beta_0]_0[\beta_0]_M \rightarrow [yes[\ ]_0]_M$ (mendo)

$[\beta_0]_K \to [\beta_0]_K[\beta_0]_K$ (div)
$[\beta_0]_K[\beta_0]_M \to [[no]_M]_K$ (mendo)
The computation stops after 2 more steps (maximum $4n+2max\{k,l\}+5$ steps in total). If there still exists a membrane 0 that is not blocked inside a J or K membrane, then the *yes* object is created inside membrane M. Otherwise, after one more step, the *no* object is created inside membrane M.

The number of membranes in the initial configuration is 7, and the number of objects is $2n+9$. The size of the working alphabet is $14n+3max\{k,l\}+15$. The number of rules in the above system is: $n+4$ rules of type (i), $3n$ rules of type (ii), $3n$ rules of type (iii), $k+n-1$ rules of type (iv), $l+n-1$ rules of type (v), $3n+2max\{k,l\}$ rules of type (vi), $4n+2max\{k,l\}+2$ rules of type (vii) and 5 rules of type $(viii)$. Hence, the size of the constructed system is $max\{\mathcal{O}(n), \mathcal{O}(max\{k,l\})\}$.

Example 2.6. Consider the Knapsack problem with $A = \{a_1, a_2, a_3\}$, $g(a_1) = 1$, $g(a_2) = 2$, $g(a_3) = 1$, $k = 2$, $r(a_1) = 2$, $r(a_2) = 1$, $r(a_3) = 3$ and $l = 3$. In this case, $n = 3$, $k = 2$, $l = 3$, and the initial configuration of the system is

$$[[\overline{\psi_3}]_M[\alpha\, e_0]_J[\overline{\alpha}]_K[d^2 d_0]_1[d^2 d_0]_2[a_1\, b_0\, c_0\, \beta_0]_0]_3$$

Graphically this is illustrated as:

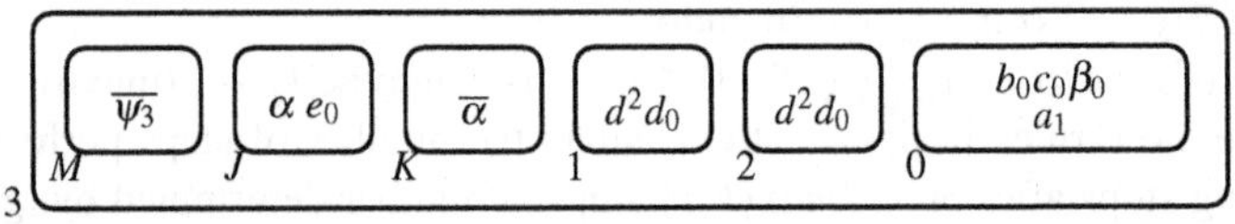

The evolution of the system is described by the following steps. The working space is generated in $n=3$ steps leading from the initial configuration 1 to configuration 4:

1. $[[\overline{\psi_3}]_M[\alpha\, e_0]_J[\overline{\alpha}]_K[d^2 d_0]_1[d^2 d_0]_2[a_1\, b_0\, c_0\, \beta_0]_0]_3$
2. $[[\overline{\psi_3}]_M[[\alpha\, e_1]_J\overline{\alpha}]_K[d\, d_0]_1^2[d\, d_0]_2^2[x_1 z_1 a_2\, b_0\, c_0\, \beta_0]_0[a_2\, b_0\, c_0\, \beta_0]_0]_3$
3. $[[\overline{\psi_3}]_M[\alpha\, e_2]_J[\overline{\alpha}]_K[d_0]_1^4[d_0]_2^4[x_1 z_1 x_2 z_2 a_3\, b_0\, c_0\, \beta_0]_0[x_1 z_1 a_3\, b_0\, c_0\, \beta_0]_0$
$[x_2 z_2 a_3\, b_0\, c_0\, \beta_0]_0[a_3\, b_0\, c_0\, \beta_0]_0]_3$
4. $[[\overline{\psi_3}]_M[[\alpha\, e_3]_J\overline{\alpha}]_K[\beta]_1^8[\psi]_2^8[x_1 z_1 x_2 z_2 x_3 z_3\, b_0\, c_0\, \overline{\beta}\, \overline{\psi}\beta_0]_0[x_1 z_1 x_2 z_2\, b_0\, c_0\, \overline{\beta}\, \overline{\psi}\beta_0]_0$
$[x_1 z_1 x_3 z_3\, b_0\, c_0 \overline{\beta}\, \overline{\psi}\beta_0]_0[x_1 z_1\, b_0\, c_0 \overline{\beta}\, \overline{\psi}\beta_0]_0[x_2 z_2 x_3 z_3\, b_0\, c_0 \overline{\beta}\, \overline{\psi}\beta_0]_0[x_2 z_2 b_0 c_0 \overline{\beta}\, \overline{\psi}\beta_0]_0$
$[x_3 z_3\, b_0\, c_0\, \overline{\beta}\, \overline{\psi}\beta_0]_0[b_0\, c_0\, \overline{\beta}\, \overline{\psi}\beta_0]_0]_3$

Graphically the working space is described by the following picture:

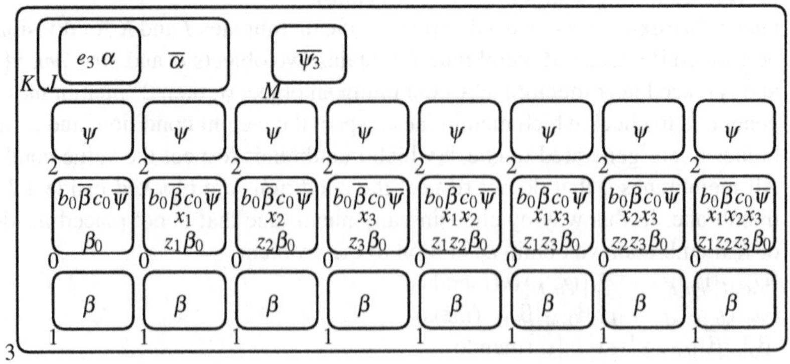

In the next steps we replace x_i by y_i and $b^{g(a_i)}$. We use y_i to mark the fact that in the membrane 0 containing it there is a subset containing a_i. The multiset of objects $b^{g(a_i)}$ is used to denote the weight of object a_i. In parallel we replace z_i by $c^{r(a_i)}$. The multiset of objects $c^{r(a_i)}$ is used to denote the value of object a_i.

5. $[[\overline{\psi_3}]_M[\alpha\, e_4]_J[\overline{\alpha}]_K[[\beta_1]_1[\psi_1]_2 x_1z_1x_2z_2x_3z_3\ b_0\ c_0\ \overline{\beta_1}\ \overline{\psi_1}]_0$
$[[\beta_1]_1[\psi_1]_2 x_1z_1x_2z_2\ b_0\ c_0\ \overline{\beta_1}\ \overline{\psi_1}\beta_0]_0[[\beta_1]_1[\psi_1]_2 x_1z_1x_3z_3\ b_0\ c_0\ \overline{\beta_1}\ \overline{\psi_1}\beta_0]_0$
$[[\beta_1]_1[\psi_1]_2 x_1z_1\ b_0\ c_0\ \overline{\beta_1}\ \overline{\psi_1}\beta_0]_0[[\beta_1]_1[\psi_1]_2 x_2z_2x_3z_3\ b_0\ c_0\ \overline{\beta_1}\ \overline{\psi_1}\beta_0]_0$
$[[\beta_1]_1[\psi_1]_2 x_2z_2\ b_0\ c_0\ \overline{\beta_1}\ \overline{\psi_1}\beta_0]_0[[\beta_1]_1[\psi_1]_2 x_3z_3\ b_0\ c_0\ \overline{\beta_1}\ \overline{\psi_1}\beta_0]_0$
$[[\beta_1]_1[\psi_1]_2 b_0\ c_0\ \overline{\beta_1}\ \overline{\psi_1}\beta_0]_0]_3$
6. $[[\overline{\psi_3}]_M[[\alpha\, e_5]_J\overline{\alpha}]_K[\beta_1]_1^8[\psi_1]_2^8[b\,c^2y_1x_2z_2x_3z_3\ b_0\ c_0\ \overline{\beta_1}\ \overline{\psi_1}\beta_0]_0$
$[b\,c^2y_1x_2z_2\ b_0\ c_0\ \overline{\beta_1}\ \overline{\psi_1}\beta_0]_0[b\,c^2y_1x_3z_3\ b_0\ c_0\ \overline{\beta_1}\ \overline{\psi_1}\beta_0]_0$
$[b\,c^2y_1b_0\ c_0\ \overline{\beta_1}\ \overline{\psi_1}\beta_0]_0[x_2z_2x_3z_3\ b_0\ c_0\ \overline{\beta_1}\ \overline{\psi_1}\beta_0]_0$
$[x_2z_2\ b_0\ c_0\ \overline{\beta_1}\ \overline{\psi_1}\beta_0]_0[x_3z_3\ b_0\ c_0\ \overline{\beta_1}\ \overline{\psi_1}\beta_0]_0[b_0\ c_0\ \overline{\beta_1}\ \overline{\psi_1}\beta_0]_0]_3$
7. $[[\overline{\psi_3}]_M[\alpha\, e_6]_J[\overline{\alpha}]_K[[\beta_2]_1[\psi_2]_2 b\,c^2y_1x_2z_2x_3z_3\ b_0\ c_0\ \overline{\beta_2}\ \overline{\psi_2}]_0$
$[[\beta_2]_1[\psi_2]_2 b\,c^2x_2z_2\ b_0\ c_0\ \overline{\beta_2}\ \overline{\psi_2}\beta_0]_0[[\beta_2]_1[\psi_2]_2 b\,c^2x_3z_3\ b_0\ c_0\ \overline{\beta_2}\ \overline{\psi_2}\beta_0]_0$
$[[\beta_2]_1[\psi_2]_2 b\,c^2\ b_0\ c_0\ \overline{\beta_2}\ \overline{\psi_2}\beta_0]_0[[\beta_2]_1[\psi_2]_2 x_2z_2x_3z_3\ b_0\ c_0\ \overline{\beta_2}\ \overline{\psi_2}\beta_0]_0$
$[[\beta_2]_1[\psi_2]_2 x_2z_2\ b_0\ c_0\ \overline{\beta_2}\ \overline{\psi_2}\beta_0]_0[[\beta_2]_1[\psi_2]_2 x_3z_3\ b_0\ c_0\ \overline{\beta_2}\ \overline{\psi_2}\beta_0]_0$
$[[\beta_2]_1[\psi_2]_2 b_0\ c_0\ \overline{\beta_2}\ \overline{\psi_2}\beta_0]_0]_3$
8. $[[\overline{\psi_3}]_M[[\alpha\, e_7]_J\overline{\alpha}]_K[\beta_2]_1^8[\psi_2]_2^8[b^3\ c^3y_1y_2x_3z_3\ b_0\ c_0\ \overline{\beta_2}\ \overline{\psi_2}\beta_0]_0$
$[b^3\ c^3y_1y_2\ b_0\ c_0\ \overline{\beta_2}\ \overline{\psi_2}\beta_0]_0[b\,c^2y_1x_3z_3\ b_0\ c_0\ \overline{\beta_2}\ \overline{\psi_2}\beta_0]_0$
$[b\,c^2y_1b_0\ c_0\ \overline{\beta_2}\ \overline{\psi_2}\beta_0]_0[b^2c\ y_2x_3z_3\ b_0\ c_0\ \overline{\beta_2}\ \overline{\psi_2}\beta_0]_0$
$[b^2c\ y_2\ b_0\ c_0\ \overline{\beta_2}\ \overline{\psi_2}\beta_0]_0[x_3z_3\ b_0\ c_0\ \overline{\beta_2}\ \overline{\psi_2}\beta_0]_0[b_0\ c_0\ \overline{\beta_2}\ \overline{\psi_2}\beta_0]_0]_3$
9. $[[\overline{\psi_3}]_M[\alpha\, e_8]_J[\overline{\alpha}]_K[[\beta_3]_1[\psi_3]_2 b^3\ c^3y_1y_2x_3z_3\ b_0\ c_0\ \overline{\beta_3}\ \overline{\psi_3}\beta_0]_0$
$[[\beta_3]_1[\psi_3]_2 b^3\ c^3y_1y_2\ b_0\ c_0\ \overline{\beta_3}\ \overline{\psi_3}\beta_0]_0[[\beta_3]_1[\psi_3]_2 b\,c^2y_1x_3z_3\ b_0\ c_0\ \overline{\beta_3}\ \overline{\psi_3}\beta_0]_0$
$[[\beta_3]_1[\psi_3]_2 b\,c^2y_1b_0\ c_0\ \overline{\beta_3}\ \overline{\psi_3}\beta_0]_0[[\beta_3]_1[\psi_3]_2 b^2c\ y_2x_3z_3\ b_0\ c_0\ \overline{\beta_3}\ \overline{\psi_3}\beta_0]_0$
$[[\beta_3]_1[\psi_3]_2 b^2c\ y_2\ b_0\ c_0\ \overline{\beta_3}\ \overline{\psi_3}\beta_0]_0[[\beta_3]_1[\psi_3]_2 x_3z_3\ b_0\ c_0\ \overline{\beta_3}\ \overline{\psi_3}\beta_0]_0$
$[[\beta_3]_1[\psi_3]_2 b_0\ c_0\ \overline{\beta_3}\ \overline{\psi_3}\beta_0]_0]_3$
10. $[[\overline{\psi_3}]_M[[\alpha\, e_9]_J\overline{\alpha}]_K[\beta_3]_1^8[\psi_3]_2^8[b^4\ c^6y_1y_2y_3\ b_0\ c_0\overline{\beta_3}\,\overline{\psi_3}\beta_0]_0[b^3\ c^3y_1y_2\ b_0\ c_0\overline{\beta_3}\,\overline{\psi_3}\beta_0]_0$
$[b^2\ c^5y_1y_3\ b_0\ c_0\ \overline{\beta_3}\ \overline{\psi_3}\beta_0]_0[b\,c^2y_1b_0\ c_0\ \overline{\beta_3}\ \overline{\psi_3}\beta_0]_0[b^3c^4\ y_2y_3\ b_0\ c_0\ \overline{\beta_3}\ \overline{\psi_3}\beta_0]_0$
$[b^2c\ y_2\ b_0\ c_0\ \overline{\beta_3}\ \overline{\psi_3}\beta_0]_0[b\,c^3y_3\ b_0\ c_0\ \overline{\beta_3}\ \overline{\psi_3}\beta_0]_0[b_0\ c_0\ \overline{\beta_3}\ \overline{\psi_3}]_0\beta_0]_3$

Graphically the working space is described by the following picture:

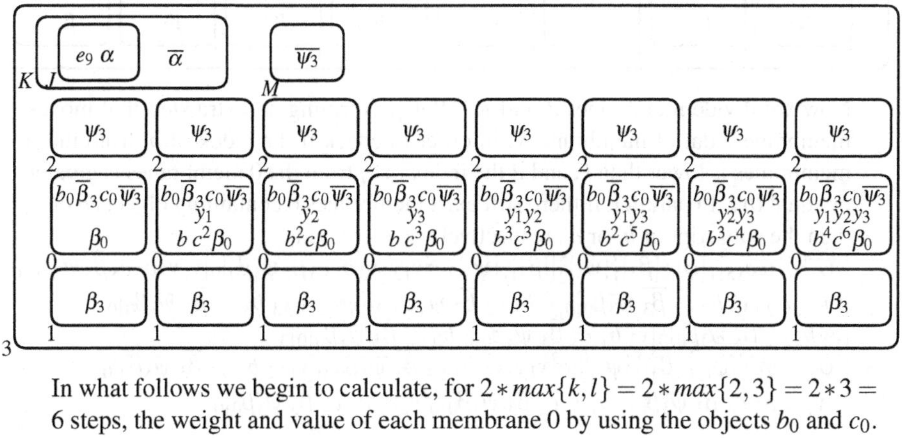

In what follows we begin to calculate, for $2*max\{k,l\} = 2*max\{2,3\} = 2*3 = 6$ steps, the weight and value of each membrane 0 by using the objects b_0 and c_0.

11. $[[\overline{\psi_3}]_M[\alpha\, e_{10}]_J[\overline{\alpha}]_K[\beta_3]_1[\psi_3]_2[[\beta_3]_1[\psi_3]_2 b^3\, c^5 y_1y_2y_3\; b_0\; c_0\; \overline{\beta_3}\;\overline{\psi_3}\beta_0]_0$
$[[\beta_3]_1[\psi_3]_2 b^2\, c^2 y_1y_2\; b_0\; c_0\; \overline{\beta_3}\;\overline{\psi_3}\beta_0]_0[[\beta_3]_1[\psi_3]_2 b\, c^4 y_1y_3\; b_0\; c_0\; \overline{\beta_3}\;\overline{\psi_3}\beta_0]_0$
$[[\beta_3]_1[\psi_3]_2 c\, y_1 b_0\; c_0\; \overline{\beta_3}\;\overline{\psi_3}\beta_0]_0[[\beta_3]_1[\psi_3]_2 b^2 c^3\; y_2y_3\; b_0\; c_0\; \overline{\beta_3}\;\overline{\psi_3}\beta_0]_0$
$[[\beta_3]_1[\psi_3]_2 b\; y_2\; b_0\; c_0\; \overline{\beta_3}\;\overline{\psi_3}\beta_0]_0[[\beta_3]_1[\psi_3]_2 c^2 y_3\; b_0\; c_0\; \overline{\beta_3}\;\overline{\psi_3}\beta_0]_0[b_0\; c_0\; \overline{\beta_3}\;\overline{\psi_3}\beta_0]_0]_3$

12. $[[\overline{\psi_3}]_M[[\alpha\, e_{11}]_J\overline{\alpha}]_K[\beta_3]_1^8[\psi_3]_2^8[b^3\, c^5 y_1y_2y_3\; b_1\; c_1\, \overline{\beta_3}\,\overline{\psi_3}\beta_0]_0[b^2\, c^2 y_1y_2\; b_1\; c_1\, \overline{\beta_3}\,\overline{\psi_3}\beta_0]_0$
$[b\, c^4 y_1y_3\; b_1\; c_1\; \overline{\beta_3}\;\overline{\psi_3}\beta_0]_0[c\; y_1 b_1\; c_1\; \overline{\beta_3}\;\overline{\psi_3}\beta_0]_0[b^2 c^3\; y_2y_3\; b_1\; c_1\; \overline{\beta_3}\;\overline{\psi_3}\beta_0]_0$
$[b\; y_2\; b_1\; c_1\; \overline{\beta_3}\;\overline{\psi_3}\beta_0]_0[c^2 y_3\; b_1\; c_1\; \overline{\beta_3}\;\overline{\psi_3}\beta_0]_0[b_0\; c_0\; \overline{\beta_3}\;\overline{\psi_3}\beta_0]_0]_3$

13. $[[\overline{\psi_3}]_M[\alpha\, e_{12}]_J[\overline{\alpha}]_K[\beta_3]_1^3[\psi_3]_2^2[[\beta_3]_1[\psi_3]_2 b^2\, c^4 y_1y_2y_3\; b_1\; c_1\; \overline{\beta_3}\;\overline{\psi_3}\beta_0]_0$
$[[\beta_3]_1[\psi_3]_2 b\; c\; y_1y_2\; b_1\; c_1\; \overline{\beta_3}\;\overline{\psi_3}\beta_0]_0[[\beta_3]_1[\psi_3]_2 c^3 y_1y_3\; b_1\; c_1\; \overline{\beta_3}\;\overline{\psi_3}\beta_0]_0$
$[[\psi_3]_2 y_1 b_1\; c_1\; \overline{\beta_3}\;\overline{\psi_3}\beta_0]_0[[\beta_3]_1[\psi_3]_2 b\; c^2\; y_2y_3\; b_1\; c_1\; \overline{\beta_3}\;\overline{\psi_3}\beta_0]_0$
$[[\beta_3]_1 y_2\; b_1\; c_1\; \overline{\beta_3}\;\overline{\psi_3}\beta_0]_0[[\psi_3]_2 c\; y_3\; b_1\; c_1\; \overline{\beta_3}\;\overline{\psi_3}\beta_0]_0[b_0\; c_0\; \overline{\beta_3}\;\overline{\psi_3}\beta_0]_0]_3$

14. $[[\overline{\psi_3}]_M[[\alpha\, e_{13}]_J\overline{\alpha}]_K[\beta_3]_1^8[\psi_3]_2^8[b^2\, c^4 y_1y_2y_3\; b_2\; c_2\; \overline{\beta_3}\;\overline{\psi_3}\beta_0]_0[b\; c\; y_1y_2\; b_2\; c_2\; \overline{\beta_3}\;\overline{\psi_3}\beta_0]_0$
$[c^3 y_1y_3\; b_2\; c_2\; \overline{\beta_3}\;\overline{\psi_3}\beta_0]_0[y_1 b_1\; c_2\; \overline{\beta_3}\;\overline{\psi_3}\beta_0]_0[b\; c^2\; y_2y_3\; b_2\; c_2\; \overline{\beta_3}\;\overline{\psi_3}\beta_0]_0$
$[y_2\; b_2\; c_1\; \overline{\beta_3}\;\overline{\psi_3}\beta_0]_0[c\; y_3\; b_1\; c_2\; \overline{\beta_3}\;\overline{\psi_3}\beta_0]_0[b_0\; c_0\; \overline{\beta_3}\;\overline{\psi_3}\beta_0]_0]_3$

15. $[[\overline{\psi_3}]_M[\alpha\, e_{14}]_J[\overline{\alpha}]_K[\beta_3]_1^5[\psi_3]_2^4[[\beta_3]_1[\psi_3]_2 b\; c^3 y_1y_2y_3\; b_2\; c_2\; \overline{\beta_3}\;\overline{\psi_3}\beta_0]_0$
$[[\beta_3]_1[\psi_3]_2\; y_1y_2\; b_2\; c_2\; \overline{\beta_3}\;\overline{\psi_3}\beta_0]_0[[\psi_3]_2 c^2 y_1y_3\; b_2\; c_2\; \overline{\beta_3}\;\overline{\psi_3}\beta_0]_0[y_1 b_1\; c_2\; \overline{\beta_3}\;\overline{\psi_3}\beta_0]_0$
$[[\beta_3]_1[\psi_3]_2 c\; y_2y_3\; b_2\; c_2\; \overline{\beta_3}\;\overline{\psi_3}\beta_0]_0[y_2\; b_2\; c_1\; \overline{\beta_3}\;\overline{\psi_3}\beta_0]_0$
$[[\psi_3]_2 y_3\; b_1\; c_2\; \overline{\beta_3}\;\overline{\psi_3}\beta_0]_0[b_0\; c_0\; \overline{\beta_3}\;\overline{\psi_3}\beta_0]_0]_3$

16. $[[\overline{\psi_3}]_M[[\alpha\, e_{15}]_J\overline{\alpha}]_K[\beta_3]_1^8[\psi_3]_2^8[b\; c^3 y_1y_2y_3\; b_3\; c_3\; \overline{\beta_3}\;\overline{\psi_3}\beta_0]_0[y_1y_2\; b_3\; c_3\; \overline{\beta_3}\;\overline{\psi_3}\beta_0]_0$
$[c^2 y_1y_3\; b_2\; c_3\, \overline{\beta_3}\,\overline{\psi_3}\beta_0]_0[y_1 b_1\; c_2\; \overline{\beta_3}\;\overline{\psi_3}\beta_0]_0[c\; y_2y_3\; b_3\; c_3\; \overline{\beta_3}\;\overline{\psi_3}\beta_0]_0[y_2\; b_2\; c_1\; \overline{\beta_3}\;\overline{\psi_3}\beta_0]_0$
$[y_3\; b_1\; c_3\; \overline{\beta_3}\;\overline{\psi_3}\beta_0]_0[b_0\; c_0\; \overline{\beta_3}\;\overline{\psi_3}\beta_0]_0]_3$

Graphically the working space is described by the following picture:

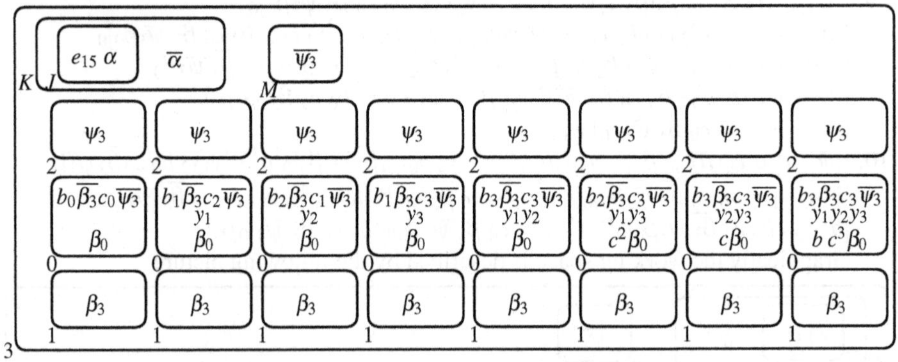

Now we divide membranes J and K, after performing an extra step that moves membrane J out of membrane K, in order to check if the index of b_i from the 0 membranes is lower than k, and if the index of c_i from the 0 membranes is greater or equal to l. In parallel, if there still exist membranes 0 containing objects b or c, then the subscript of b_i or c_i, respectively, is increased.

17. $[[\overline{\psi_3}]_M[\alpha e_{15}]_J[\varphi]_K[\beta_3]_1^7[\psi_3]_2^5[[\beta_3]_1[\psi_3]_2 c^2 y_1y_2y_3 b_3 c_3\overline{\beta_3}\,\overline{\psi_3}\beta_0]_0[y_1y_2 b_3 c_3\overline{\beta_3}\,\overline{\psi_3}\beta_0]_0$
$[[\psi_3]_2 c\; y_1y_3\; b_2\; c_3\; \overline{\beta_3}\;\overline{\psi_3}\beta_0]_0[y_1 b_1\; c_2\; \overline{\beta_3}\;\overline{\psi_3}\beta_0]_0[[\psi_3]_2 y_2y_3\; b_3\; c_3\; \overline{\beta_3}\;\overline{\psi_3}\beta_0]_0$
$[y_2\; b_2\; c_1\; \overline{\beta_3}\;\overline{\psi_3}\beta_0]_0[y_3\; b_1\; c_3\; \overline{\beta_3}\;\overline{\psi_3}\beta_0]_0[b_0\; c_0\; \overline{\beta_3}\;\overline{\psi_3}\beta_0]_0]_3$

18. $[[\overline{\psi_3}]_M[\gamma_1]_J^2[\varphi_1]_K^2[\beta_3]_1^8[\psi_3]_2^8[c^2 y_1y_2y_3\; b_4\; c_4\; \overline{\beta_3}\;\overline{\psi_3}\beta_0]_0[y_1y_2\; b_3\; c_3\; \overline{\beta_3}\;\overline{\psi_3}\beta_0]_0$
$[c\; y_1y_3\; b_2\; c_4\; \overline{\beta_3}\;\overline{\psi_3}\beta_0]_0[y_1 b_1\; c_2\; \overline{\beta_3}\;\overline{\psi_3}\beta_0]_0[y_2y_3\; b_3\; c_4\; \overline{\beta_3}\;\overline{\psi_3}\beta_0]_0$

$[y_2\, b_2\, c_1\, \overline{\beta_3}\, \overline{\psi_3}\beta_0]_0[y_3\, b_1\, c_3\, \overline{\beta_3}\, \overline{\psi_3}\beta_0]_0[b_0\, c_0\, \overline{\beta_3}\, \overline{\psi_3}\beta_0]_0]_3$

19. $[[\overline{\psi_3}]_M[\gamma_2]^4_J[\varphi_2]^4_K[\beta_3]^8_1[\psi_3]^7_2[[\psi_3]_2 c\, y_1y_2y_3\, b_4\, c_4\, \overline{\beta_3}\, \overline{\psi_3}\beta_0]_0[y_1y_2\, b_3\, c_3\, \overline{\beta_3}\, \overline{\psi_3}\beta_0]_0$
$[[\psi_3]_2 y_1y_3\, b_2\, c_4\, \overline{\beta_3}\, \overline{\psi_3}\beta_0]_0[y_1b_1\, c_2\, \overline{\beta_3}\, \overline{\psi_3}\beta_0]_0[y_2y_3\, b_3\, c_4\, \overline{\beta_3}\, \overline{\psi_3}\beta_0]_0$
$[y_2\, b_2\, c_1\, \overline{\beta_3}\, \overline{\psi_3}\beta_0]_0[y_3\, b_1\, c_3\, \overline{\beta_3}\, \overline{\psi_3}\beta_0]_0[b_0\, c_0\, \overline{\beta_3}\, \overline{\psi_3}\beta_0]_0]_3$

20. $[[\overline{\psi_3}]_M[\beta_3]^8_J[\psi_3]^8_K[\beta_3]^8_1[\psi_3]^8_2[c\, y_1y_2y_3\, b_4\, c_5\, \overline{\beta_3}\, \overline{\psi_3}\beta_0]_0[y_1y_2\, b_3\, c_3\, \overline{\beta_3}\, \overline{\psi_3}\beta_0]_0$
$[y_1y_3\, b_2\, c_5\, \overline{\beta_3}\, \overline{\psi_3}\beta_0]_0[y_1b_1\, c_2\, \overline{\beta_3}\, \overline{\psi_3}\beta_0]_0[y_2y_3\, b_3\, c_4\, \overline{\beta_3}\, \overline{\psi_3}\beta_0]_0$
$[y_2\, b_2\, c_1\, \overline{\beta_3}\, \overline{\psi_3}\beta_0]_0[y_3\, b_1\, c_3\, \overline{\beta_3}\, \overline{\psi_3}\beta_0]_0[b_0\, c_0\, \overline{\beta_3}\, \overline{\psi_3}\beta_0]_0]_3$

The next steps are used to generate a *yes* object in membrane M. We present only the final configuration obtained after $4n+2max\{k,l\}+5 = 4*3+2*3+5 = 23$ steps.

24. $[[\beta_3]^5_J[\psi_3]^4_K[\overline{\beta_0}]^2_K[\beta_3]^8_1[\psi_3]^8_2[[c\, y_1y_2y_3\, c_5\, \overline{\psi_3}\beta_0]_0]_J[[y_1y_2\, c_3\, \overline{\psi_3}\beta_0]_0]_J$
$[yes[y_1y_3\, b_2\, c_5\, \overline{\beta_3}\, \overline{\psi_3}]_0\beta_0]_M[[y_1b_1\, \overline{\beta_3}\beta_0]_0]_K[[y_2y_3\, c_4\, \overline{\psi_3}\beta_0]_0]_J$
$[[y_2\, b_2\, \overline{\beta_3}\beta_0]_0]_K[y_3\, b_1\, c_3\, \overline{\beta_3}\, \overline{\psi_3}\beta_0]_0[[b_0\, \overline{\beta_3}\beta_0]_0]_K]_3$

Exercise 2.9. Solve the Knapsack problem using other classes of mobile membranes.

2.5.6 2-Partition Problem

The problem can be enounced as follows: given a finite set A, a weight function $g : A \rightarrow \mathbb{N}$, decide whether or not there exists a partition of A into two subsets such that they have the same weight.

Consider $A = \{a_1,\dots,a_n\}$, and a system of mutual mobile membranes having the initial configuration

$$[[\overline{\alpha}]_M[\alpha\, e_0]_J[\overline{\alpha}]_K[d^{n-1}d_0]_1[d^{n-1}d_0]_2[a_1\, b_0\, c_0\, \beta_0]_0]_3$$

and working over the alphabet

$$V = \{yes, no, b, c, d, d_0, \alpha, \alpha_1, \overline{\alpha}, \beta, \overline{\beta}, \psi, \overline{\psi}\}$$
$$\cup\{a_i, x_i, y_i, z_i, t_i, \psi_i, \overline{\psi}_i \mid 1 \le i \le n\}$$
$$\cup\{\beta_i, \overline{\beta}_i \mid 0 \le i \le n\} \cup \{c_i, b_i \mid 0 \le i \le g(A)\} \cup \{e_i \mid 0 \le i \le 4n+g(a)+1\}$$

In addition to mutual endocytosis and mutual exocytosis rules, elementary division rules are used to generate all the possible subsets. The system of mutual mobile membranes solving the 2-partition problem uses the rules:

(i) $[a_i]_0 \rightarrow [x_i a_{i+1}]_0[z_i a_{i+1}]_0$, for $1 \le i \le n-1$ (div)
$[a_n]_0 \rightarrow [x_n\beta\, \overline{\psi}]_0[z_n\beta\, \overline{\psi}]_0$ (div)
$[d]_j \rightarrow [\ ]_j[\ ]_j$, for $1 \le j \le 2$ (div)
$[d_0]_1 \rightarrow [\beta]_1[\beta]_1$ (div)
$[d_0]_2 \rightarrow [\psi]_2[\psi]_2$ (div)
The first two rules generate 2^n membranes labelled by 0 containing all the possible subsets over variables $\{x_1,\dots,x_n\}$. The complementary subset is formed over variables $\{z_1,\dots,z_n\}$, in a such a manner that x_i and z_i do not exist at the same time inside a membrane, and the total number of x_i and z_j from each membrane after the generation stage is n. The objects x_i and z_i are used to introduce objects

that represent the weight of the objects a_i. In each membrane are placed also two symbols $\overline{\beta}$ and $\overline{\psi}$. The next two rules generate 2^n membranes labelled by 1 each containing an object β and 2^n membranes labelled by 2 each containing an object ψ. The symbols β, $\overline{\beta}$, ψ and $\overline{\psi}$ are used in the mobility of membranes 1 and 2.

(ii) $[\beta]_1[\overline{\beta}]_0 \to [[\beta_1]_1\overline{\beta_1}]_0$ (mendo)
$[\beta_i]_1[\overline{\beta_i}]_0 \to [[\beta_{i+1}]_1\overline{\beta_{i+1}}]_0$, for $1 \le i \le n-1$ (mendo)
$[[\beta_i]_1 x_i \overline{\beta_i}]_0 \to [\beta_i]_1[y_i\, b^{g(a_i)}\overline{\beta_i}]_0$, for $1 \le i \le n$ (mexo)
$[[\beta_i]_1\overline{\beta_i}]_0 \to [\beta_i]_1[\overline{\beta_i}]_0$, for $1 \le i \le n$ (mexo)
These rules are used to replace x_i by $y_i\, b^{g(a_i)}$, for $1 \le i \le n$. If x_i is not present in a membrane 0, then the indexes of β and $\overline{\beta}$ are incremented. After applying these rules, each membrane 0 contains a number of objects b equal to the weight of one partition of A, and also objects y_i used to remember which objects from the first partition are contained in this membrane.

(iii) $[\psi]_2[\overline{\psi}]_0 \to [[\psi_1]_2\overline{\psi_1}]_0$ (mendo)
$[\psi_i]_2[\overline{\psi_i}]_0 \to [[\psi_{i+1}]_2\overline{\psi_{i+1}}]_0$, for $1 \le i \le n-1$ (mendo)
$[[\psi_i]_2 z_i \overline{\psi_i}]_0 \to [\psi_i]_2[t_i c^{g(a_i)}\overline{\psi_i}]_0$, for $1 \le i \le n$ (mexo)
$[[\psi_i]_2\overline{\psi_i}]_0 \to [\psi_i]_2[\overline{\psi_i}]_0$, for $1 \le i \le n$ (mexo)
In parallel with (ii), these rules are used to replace z_i by $t_i\, c^{g(a_i)}$, for $1 \le i \le n$. If z_i is not present in a membrane 0, then the indexes of ψ and $\overline{\psi}$ are incremented. After applying these rules, each membrane 0 contains a number of objects c equal to the weight of the other partition of A, and also objects t_i used to remember which objects from the other partition are contained in this membrane.

(iv) $[\beta_n]_1[b\,\overline{\beta_n}]_0 \to [[\beta_n]_1\overline{\beta_n}]_0$ (mendo)
$[[\beta_n]_1 b_i\, \overline{\beta_n}]_0 \to [\beta_n]_1[b_{i+1}\, \overline{\beta_n}]_0$, for $0 \le i$ (mexo)
These rules are used to calculate the weights of the first partition of A, by using the objects b_0 that appear in all 0 membranes. If an object b is present, then the index of b_i, placed in the same membrane as b, is incremented.

(v) $[\psi_n]_2[c\,\overline{\psi_n}]_0 \to [[\psi_n]_2\overline{\psi_n}]_0$ (mendo)
$[[\psi_n]_2 c_i\, \overline{\psi_n}]_0 \to [\psi_n]_2[c_{i+1}\, \overline{\psi_n}]_0$, for $0 \le i$ (mexo)
In parallel with the rules (iv), these rules are used to calculate the weights of the other partition of A, by using the objects c_0 that appear in all 0 membranes. If an object c is present, then the index of c_i, placed in the same membrane as c, is incremented.

(vi) $[\alpha\, e_i]_J[\overline{\alpha}]_K \to [[\alpha\, e_{i+1}]_J\overline{\alpha}]_K$ (mendo)
$[[\alpha\, e_i]_J\overline{\alpha}]_K \to [\alpha\, e_{i+1}]_J[\overline{\alpha}]_K$, for $0 \le i < 3n + g(A) - 1$ (mexo)
$[[\alpha\, e_i]_J\overline{\alpha}]_K \to [\alpha\, e_{i+1}]_J[\alpha\, e_{i+1}]_K$, for $i = 3n + g(A) - 1$ (mexo)
We use these rules in parallel to calculate the number of steps performed. The weight of the set A is given by $g(A) = \Sigma_{a \in A}\, g(a)$. The counting stops after $3n + g(A) + 1$ steps, a number determined by: generating space (n steps), replacing x_i and z_i by corresponding weights ($2n$ steps) and calculating weights ($g(A)$ steps).

(vii) $[[\alpha\, e_{3n+g(A)}]_J\overline{\alpha}]_K \to [\alpha\, e_{3n+g(A)}]_J[\alpha\, e_{3n+g(A)}]_K$ (mexo)
$[\alpha\, e_{3n+g(A)}]_J \to [\gamma_1]_J[\gamma_1]_J$ (div)
$[\gamma_i]_J \to [\gamma_{i+1}]_J[\gamma_{i+1}]_J$, for $1 \le i \le n-2$ (div)
$[\gamma_{n-1}]_J \to [\beta_n]_J[\beta_n]_J$ (div)

$[\beta_n]_J[\overline{\beta_n}\,\overline{\psi_n} b_i c_j]_0 \to [[\,]_J]_0$, for $i \neq j$ (mendo)
One extra step, to prepare the membrane J for division, is performed if needed. We generate 2^n membranes J to check which membranes respect the weight condition. If membrane J contains an object $e_{3n+g(A)}$, and is placed near membrane K, then 2^n membranes are generated (n steps) in order to check in 2 steps which membranes contain c_i and b_i.

(viii) $[\alpha\, e_i]_K[\overline{\alpha}]_M \to [[\alpha\, e_{i+1}]_K\overline{\alpha}]_M$ for $3n+g(A) \leq i \leq 4n+g(A)$ (mendo)
$[[\alpha\, e_i]_K\overline{\alpha}]_M \to [\underline{\alpha}\, e_{i+1}]_K[\overline{\alpha}]_M$, for $3n+g(A) \leq i < 4n+g(A)$ (mexo)
$[[\underline{\alpha}\, e_i]_K\overline{\alpha}]_M \to [\overline{\beta_0}\beta_0]_K[\alpha_1]_M$, for $i = 4n+g(A)$ or $i = 4n+g(A)+1$ (mexo)
$[\overline{\beta_0}]_K[\beta_0]_0 \underline{\to} [\underline{yes}[\,]_0]_K$ (mendo)
$[\alpha_1]_M \underline{\to} [\beta_0]_M[\overline{\beta_0}]_M$ (div)
$[\beta_0]_K[\overline{\beta_0}]_M \to [[no]_K]_M$ (mendo)
The above rules are used in parallel to calculate the number of steps performed. The number $4n+g(A)+1$ is determined by: generating space (n steps), replacing x_i and z_i by corresponding weights ($2n$ steps) and calculating weights ($g(A)$ steps), generating J membranes ($n-1$ steps), preparing division of J (1 step), blocking all membranes 0 that do not satisfy conditions (1 step). If there still exists a membrane 0 that is not inside a membrane J, then the object *yes* is created inside membrane K. Otherwise, after one more step, the *no* object is created inside membrane K. The computation stops after maximum $4n+g(A)+3$ steps, with the answer placed inside membrane K.

The number of membranes in the initial configuration is 6, and the number of objects is $2n+5$. The size of the working alphabet is $13n+3g(A)+15$. The number of rules in the above system is: $n+4$ rules of type (i), $3n$ rules of type (ii), $3n$ rules of type (iii), $2g(A)$ rules of type (iv), $2g(A)$ rules of type (v), $3n+g(A)$ rules of type (vi), $n+2$ rules of type (vii) and $2n+5$ rules of type $(viii)$. Hence, the size of the constructed system of mutual mobile membranes is $max\{\mathcal{O}(n), \mathcal{O}(g(A))\}$.

Example 2.7. Consider the 2-partition problem with $A = \{a_1, a_2, a_3\}$, $g(a_1) = 1$, $g(a_2) = 3$ and $g(a_3) = 2$. In this case, $n = 3$, $g(A) = 6$, and the initial configuration of the system of mutual mobile membranes

$$[[\overline{\alpha}]_M[\alpha\, e_0]_J[\overline{\alpha}]_K[d^2 d_0]_1[d^2 d_0]_2[a_1 b_0\, c_0\, \beta_0]_0]_3$$

Graphically this is illustrated as:

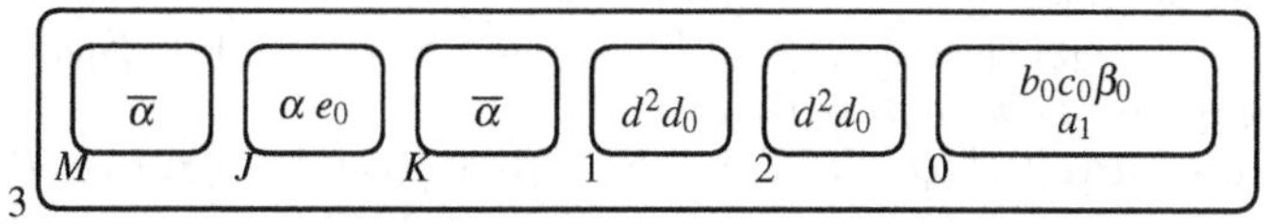

The evolution of the system is described by the following steps. The working space is generated in $n = 3$ steps leading from the initial configuration 1 to configuration 4.

1. $[[\overline{\alpha}]_M[\alpha\, e_0]_J[\overline{\alpha}]_K[d^2 d_0]_1[d^2 d_0]_2[a_1 b_0\, c_0\, \beta_0]_0]_3$
2. $[[\overline{\alpha}]_M[[\alpha\, e_1]_J\overline{\alpha}]_K[d\, d_0]_1^2[d\, d_0]_2^2[x_1 a_2\, b_0\, c_0\, \beta_0]_0[z_1 a_2\, b_0\, c_0\, \beta_0]_0]_3$
3. $[[\overline{\alpha}]_M[\alpha\, e_2]_J[\overline{\alpha}]_K[d_0]_1^4[d_0]_2^4[x_1 x_2 a_3\, b_0\, c_0\, \beta_0]_0[x_1 z_2 a_3\, b_0\, c_0\, \beta_0]_0$
$[z_1 x_2 a_3\, b_0\, c_0\, \beta_0]_0[z_1 z_2 a_3\, b_0\, c_0\, \beta_0]_0]_3$

4. $[[\overline{\alpha}]_M[[\alpha\, e_3]_J\overline{\alpha}]_K[\beta]_1^8[\psi]_2^8[x_1x_2x_3\, b_0\, c_0\, \overline{\beta}\, \overline{\psi}\, \beta_0]_0[x_1x_2z_3\, b_0\, c_0\, \overline{\beta}\, \overline{\psi}\, \beta_0]_0$
$[x_1z_2x_3\, b_0\, c_0\, \overline{\beta}\, \overline{\psi}\, \beta_0]_0[x_1z_2z_3\, b_0\, c_0\, \overline{\beta}\, \overline{\psi}\, \beta_0]_0[z_1x_2x_3\, b_0\, c_0\, \overline{\beta}\, \overline{\psi}\, \beta_0]_0$
$[z_1x_2z_3\, b_0\, c_0\, \overline{\beta}\, \overline{\psi}\, \beta_0]_0[z_1z_2x_3\, b_0\, c_0\, \overline{\beta}\, \overline{\psi}\, \beta_0]_0[z_1z_2z_3\, b_0\, c_0\, \overline{\beta}\, \overline{\psi}\, \beta_0]_0]_3$
Graphically the working space is described by the following picture:

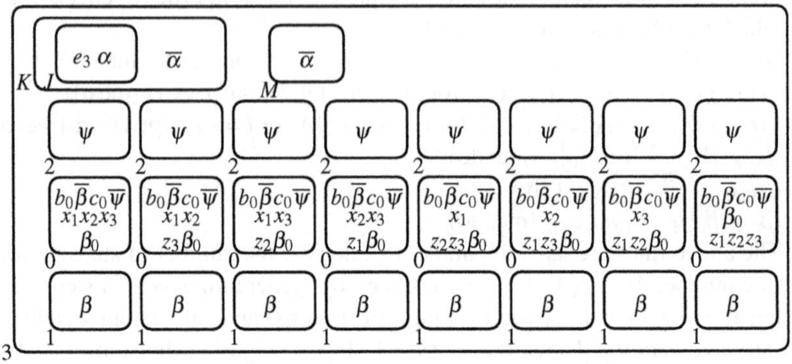

We notice from this image that each partition appears twice, but this does not increase the number of steps performed. In the next steps we replace x_i by y_i and $b^{g(a_i)}$. We use y_i to mark the fact that in the membrane 0 containing it there is a subset containing a_i. The multiset of objects $b^{g(a_i)}$ is used to denote the weight of object a_i. In parallel we replace z_j by t_j. We use t_j to mark the fact that in the membrane 0 containing it there is a subset containing a_j.

5. $[[\overline{\alpha}]_M[\alpha\, e_4]_J[\overline{\alpha}]_K[[\beta_1]_1[\psi_1]_2x_1x_2x_3\, b_0\, c_0\, \overline{\beta_1}\, \overline{\psi_1}\beta_0]_0$
$[[\beta_1]_1[\psi_1]_2x_1x_2z_3\, b_0\, c_0\, \overline{\beta_1}\, \overline{\psi_1}\beta_0]_0[[\beta_1]_1[\psi_1]_2x_1z_2x_3\, b_0\, c_0\, \overline{\beta_1}\, \overline{\psi_1}\beta_0]_0$
$[[\beta_1]_1[\psi_1]_2x_1z_2z_3\, b_0\, c_0\, \overline{\beta_1}\, \overline{\psi_1}\beta_0]_0[[\beta_1]_1[\psi_1]_2z_1x_2x_3\, b_0\, c_0\, \overline{\beta_1}\, \overline{\psi_1}\beta_0]_0$
$[[\beta_1]_1[\psi_1]_2z_1x_2z_3\, b_0\, c_0\, \overline{\beta_1}\, \overline{\psi_1}\beta_0]_0[[\beta_1]_1[\psi_1]_2z_1z_2x_3\, b_0\, c_0\, \overline{\beta_1}\, \overline{\psi_1}\beta_0]_0$
$[[\beta_1]_1[\psi_1]_2z_1z_2z_3\, b_0\, c_0\, \overline{\beta_1}\, \overline{\psi_1}\beta_0]_0]_3$
6. $[[\overline{\alpha}]_M[[\alpha\, e_5]_J\overline{\alpha}]_K[\beta_1]_1^8[\psi_1]_2^8[b\, y_1x_2x_3\, b_0\, c_0\, \overline{\beta_1}\, \overline{\psi_1}\beta_0]_0[b\, y_1x_2z_3\, b_0\, c_0\, \overline{\beta_1}\, \overline{\psi_1}\beta_0]_0$
$[by_1z_2x_3\, b_0\, c_0\, \overline{\beta_1}\, \overline{\psi_1}\beta_0]_0[b\, y_1z_2z_3\, b_0\, c_0\, \overline{\beta_1}\, \overline{\psi_1}\beta_0]_0[c\, t_1x_2x_3\, b_0\, c_0\, \overline{\beta_1}\, \overline{\psi_1}\beta_0]_0$
$[c\, t_1x_2z_3\, b_0\, c_0\, \overline{\beta_1}\, \overline{\psi_1}\beta_0]_0[c\, t_1z_2x_3\, b_0\, c_0\, \overline{\beta_1}\, \overline{\psi_1}\beta_0]_0[c\, t_1z_2z_3\, b_0\, c_0\, \overline{\beta_1}\, \overline{\psi_1}\beta_0]_0]_3$
7. $[[\overline{\alpha}]_M[\alpha\, e_6]_J[\overline{\alpha}]_K[[\beta_2]_1[\psi_2]_2b\, y_1x_2x_3\, b_0\, c_0\, \overline{\beta_2}\, \overline{\psi_2}\beta_0]_0$
$[[\beta_2]_1[\psi_2]_2b\, y_1x_2z_3\, b_0\, c_0\, \overline{\beta_2}\, \overline{\psi_2}\beta_0]_0[[\beta_2]_1[\psi_2]_2by_1z_2x_3\, b_0\, c_0\, \overline{\beta_2}\, \overline{\psi_2}\beta_0]_0$
$[[\beta_2]_1[\psi_2]_2b\, y_1z_2z_3\, b_0\, c_0\, \overline{\beta_2}\, \overline{\psi_2}\beta_0]_0[[\beta_2]_1[\psi_2]_2c\, t_1x_2x_3\, b_0\, c_0\, \overline{\beta_2}\, \overline{\psi_2}\beta_0]_0$
$[[\beta_2]_1[\psi_2]_2c\, t_1x_2z_3\, b_0\, c_0\, \overline{\beta_2}\, \overline{\psi_2}\beta_0]_0[[\beta_2]_1[\psi_2]_2c\, t_1z_2x_3\, b_0\, c_0\, \overline{\beta_2}\, \overline{\psi_2}\beta_0]_0$
$[[\beta_2]_1[\psi_2]_2c\, t_1z_2z_3\, b_0\, c_0\, \overline{\beta_2}\, \overline{\psi_2}\beta_0]_0]_3$
8. $[[\overline{\alpha}]_M[[\alpha\, e_7]_J\overline{\alpha}]_K[\beta_2]_1^8[\psi_2]_2^8[b^4\, y_1y_2x_3\, b_0\, c_0\, \overline{\beta_2}\, \overline{\psi_2}\beta_0]_0[b^4\, y_1y_2z_3\, b_0\, c_0\, \overline{\beta_2}\, \overline{\psi_2}\beta_0]_0$
$[b\, c^3y_1t_2x_3\, b_0\, c_0\, \overline{\beta_2}\, \overline{\psi_2}\beta_0]_0[b\, c^3y_1t_2z_3\, b_0\, c_0\, \overline{\beta_2}\, \overline{\psi_2}\beta_0]_0[b^3c\, t_1y_2x_3\, b_0\, c_0\, \overline{\beta_2}\, \overline{\psi_2}\beta_0]_0$
$[b^3c\, t_1y_2z_3\, b_0\, c_0\, \overline{\beta_2}\, \overline{\psi_2}\beta_0]_0[c^4\, t_1t_2x_3\, b_0\, c_0\, \overline{\beta_2}\, \overline{\psi_2}\beta_0]_0[c^4\, t_1t_2z_3\, b_0\, c_0\, \overline{\beta_2}\, \overline{\psi_2}\beta_0]_0]_3$
9. $[[\overline{\alpha}]_M[\alpha\, e_8]_J[\overline{\alpha}]_K[[\beta_3]_1[\psi_3]_2b^4\, y_1y_2x_3\, b_0\, c_0\, \overline{\beta_3}\, \overline{\psi_3}\beta_0]_0$
$[[\beta_3]_1[\psi_3]_2b^4\, y_1y_2z_3\, b_0\, c_0\, \overline{\beta_3}\, \overline{\psi_3}\beta_0]_0[[\beta_3]_1[\psi_3]_2b\, c^3y_1t_2x_3\, b_0\, c_0\, \overline{\beta_3}\, \overline{\psi_3}\beta_0]_0$
$[[\beta_3]_1[\psi_3]_2b\, c^3y_1t_2z_3\, b_0\, c_0\, \overline{\beta_3}\, \overline{\psi_3}\beta_0]_0[[\beta_3]_1[\psi_3]_2b^3c\, t_1y_2x_3\, b_0\, c_0\, \overline{\beta_3}\, \overline{\psi_3}\beta_0]_0$
$[[\beta_3]_1[\psi_3]_2b^3c\, t_1y_2z_3\, b_0\, c_0\, \overline{\beta_3}\, \overline{\psi_3}\beta_0]_0[[\beta_3]_1[\psi_3]_2c^4\, t_1t_2x_3\, b_0\, c_0\, \overline{\beta_3}\, \overline{\psi_3}\beta_0]_0$
$[[\beta_3]_1[\psi_3]_2c^4\, t_1t_2z_3\, b_0\, c_0\, \overline{\beta_3}\, \overline{\psi_3}\beta_0]_0]_3$

10. $[[\overline{\alpha}]_M[[\alpha\, e_9]_J\overline{\alpha}]_K[\beta_3]_1^8[\psi_3]_2^8[b^6\, y_1y_2y_3\, b_0\, c_0\, \overline{\beta_3}\, \overline{\psi_3}\beta_0]_0[b^4c^2\, y_1y_2t_3\, b_0\, c_0\, \overline{\beta_3}\, \overline{\psi_3}\beta_0]_0$
$[b^3\, c^3y_1t_2y_3\, b_0\, c_0\, \overline{\beta_3}\, \overline{\psi_3}\beta_0]_0[b\, c^5y_1t_2t_3\, b_0\, c_0\, \overline{\beta_3}\, \overline{\psi_3}\beta_0]_0[b^5c\, t_1y_2y_3\, b_0\, c_0\, \overline{\beta_3}\, \overline{\psi_3}\beta_0]_0$
$[b^3c^3\, t_1y_2t_3\, b_0\, c_0\, \overline{\beta_3}\, \overline{\psi_3}\beta_0]_0[b^2c^4\, t_1t_2y_3\, b_0\, c_0\, \overline{\beta_3}\, \overline{\psi_3}\beta_0]_0[c^6\, t_1t_2t_3\, b_0\, c_0\, \overline{\beta_3}\, \overline{\psi_3}\beta_0]_0]_3$
Graphically the working space is described by the following picture:

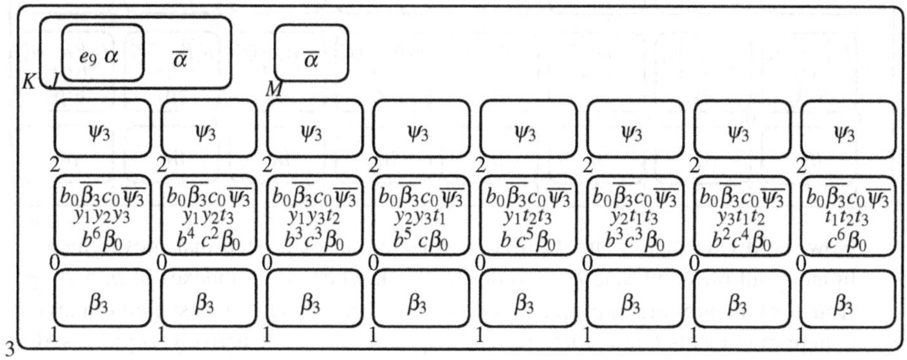

In what follows we calculate in $g(A) = 1+3+2 = 6$ steps, the weights of the subsets (over $\{y_1,y_2,y_3\}$) and complementary subsets (over $\{t_1,t_2,t_3\}$) from each membrane 0 by using the objects b_0 and c_0.

11. $[[\overline{\alpha}]_M[\alpha\, e_{10}]_J[\overline{\alpha}]_K[\beta_3]_1[\psi_3]_2[[\beta_3]_1b^5\, y_1y_2y_3\, b_0\, c_0\, \overline{\beta_3}\, \overline{\psi_3}\beta_0]_0$
$[[\beta_3]_1[\psi_3]_2b^3c\, y_1y_2t_3\, b_0\, c_0\, \overline{\beta_3}\, \overline{\psi_3}\beta_0]_0[[\beta_3]_1[\psi_3]_2b^2\, c^2y_1t_2y_3\, b_0\, c_0\, \overline{\beta_3}\, \overline{\psi_3}\beta_0]_0$
$[[\beta_3]_1[\psi_3]_2c^4y_1t_2t_3\, b_0\, c_0\, \overline{\beta_3}\, \overline{\psi_3}\beta_0]_0[[\beta_3]_1[\psi_3]_2b^4t_1y_2y_3\, b_0\, c_0\, \overline{\beta_3}\, \overline{\psi_3}\beta_0]_0$
$[[\beta_3]_1[\psi_3]_2b^2c^2\, t_1y_2t_3\, b_0\, c_0\, \overline{\beta_3}\, \overline{\psi_3}\beta_0]_0[[\beta_3]_1[\psi_3]_2b\, c^3\, t_1t_2y_3\, b_0\, c_0\, \overline{\beta_3}\, \overline{\psi_3}\beta_0]_0$
$[[\psi_3]_2c^5\, t_1t_2t_3\, b_0\, c_0\, \overline{\beta_3}\, \overline{\psi_3}\beta_0]_0]_3$
12. $[[\overline{\alpha}]_M[[\alpha\, e_{11}]_J\overline{\alpha}]_K[\beta_3]_1^8[\psi_3]_2^8[b^5\, y_1y_2y_3\, b_1\, c_0\, \overline{\beta_3}\, \overline{\psi_3}\beta_0]_0[b^3c\, y_1y_2t_3\, b_1\, c_1\, \overline{\beta_3}\, \overline{\psi_3}\beta_0]_0$
$[b^2\, c^2y_1t_2y_3\, b_1\, c_1\, \overline{\beta_3}\, \overline{\psi_3}\beta_0]_0[c^4y_1t_2t_3\, b_1\, c_1\, \overline{\beta_3}\, \overline{\psi_3}\beta_0]_0[b^4t_1y_2y_3\, b_1\, c_1\, \overline{\beta_3}\, \overline{\psi_3}\beta_0]_0$
$[b^2c^2\, t_1y_2t_3\, b_1\, c_1\, \overline{\beta_3}\, \overline{\psi_3}\beta_0]_0[b\, c^3\, t_1t_2y_3\, b_1\, c_1\, \overline{\beta_3}\, \overline{\psi_3}\beta_0]_0[c^5\, t_1t_2t_3\, b_0\, c_1\, \overline{\beta_3}\, \overline{\psi_3}\beta_0]_0]_3$
13. $[[\overline{\alpha}]_M[\alpha\, e_{12}]_J[\overline{\alpha}]_K[\beta_3]_1^2[\psi_3]_2^2[[\beta_3]_1b^4\, y_1y_2y_3\, b_1\, c_0\, \overline{\beta_3}\, \overline{\psi_3}\beta_0]_0$
$[[\beta_3]_1[\psi_3]_2b^2y_1y_2t_3\, b_1\, c_1\, \overline{\beta_3}\, \overline{\psi_3}\beta_0]_0[[\beta_3]_1[\psi_3]_2b\, c\, y_1t_2y_3\, b_1\, c_1\, \overline{\beta_3}\, \overline{\psi_3}\beta_0]_0$
$[[\psi_3]_2c^3y_1t_2t_3\, b_1\, c_1\, \overline{\beta_3}\, \overline{\psi_3}\beta_0]_0[[\beta_3]_1b^3t_1y_2y_3\, b_1\, c_1\, \overline{\beta_3}\, \overline{\psi_3}\beta_0]_0$
$[[\beta_3]_1[\psi_3]_2b\, c\, t_1y_2t_3\, b_1\, c_1\, \overline{\beta_3}\, \overline{\psi_3}\beta_0]_0[[\beta_3]_1[\psi_3]_2c^2\, t_1t_2y_3\, b_1\, c_1\, \overline{\beta_3}\, \overline{\psi_3}\beta_0]_0$
$[[\psi_3]_2c^4\, t_1t_2t_3\, b_0\, c_1\, \overline{\beta_3}\, \overline{\psi_3}\beta_0]_0]_3$
14. $[[\overline{\alpha}]_M[[\alpha\, e_{13}]_J\overline{\alpha}]_K[\beta_3]_1^8[\psi_3]_2^8[b^4\, y_1y_2y_3\, b_2\, c_0\, \overline{\beta_3}\, \overline{\psi_3}\beta_0]_0[b^2y_1y_2t_3\, b_2\, c_2\, \overline{\beta_3}\, \overline{\psi_3}\beta_0]_0$
$[b\, c\, y_1t_2y_3\, b_2\, c_2\, \overline{\beta_3}\, \overline{\psi_3}\beta_0]_0[c^3y_1t_2t_3\, b_1\, c_2\, \overline{\beta_3}\, \overline{\psi_3}\beta_0]_0[b^3t_1y_2y_3\, b_2\, c_1\, \overline{\beta_3}\, \overline{\psi_3}\beta_0]_0$
$[b\, c\, t_1y_2t_3\, b_2\, c_2\, \overline{\beta_3}\, \overline{\psi_3}\beta_0]_0[c^2\, t_1t_2y_3\, b_2\, c_2\, \overline{\beta_3}\, \overline{\psi_3}\beta_0]_0[c^4\, t_1t_2t_3\, b_0\, c_2\, \overline{\beta_3}\, \overline{\psi_3}\beta_0]_0]_3$
15. $[[\overline{\alpha}]_M[\alpha\, e_{14}]_J[\overline{\alpha}]_K[\beta_3]_1^3[\psi_3]_2^3[[\beta_3]_1b^3\, y_1y_2y_3\, b_2\, c_0\, \overline{\beta_3}\, \overline{\psi_3}\beta_0]_0$
$[[\beta_3]_1b\, y_1y_2t_3\, b_2\, c_2\, \overline{\beta_3}\, \overline{\psi_3}\beta_0]_0[[\beta_3]_1[\psi_3]_2y_1t_2y_3\, b_2\, c_2\, \overline{\beta_3}\, \overline{\psi_3}\beta_0]_0$
$[[\psi_3]_2c^2y_1t_2t_3\, b_1\, c_2\, \overline{\beta_3}\, \overline{\psi_3}\beta_0]_0[[\beta_3]_1b^2t_1y_2y_3\, b_2\, c_1\, \overline{\beta_3}\, \overline{\psi_3}\beta_0]_0$
$[[\beta_3]_1[\psi_3]_2t_1y_2t_3\, b_2\, c_2\, \overline{\beta_3}\, \overline{\psi_3}\beta_0]_0[[\psi_3]_2c\, t_1t_2y_3\, b_2\, c_2\, \overline{\beta_3}\, \overline{\psi_3}\beta_0]_0$
$[[\psi_3]_2c^3\, t_1t_2t_3\, b_0\, c_2\, \overline{\beta_3}\, \overline{\psi_3}\beta_0]_0]_3$
16. $[[\overline{\alpha}]_M[[\alpha\, e_{15}]_J\overline{\alpha}]_K[\beta_3]_1^8[\psi_3]_2^8[b^3\, y_1y_2y_3\, b_3\, c_0\, \overline{\beta_3}\, \overline{\psi_3}\beta_0]_0[b\, y_1y_2t_3\, b_3\, c_2\, \overline{\beta_3}\, \overline{\psi_3}\beta_0]_0$
$[y_1t_2y_3\, b_3\, c_3\, \overline{\beta_3}\, \overline{\psi_3}\beta_0]_0[c^2y_1t_2t_3\, b_1\, c_3\, \overline{\beta_3}\, \overline{\psi_3}\beta_0]_0[b^2t_1y_2y_3\, b_3\, c_1\, \overline{\beta_3}\, \overline{\psi_3}\beta_0]_0$
$[t_1y_2t_3\, b_3\, c_3\, \overline{\beta_3}\, \overline{\psi_3}\beta_0]_0[c\, t_1t_2y_3\, b_2\, c_3\, \overline{\beta_3}\, \overline{\psi_3}\beta_0]_0[c^3\, t_1t_2t_3\, b_0\, c_3\, \overline{\beta_3}\, \overline{\psi_3}\beta_0]_0]_3$
Graphically the working space is described by the following picture:

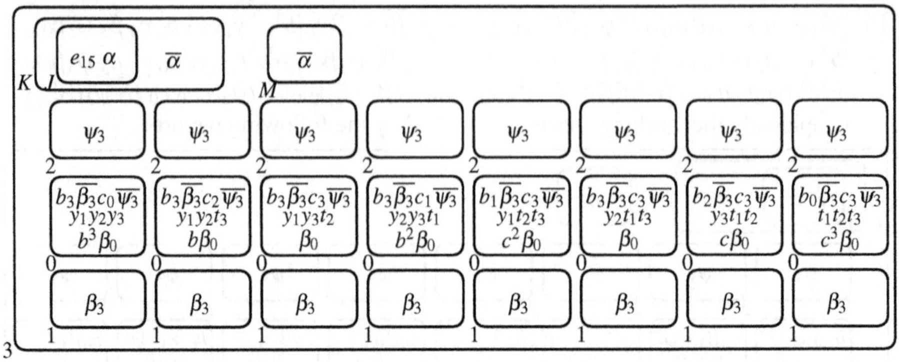

Now we divide membrane J, after performing an extra step that moves membrane J out of membrane K, in order to check whether the indexes of b_i and c_j from the 0 membranes are equal or not. In parallel, if there still exist membranes 0 containing objects b or c, then the subscript of b_i or c_i, respectively, is increased.

17. $[[\overline{\alpha}]_M[\alpha\, e_{15}]_J[\alpha\, e_{15}]_K[\beta_3]_1^5[\psi_3]_2^5[[\beta_3]_1 b^2\, y_1y_2y_3\, b_3\, c_0\, \overline{\beta_3}\,\overline{\psi_3}\beta_0]_0$
$[[\beta_3]_1 y_1y_2t_3\, b_3\, c_2\, \overline{\beta_3}\,\overline{\psi_3}\beta_0]_0[y_1t_2y_3\, b_3\, c_3\, \overline{\beta_3}\,\overline{\psi_3}\beta_0]_0[[\psi_3]_2 c\, y_1t_2t_3\, b_1\, c_3\, \overline{\beta_3}\,\overline{\psi_3}\beta_0]_0$
$[[\beta_3]_1 b\, t_1y_2y_3\, b_3\, c_1\, \overline{\beta_3}\,\overline{\psi_3}\beta_0]_0[t_1y_2t_3\, b_3\, c_3\, \overline{\beta_3}\,\overline{\psi_3}\beta_0]_0[[\psi_3]_2 t_1t_2y_3\, b_2\, c_3\, \overline{\beta_3}\,\overline{\psi_3}\beta_0]_0$
$[[\psi_3]_2 c^2\, t_1t_2t_3\, b_0\, c_3\, \overline{\beta_3}\,\overline{\psi_3}\beta_0]_0]_3$
18. $[[\overline{\alpha}[\alpha\, e_{16}]_K]_M[\gamma_1]_J^2[\beta_3]_1^8[\psi_3]_2^8[b^2\, y_1y_2y_3\, b_4\, c_0\, \overline{\beta_3}\,\overline{\psi_3}\beta_0]_0[y_1y_2t_3\, b_4\, c_2\, \overline{\beta_3}\,\overline{\psi_3}\beta_0]_0$
$[y_1t_2y_3\, b_3\, c_3\, \overline{\beta_3}\,\overline{\psi_3}\beta_0]_0[c\, y_1t_2t_3\, b_1\, c_4\, \overline{\beta_3}\,\overline{\psi_3}\beta_0]_0[b\, t_1y_2y_3\, b_4\, c_1\, \overline{\beta_3}\,\overline{\psi_3}\beta_0]_0$
$[t_1y_2t_3\, b_3\, c_3\, \overline{\beta_3}\,\overline{\psi_3}\beta_0]_0[t_1t_2y_3\, b_2\, c_4\, \overline{\beta_3}\,\overline{\psi_3}\beta_0]_0[c^2\, t_1t_2t_3\, b_0\, c_4\, \overline{\beta_3}\,\overline{\psi_3}\beta_0]_0]_3$
19. $[[\overline{\alpha}]_M[\alpha\, e_{17}]_K[\gamma_2]_J^4[\overline{\alpha}]_K[\beta_3]_1^6[\psi_3]_2^6[[\beta_3]_1 b\, y_1y_2y_3\, b_4\, c_0\, \overline{\beta_3}\,\overline{\psi_3}\beta_0]_0$
$[y_1y_2t_3\, b_4\, c_2\, \overline{\beta_3}\,\overline{\psi_3}\beta_0]_0[y_1t_2y_3\, b_3\, c_3\, \overline{\beta_3}\,\overline{\psi_3}\beta_0]_0[[\psi_3]_2 y_1t_2t_3\, b_1\, c_4\, \overline{\beta_3}\,\overline{\psi_3}\beta_0]_0$
$[[\beta_3]_1 t_1y_2y_3\, b_4\, c_1\, \overline{\beta_3}\,\overline{\psi_3}\beta_0]_0[t_1y_2t_3\, b_3\, c_3\, \overline{\beta_3}\,\overline{\psi_3}\beta_0]_0[t_1t_2y_3\, b_2\, c_4\, \overline{\beta_3}\,\overline{\psi_3}\beta_0]_0$
$[[\psi_3]_2 c\, t_1t_2t_3\, b_0\, c_4\, \overline{\beta_3}\,\overline{\psi_3}\beta_0]_0]_3$
20. $[[\overline{\alpha}[\alpha\, e_{18}]_K]_M[\beta_3]_J^8[\overline{\alpha}]_K[\beta_3]_1^8[\psi_3]_2^8[b\, y_1y_2y_3\, b_5\, c_0\, \overline{\beta_3}\,\overline{\psi_3}\beta_0]_0[y_1y_2t_3\, b_4\, c_2\, \overline{\beta_3}\,\overline{\psi_3}\beta_0]_0$
$[y_1t_2y_3\, b_3\, c_3\, \overline{\beta_3}\,\overline{\psi_3}\beta_0]_0[y_1t_2t_3\, b_1\, c_5\, \overline{\beta_3}\,\overline{\psi_3}\beta_0]_0[t_1y_2y_3\, b_5\, c_1\, \overline{\beta_3}\,\overline{\psi_3}\beta_0]_0$
$[t_1y_2t_3\, b_3\, c_3\, \overline{\beta_3}\,\overline{\psi_3}\beta_0]_0[t_1t_2y_3\, b_2\, c_4\, \overline{\beta_3}\,\overline{\psi_3}\beta_0]_0[c\, t_1t_2t_3\, b_0\, c_5\, \overline{\beta_3}\,\overline{\psi_3}\beta_0]_0]_3$
21. $[[\alpha_1]_M[\beta_0\overline{\beta_0}]_K[\beta_3]_J^8[\overline{\alpha}]_K[\beta_3]_1^8[\psi_3]_2^8[b\, y_1y_2y_3\beta_0[\ ]_J]_0[y_1y_2t_3\beta_0[\ ]_J]_0$
$[y_1t_2y_3\, b_3\, c_3\, \overline{\beta_3}\,\overline{\psi_3}\beta_0]_0[y_1t_2t_3\, \beta_0[\ ]_J]_0[t_1y_2y_3\, \beta_0[\ ]_J]_0[t_1y_2t_3\, b_3\, c_3\, \overline{\beta_3}\,\overline{\psi_3}\beta_0]_0$
$[t_1t_2y_3\, \beta_0[\ ]_J]_0[c\, t_1t_2t_3\, \beta_0[\ ]_J]_0]_3$
The next step is used to generate a *yes* object in membrane M. The computation stops after $4n+g(A)+2=4*3+6+2=20$ steps.
22. $[[\overline{\beta_0}]_M^2[\beta_3]_J^8[\overline{\alpha}]_K[\beta_3]_1^8[\psi_3]_2^8[b\, y_1y_2y_3\beta_0[\ ]_J]_0[y_1y_2t_3\beta_0[\ ]_J]_0[y_1t_2y_3\, b_3\, c_3\, \overline{\beta_3}\,\overline{\psi_3}\beta_0]_0$
$[y_1t_2t_3\, \beta_0[\ ]_J]_0[t_1y_2y_3\, \beta_0[\ ]_J]_0[yes\, \beta_0[t_1y_2t_3\, b_3\, c_3\, \overline{\beta_3}\,\overline{\psi_3}]_0]_K$
$[t_1t_2y_3\, \beta_0[\ ]_J]_0[c\, t_1t_2t_3\, \beta_0[\ ]_J]_0]_3$

Exercise 2.10. Solve the 2-Partition problem using other classes of mobile membranes.

2.6 Decidability Results

In [11] we investigate the problem of reaching a configuration from another configuration in a special class of systems of mobile membranes. We prove that reachability can be decided by reducing it to the reachability problem of a version of pure and public ambient calculus without the capability open. The relationship between mobile ambients and mobile membranes is presented in Section 3.3.

Reachability is the problem of deciding whether a system may reach a given configuration during its execution. This is one of the most critical properties in the verification of systems; most of the safety properties of computing systems can be reduced to the problem of checking whether a system may reach an "unintended state".

In what follows we investigate the problem of reaching a certain configuration in systems of mobile membranes starting from a given configuration. We prove that reachability in systems of mobile membranes can be decided by reducing it to the reachability problem of a version of pure and public ambient calculus from which the open capability has been removed. It is proven in [28] that reachability for this fragment of ambient calculus is decidable by reducing it to marking reachability for Petri nets, which is proven to be decidable in [115]. The reachability problem is investigated in [82] for other classes of P systems, namely for extensions of PB systems with volatile membranes.

When working with Petri nets, reachability is a property of general interest. Given a net with initial marking ω_0, we say that the marking ω is reachable if there exists a sequence of firings $\omega_0 \to \omega_1 \to \dots \omega_n = \omega$ of the net. The reachability problems is decidable in Petri nets, even if they tend to have a very large complexity in practice. A good survey of the known decidability issues for Petri nets is given in [83].

2.6.1 Mobile Membranes with Replication

Since we use reduction to mobile ambients, we construct a class of systems of mobile membranes in which replication from mobile ambients is expressed explicitly by duplicating objects or membranes in systems of mobile membranes.

Definition 2.13. A system of n mobile membranes with replication rules is a structure

$$\Pi = (V \cup \overline{V}, H \cup \overline{H}, \mu, w_1, \dots, w_n, R), \text{ where:}$$

1. $n \geq 1$ represents the initial degree of the system;
2. $V \cup \overline{V}$ is an alphabet (its elements are called objects), where $V \cap \overline{V} = \emptyset$;
3. $H \cup \overline{H}$ is a finite set of labels for membranes, where $H \cap \overline{H} = \emptyset$;
4. $\mu \subseteq H \times H$ describes the membrane structure, such that $(i,j) \in \mu$ denotes that the membrane labelled by j is contained in the membrane labelled by i; we distinguish the external membrane (usually called the "skin" membrane) and several

internal membranes; a membrane without any other membrane inside it is said to be elementary;
5. $w_1, w_2, \ldots, w_n$ are multisets of objects from $V \cup \overline{V}$ placed in the n membranes of the system;
6. R is a finite set of developmental rules, of the following forms:

   a. $[\overline{a}\!\downarrow \rightarrow \overline{a}\!\downarrow a\!\downarrow]_h$, for $h \in H$, $a\!\downarrow \in V$, $\overline{a}\!\downarrow \in \overline{V}$; replication rule
   The objects $\overline{a}\!\downarrow$ are used to create new objects $a\!\downarrow$ without being consumed.
   b. $[\overline{a}\!\downarrow a\!\downarrow \rightarrow \overline{a}\!\downarrow]_h$, for $h \in H$, $a\!\downarrow \in V$, $\overline{a}\!\downarrow \in \overline{V}$; consumption rule
   The objects $a\!\downarrow$ are consumed.
   c. $[\overline{a}\!\uparrow \rightarrow \overline{a}\!\uparrow a\!\uparrow]_h$, for $h \in H$, $a\!\uparrow \in V$, $\overline{a}\!\uparrow \in \overline{V}$; replication rule
   The objects $\overline{a}\!\uparrow$ are used to create new objects $a\!\uparrow$ without being consumed.
   d. $[\overline{a}\!\uparrow a\!\uparrow \rightarrow \overline{a}\!\uparrow]_h$, for $h \in H$, $a\!\uparrow \in V$, $\overline{a}\!\uparrow \in \overline{V}$; consumption rule
   The objects $a\!\uparrow$ are consumed.
   e. $[\,a\!\downarrow\,]_h\,[\,]_a \rightarrow [\,[\,]_h\,]_a$, for $a, h \in H, a\!\downarrow \in V$; endocytosis
   An elementary membrane labelled h (containing an object $a\!\downarrow$) enters the adjacent membrane labelled a. The labels h and a remain unchanged during this process; however the object $a\!\downarrow$ is consumed during the operation. Membrane a is not necessarily elementary.
   f. $[\,[\,a\!\uparrow\,]_h\,]_a \rightarrow [\,]_h\,[\,]_a$, for $a, h \in H, a\!\uparrow \in V$; exocytosis
   An elementary membrane labelled h (containing an object $a\!\uparrow$) is sent out of a membrane labelled a. The labels of the two membranes remain unchanged; the object $a\!\uparrow$ of membrane h is consumed during the operation. Membrane a is not necessarily elementary.
   g. $[\,]_{\overline{h}} \rightarrow [\,]_{\overline{h}}[\,]_h$ for $h \in H, \overline{h} \in \overline{H}$ division rules
   An elementary membrane labelled $\overline{h}$ is divided into two membranes labelled by $\overline{h}$ and h and having the same objects.

$V \cap \overline{V} = \emptyset$ states that objects from $\overline{V}$ can participate only in the rules of type (a) – (d). Similarly, $H \cap \overline{H} = \emptyset$ states that membranes having labels from the set $\overline{H}$ can participate only in rules of type (g).

The rules are applied using the following principles:

1. In biological systems molecules are divided into classes of different types. We make the same decision here and split the objects into four classes: $a\downarrow$ - objects which control the *endocytosis*, $a\uparrow$ - objects which control the *exocytosis*, and $\overline{a}\downarrow$, $\overline{a}\uparrow$ - objects which produce new objects of the first two classes without being consumed.
2. All the rules are applied in parallel, non-deterministically choosing the rules, the membranes, and the objects in such a way that the parallelism is maximal; this means that in each step we apply a set of rules such that no further rule, no further membranes and objects can evolve at the same time.
3. A membrane a from each rule of type (e) and (f) is said to be *passive*, while membrane h is said to be *active*. In any step of a computation, any object and any active membrane can be involved in at most one rule, but the passive membranes

are not considered involved in the use of rules (hence they can be used by several rules at the same time).
4. When a membrane is moved across another membrane, by endocytosis or exocytosis, its whole content (its objects) is moved.
5. If a membrane is divided, then its content is replicated in the two new copies.
6. The skin membrane can never be divided.

According to these rules, we get transitions among the configurations of the system. For two systems of mobile membranes M and N, we say that M reduces to N if there is a sequence of rules applicable in the system of mobile membranes M in order to obtain the system of mobile membranes N.

In what follows we prove that the problem of reaching a configuration starting from a certain configuration is decidable for the systems of mobile membranes from Definition 2.13.

Theorem 2.21. *For two arbitrary systems of mobile membranes with replication rules M_1 and M_2, it is decidable whether M_1 reduces to M_2.*

The main steps of the proof are as follows:

1. systems of mobile membranes are reduced to pure and public mobile ambients without the capability *open*;
2. the reachability problem for two arbitrary systems of mobile membranes is expressed as the reachability problem for the corresponding mobile ambients.
3. the reachability problem is decidable for a fragment of pure and public mobile ambients without the capability *open*.

The rest of this section is devoted to the proof of Theorem 2.21.

2.6.2 From Mobile Membranes to Mobile Ambients

We use the following translation steps:

1. any object $a\downarrow$ is translated into a capability *in a*;
2. any object $a\uparrow$ is translated into a capability *out a*;
3. any object $\overline{a}\downarrow$ is translated into a replication !*in a*
4. any object $\overline{a}\uparrow$ is translated into a replication !*out a*
5. a membrane h is translated into an ambient h
6. an elementary membrane $\overline{h}$ is translated into a replication $!h[\,]$ while all the objects inside membrane h are translated into capabilities in ambient h using the above steps.

A correspondence exists between the rules of the systems of mobile membranes and the reduction rules of the mobile ambients as follows:

- rule (c) corresponds to rule **(In)**;

- rule (d) corresponds to rule **(Out)**;
- rules $(a),(b),(e)$ correspond to instances of rule **(Repl)**.

The rule **(Repl)** from mobile ambients has the form $A \Rightarrow_{amb} !A \mid A$. If we start with a system of mobile membranes M, we denote by $\mathscr{T}(M)$ the mobile ambient obtained using the above translation steps. For example, starting from the system of mobile membranes $M = [m{\downarrow}\ m{\uparrow}]_n[\ ]_m$ we obtain $\mathscr{T}(M) = n[in\ m \mid out\ m] \mid m[\ ]$.

Proposition 2.2. *For two systems of mobile membranes M and N, M reduces to N by applying one rule if and only if $\mathscr{T}(M)$ reduces to $\mathscr{T}(N)$ by applying only one reduction rule.*

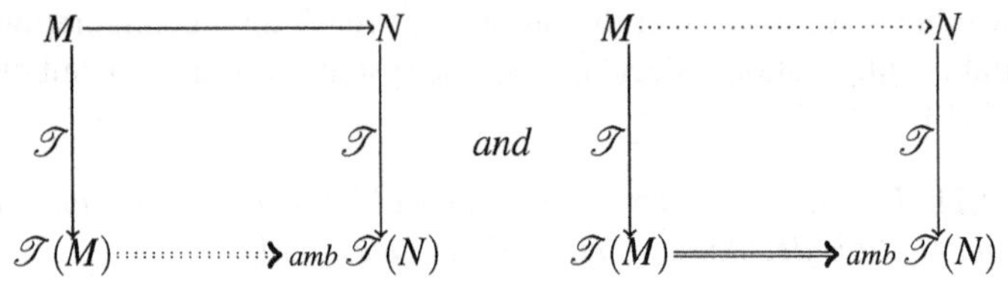

Proof (Sketch). Since M reduces to N by applying one rule, then one of the rules of type $(a),\ldots,(e)$ is applied. We treat only the case when a rule of type (a) is applied, the others being treated in a similar manner.

If a rule $\overline{a}{\downarrow} \to \overline{a}{\downarrow}\ a{\downarrow}$ is applied, only one object from the system of mobile membranes M is used (namely $\overline{a}{\downarrow}$) to create a new object $a{\downarrow}$, thus obtaining the system of mobile membranes N. By translating the system of mobile membranes M into $\mathscr{T}(M)$, we have that $\overline{a}{\downarrow}$ is translated into $!in\ a$. By applying the reduction rule corresponding to (a) (namely the rule **(Repl)**) to $!in\ a$, we have that $!in\ a \Rightarrow_{amb} in\ a \mid !in\ a$, and so a new capability $in\ a$ is created. We observe that $\mathscr{T}(\overline{a}{\downarrow}\ a{\downarrow}) = !in\ a \mid in\ a$, which means that the obtained mobile ambient is $\mathscr{T}(N)$ (in fact it is structurally congruent to $\mathscr{T}(N)$). □

According to Proposition 2.2 the reachability problem for systems of mobile membranes can be reduced to a similar problem for mobile ambients.

2.6.3 From Mobile Ambients to Petri Nets

After translating the systems of mobile membranes into a fragment of mobile ambients, we present the algorithm used in [28] to translate this fragment of mobile ambients into a fragment of Petri nets which is known to be decidable from [115]. The fragment of mobile ambients used here is a subset of the fragment of mobile ambients used in [28] and the difference is provided by the extra rule $!A \Rightarrow_{amb} !A \mid !A$ used in [28].

We observe that applying a reduction rule over a process either increases the number of ambients or leaves it unchanged. The only reduction rule which increases the number of ambients when applied is the rule **(Repl)**, while the other reduction rules leave the number of ambients unchanged. If we reach process B starting from process A, then the number of ambients of process B is known. Therefore, we can use this information to know how many times the reduction rule **(Repl)** is applied to replicate ambients. A similar argument does not hold for capabilities as they can be consumed by the reduction rules **(In)** and **(Out)**.

An ambient context $\mathscr{C}$ is a process in which some holes may occur (denoted by $\Box$). Using the ambient contexts, we split a process into two parts: one is a context containing ambients, whereas the other is a process without ambients. In order to uniquely identify all the occurrences of replication, ambient, capability or hole $\Box$ within an ambient context or a process, we introduce a labelling system. Using a countable set of labels, we say that a process A or an ambient context $\mathscr{C}$ is well-labelled if any label occurs at most once in A or $\mathscr{C}$. We denote by $Amb(\mathscr{C})$ the multiset of ambients occurring in an ambient context $\mathscr{C}$. We say that two processes are label-free-equivalent if after removing all the labels from the two processes, they are structurally congruent.

2.6.3.1 I) Labelled Transition System.

For the reachability problem $A \Rightarrow^* B$, we denote by $\mathscr{C}_A$ a well-labelled ambient context, and by θ_A a mapping from the set of holes in $\mathscr{C}_A$ to some labelled processes without replicable ambients such that $\theta_A(\mathscr{C}_A)$ is well-labelled, and $\theta_A(\mathscr{C}_A) = A$ where labels are ignored.

A labelled transition system $L_{A,B}$ describes all possible reductions for a context $\mathscr{C}_A$: this includes reductions of replications and capabilities contained in $\mathscr{C}_A$ and in the processes associated with the holes of the context. The states of the labelled transition system $L_{A,B}$ are associative-commutative equivalent classes of ambient contexts and, for simplicity, we often identify a state as one of the representatives of its class.

We define a mapping $\theta_{L_{A,B}}$ which extends the mapping θ_A. Initially, $L_{A,B}$ contains (the equivalence class of) $\mathscr{C}_A$ as a unique state, and we have $\theta_{L_{A,B}} = \theta_A$. We present in what follows the construction steps of $\theta_{L_{A,B}}$, where *cap* stands for *in* or *out*:

1. For any ambient context $\mathscr{C}$ from $L_{A,B}$ and for any labelled capability $cap^w n$ in $\mathscr{C}$, if this capability can be executed using one of the rules **(In)** or **(Out)** leading to some ambient context $\mathscr{C}'$, then a state $\mathscr{C}'$ and a transition from $\mathscr{C}$ to $\mathscr{C}'$ labelled by $cap^w n$ are added to $L_{A,B}$.
2. For any ambient context $\mathscr{C}$ from $L_{A,B}$ and for any labelled replication $!^w$ in $\mathscr{C}$ such that the reduction rule **(Repl)** is applied, we define the ambient context $\mathscr{C}'$ as follows: $\mathscr{C}'$ is identical to $\mathscr{C}$ except that the subcontext $!^w\mathscr{C}_a$ in $\mathscr{C}$ is replaced by $!^w\mathscr{C}_a \mid \gamma(\mathscr{C}_a)$ in $\mathscr{C}'$; the mapping γ relabels $\mathscr{C}_a$ with fresh labels, such that $\mathscr{C}'$ is well-labelled. If $Amb(\mathscr{C}') \subseteq Amb(B)$, then state $\mathscr{C}'$ and a transition from $\mathscr{C}$ to $\mathscr{C}'$ labelled by $!^w$ is added to $L_{A,B}$. Additionally, we define $\theta'_{L_{A,B}}$ as an extension of $\theta_{L_{A,B}}$ such that for all $\Box^{w'}$ in $\mathscr{C}_a$ we have:

   (i) $\theta'_{L_{A,B}}(\gamma(\Box^{w'}))$ and $\theta_{L_{A,B}}(\Box^{w'})$ are label-free-equivalent,

(ii) labels in $\theta'_{L_{A,B}}(\gamma(\Box^{w'}))$ are fresh in the currently built transition system $L_{A,B}$,
(iii) $\theta'_{L_{A,B}}(\gamma(\Box^{w'}))$ is well-labelled.

Finally, we set $\theta_{L_{A,B}}$ to be $\theta'_{L_{A,B}}$.

3. For any ambient context $\mathscr{C}$ from $L_{A,B}$, for any labelled hole $\Box^w$ in $\mathscr{C}$ and for any capability $\mathsf{cap}^w\mathsf{n}$ in the process $\theta_{L_{A,B}}(\Box^w)$, we consider the ambient context $\mathscr{C}_m$ identical to $\mathscr{C}$ except that $\Box^w$ in $\mathscr{C}$ has been replaced by $\Box^w \mid cap^w n$ in $\mathscr{C}_m$. If the capability $cap^w n$ can be consumed in $\mathscr{C}_m$ using one of the rules **(In)** or **(Out)** leading to an ambient context $\mathscr{C}'$, then state $\mathscr{C}'$ and a transition from $\mathscr{C}$ to $\mathscr{C}'$ labelled by $cap^w n$ are added to transition system $L_{A,B}$.
4. For any ambient context $\mathscr{C}$ from $L_{A,B}$ and for any labelled hole $\Box^w$ in $\mathscr{C}$ associated by $\theta_{L_{A,B}}$ with a process of the form $!^{w'}A'$, if a replication $!^{w'}$ can be reduced in process $\theta_{L_{A,B}}(\mathscr{C})$ using rule **(Repl)**, then a transition from $\mathscr{C}$ to itself labelled by $!^{w''}$ is added to $L_{A,B}$ for any replication $!^{w''}$ in $\theta_{L_{A,B}}(\Box^w)$.

In the second step, the reduction of a replication contained in the ambient context by means of the rule **(Repl)** is done only when the number of ambients in the resulting process is smaller than the number of ambients in the target process B, namely $Amb(\mathscr{C}') \subseteq Amb(B)$. This requirement is crucial as it implies that the transition system $L_{A,B}$ has only finitely many states.

As an example, we give in Figure 2.6 the labelled transition system associated with the process $n[!^1 in\ m.!^2 out\ m] \mid m[\,]$ (we omit in this process unnecessary labels). We use the labelled replications $!^1$ and $!^2$ to distinguish between different replication operators which appear in this process.

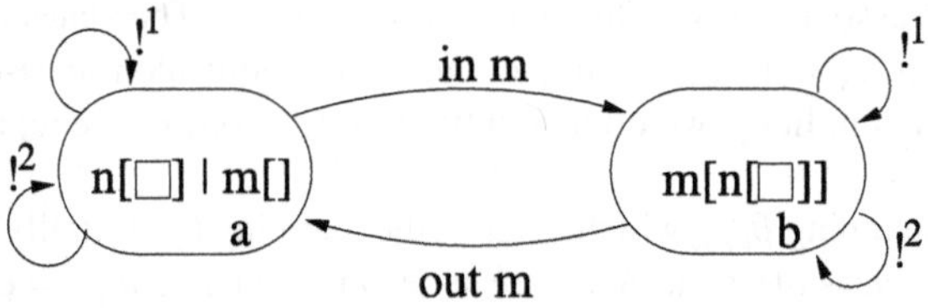

Fig. 2.6 A Labelled Transition System for the Process $n[!^1 in\ m.!^2 out\ m] \mid m[\,]$

We observe that the labelled transitions in $L_{A,B}$ for replications and capabilities of an ambient context correspond to the reductions performed over processes. As shown in steps 3 and 4, the transitions applied for any capabilities or replications associated with the holes are independent of the fact that they are effectively available to perform a transition (at this point).

2.6.3.2 II) From Processes Without Ambients to Petri Nets.

In what follows we show how to build a Petri net from a labelled process without ambients. We denote by $\mathscr{E}(E)$ the set of all multisets which can be built with elements from the set E.

We recall that a Petri net is given by a 5-tuple $(\mathscr{P}, \mathscr{P}_i, \mathscr{T}, Pre, Post)$, where

- $\mathscr{P}$ is a finite set of *places*;
- $\mathscr{P} \subseteq \mathscr{P}_i$ is a set of initial places;
- $\mathscr{T}$ is a finite set of transitions;
- $Pre, Post : \mathscr{T} \to \mathscr{E}(\mathscr{P})$ are mappings from transitions to multisets of places.

We say that an ambient-free process is *rooted* if it is of the form $cap^w n.A'$ or of the form $!^w A'$. We define the Petri net $PN_{A'}$ associated with a rooted process A' as follows: the places of $PN_{A'}$ are precisely the rooted subprocesses of A', and A' itself is the unique initial place; the transitions are defined as the set of all capabilities $in^w n$, $out^{w'} n$ and replications $!^w$ occurring in A'. Finally, *Pre* and *Post* are defined for all transitions as follows:

- $Pre(cap^w n) = \{cap^w n\}$ and $Post(cap^w n) = \emptyset$ if $cap^w n$ is a place in $PN_{A'}$.
- $Pre(cap^w n) = \{cap^w n.(A_1 \mid \ldots \mid A_k)\}$ and $Post(cap^w n) = \{A_1 \mid \ldots \mid A_k\}$ if $cap^w n.(A_1 \mid \ldots \mid A_k)$ is a place in $PN_{A'}$ ($A_1 \mid \ldots \mid A_k$ being rooted processes).
- $Pre(!^w) = \{!^w A'\}$ and $Post(!^w) = \{!^w A', A'\}$ if $!^w A'$ is a place in $PN_{A'}$.

For $!^1 in\ m.!^2 out\ m$, we obtain the Petri net given in Figure 2.7.

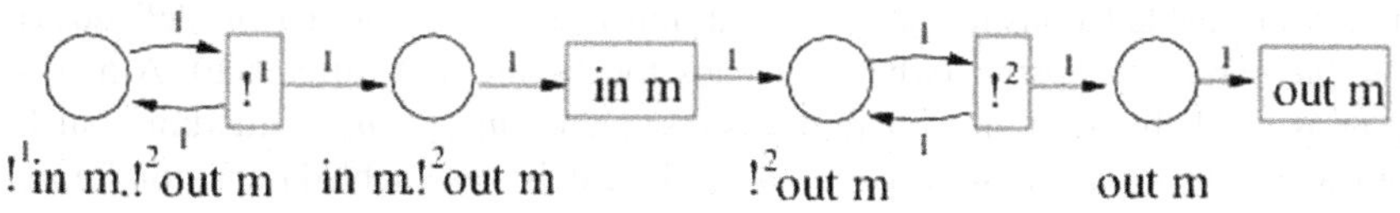

Fig. 2.7 A Petri Net for the Process $!^1 in\ m.!^2 out\ m$

We denote by $PN_{\Box^w}$ the Petri net $PN(\theta_{L_{A,B}}(\Box^w))$, that is, the Petri net corresponding to the rooted ambient-free process associated with $\Box^w$ by $\theta_{L_{A,B}}$. In what follows we show how to combine the transition system $L_{A,B}$ and the Petri nets $PN_{\Box^w}$ into one single Petri net.

2.6.3.3 III) Combining the Transition Systems and Petri Nets.

We first turn the labelled transition system $L_{A,B}$ into a Petri net

$$PN_L = (\mathscr{P}_L, \mathscr{P}_L^i, \mathscr{T}_L, Pre_L, Post_L), \text{ where}$$

- $\mathscr{P}_L$ is a set of states of $L_{A,B}$;
- $\mathscr{P}_L^i$ is a singleton set containing the state corresponding to the ambient context $\mathscr{C}_A$ of A;
- $\mathscr{T}_L$ is the set of transitions of the form (s, l, s'), with
  - s and s' states from $L_{A,B}$,
  - a transition l from s to s' in $L_{A,B}$;

- $Pre(t) = s$ and $Post(t) = \{s'\}$ for all transitions $t = (s, l, s')$.

We define a Petri net $PN_{A,B} = (\mathscr{P}_{A,B}, \mathscr{P}^i_{A,B}, \mathscr{T}_{A,B}, Pre_{A,B}, Post_{A,B})$ by

- places (initial places) of $PN_{A,B}$ are the union of places (initial places) of PN_L and of each of the Petri nets $PN_{\Box^w}$ (for $\Box^w$ occurring in one of the states of $L_{A,B}$);
- transitions of $PN_{A,B}$ are precisely the transitions of PN_L;
- the mappings $Pre_{A,B}$ and $Post_{A,B}$ are defined for all transitions $t = (a, f, b)$ as:

  (i) $Pre_{A,B}(t) = \{a\}$ and $Post_{A,B}(t) = \{b\}$ if f does not occur as a transition in any $PN_{\Box^w}$ (for $\Box^w$ occurring in one of the states of $L_{A,B}$),
  (ii) if f is a transition of $PN_{\Box^w}$, then $Pre_{A,B}(t) = \{a\} \cup Pre_{\Box^w}(f)$ and $Post_{A,B}(t) = \{b\} \cup Post_{\Box^w}(f)$, where $Pre_{\Box^w}$ and $Post_{\Box^w}$ are the mappings Pre and $Post$ of $PN_{\Box^w}$, respectively.

2.6.4 Deciding Reachability

We recall that for a Petri net $PN = (\mathscr{P}, \mathscr{P}^i, \mathscr{T}, Pre, Post)$, a marking m is a multiset from $\mathscr{E}(P)$. A transition t is enabled by a marking m if $Pre(t) \subseteq m$. Executing an enabled transition t for a marking m gives a marking m' defined as $m' = (m \setminus Pre(t)) \cup Post(t)$ (where $\setminus$ stands for the multiset difference). A marking m' is reachable from m if there exists a sequence $m_0, \ldots, m_k$ of markings such that $m_0 = m$, $m_k = m'$ and for each m_i, m_{i+1}, there exists an enabled transition for m_i whose execution gives m_{i+1}.

Theorem 2.22 ([115]). *For all Petri nets P, for all markings m, m' of P, one can decide whether m' is reachable from m.*

For the reachability problem $A \Rightarrow^* B$ over ambients, we consider the Petri net $PN_{A,B}$ and the initial marking m_A defined as $m_A = \mathscr{P}^i_{A,B}$. Figure 2.8 depicts the initial marking for process $n[!^1 in\ m.!^2 out\ m] \mid m[\,]$ as a combination of the labelled transition system of Figure 2.6 and the Petri net of Figure 2.7.

It should be noticed that for any marking m reachable from m_A, m contains exactly one occurrence of a place from $\mathscr{P}_L$. Roughly speaking, to any reachable marking corresponds exactly one ambient context. Moreover, the execution of one transition in the Petri net $PN_{A,B}$ simulates a reduction from $\Rightarrow_{amb}$.

We define now the set $\mathscr{M}_B$ of markings of $PN_{A,B}$ corresponding to B. Intuitively, a marking m belongs to $\mathscr{M}_B$ if m contains exactly one occurrence $\mathscr{C}$ of a place from $\mathscr{P}_L$ (that is, representing some ambient context) and in the context $\mathscr{C}$, the holes can be replaced with processes without ambients to obtain B. Each of the processes without replication must correspond to a marking of the sub-Petri net associated with the hole it fills up. $\mathscr{M}_B$ is defined as the set of markings m for $PN_{A,B}$ satisfying:

(i) there exists exactly one ambient context $\mathscr{C}_m$ in m;

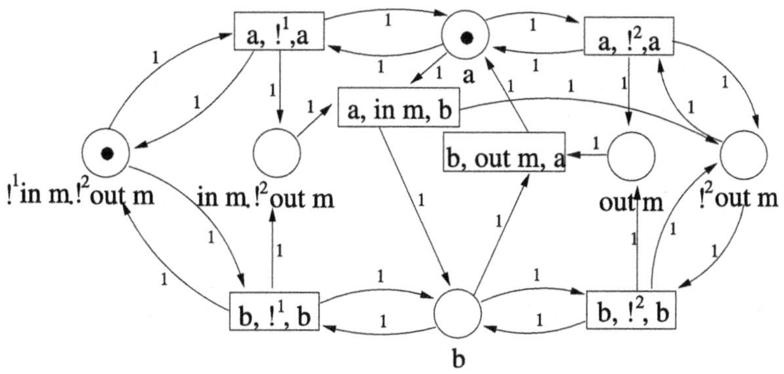

Fig. 2.8 The Petri Net for the Labelled Process $n[!^1 in\ m.!^2 out\ m] \mid m[\,]$

(ii) $\sigma_m(\mathscr{C}_m)$ and B are label-free-equivalent, for any substitution σ_m from holes $\square^w$ occurring in $\mathscr{C}_m$ to processes without ambients defined as $\sigma_m(\square_m) = P_1 \mid \ldots \mid P_k$ for $\{P_1, \ldots, P_k\}$ the multiset corresponding to the restriction of m to the places of $PN_{\square^w}$;

(iii) for all holes $\square^w$ occurring in a state of the transition system $L_{A,B}$ but not in $\mathscr{C}_m$, the restriction of m to places of $PN_{\square^w}$ is precisely the set of initial places of $PN_{\square^w}$.

We adapt the results presented in [28] to our restricted fragment of mobile ambients.

Proposition 2.3. *For a Petri net $PN_{A,B}$, there are only finitely many markings corresponding to a process B, and the set $\mathscr{M}_B$ can be computed.*

The translation correctness is ensured by the following result.

Proposition 2.4. *For all processes A, B we have that $A \Rightarrow_{amb} B$ if and only if there exists a marking from $\mathscr{M}_B$ such that m_B is reachable from m_A in $PN_{A,B}$.*

Using Proposition 2.4 and Theorem 2.22, we can decide whether an ambient A can be reduced to an ambient B.

Theorem 2.23. *For two arbitrary ambients A and B from our restricted fragment, it is decidable whether A reduces to B.*

Chapter 3
Encodings

Abstract The difference between the two research areas (process algebra and membrane computing) is the fact that process algebra represents a tool for the high-level description of interactions, communications, and synchronizations between a collection of independent agents or processes, providing also algebraic laws that allow process descriptions to be manipulated and analyzed, and permit formal reasoning about equivalences between processes (e.g., using bisimulation), while membrane computing uses techniques from languages, automata, complexity, and dynamical systems. In this chapter we establish several links between these two fields in order to be able to use techniques from one area in the other one. We consider our encodings as the first efforts towards bridging the gap between process calculi and mobile membranes.

3.1 $D\pi$ into $tD\pi$

In order to compare the expressive power of $tD\pi$ we use a method of *embeddings among languages* introduced in [148]. The method is based on a tuple composed of a set of process expressions $\mathscr{P}$, a partial operation over $\mathscr{P}$ (in process calculi we choose the parallel composition operator) and an observational equivalence. To compare two formalisms by looking at their sets of syntactic expressions (languages) L_1 and L_2, we are required to identify the corresponding *algebraic languages* $(\mathscr{P}; |; \simeq)$ respectively $(\mathscr{P}'; |'; \simeq')$. The method was refined in [81] by defining several constraints for the *coder* and *decoder* functions; the resulting method is called *modular embedding*. The coder translates an expression of the first language into an expression of the second language. The decoder translates back the observational outcomes of the second language into observables of the first language. We adapt this method and use it to show that $tD\pi$ is more expressive than the underlining $D\pi$. The method was shown to be useful [147] first in proving some negative results (languages not-embedded) for a set of concurrent programming languages. The study was based on properties like compositionality of the

B. Aman, G. Ciobanu, *Mobility in Process Calculi and Natural Computing*, Natural Computing Series, DOI 10.1007/978-3-642-24867-2_3,

observational equivalence, interference freedom or composition with hiding of the languages. Later in [148] some more complicated positive results were given in the context of the same set of concurrent programming languages.

Embeddings have also been used to compare the expressivity of some timed coordination languages in [97]. Using modular embedding in [33], the authors compare the expressive power of three classes of quite different coordination languages. Embeddings are extended to *architectural embeddings* in [29] in order to be able to compare coordination architectures.

3.1.1 Algebraic Languages

In our case, for the message passing calculi $D\pi$ and $tD\pi$ the first two components are simple to identify. $\mathscr{P}$ and $\mathscr{P}'$ are the sets of syntactic terms denoting process expressions defined by the syntax of each calculus. The second component of the algebraic language represents a set of operations over the set of process expressions $\mathscr{P}$. We confine our presentation to $tD\pi$ and consequently we take as the sole operation over $\mathscr{P}$ the parallel composition of processes (we do not have summation). Not so simple is the choice of the *semantic equivalence relation* $\simeq$ over $\mathscr{P}$.

The observational equivalence of the algebraic language is defined as the *kernel*[1] of a function $\mathscr{O}$. Such a function $\mathscr{O}$ could be the *interpretation* of the process expressions of our calculi. $\mathscr{O}_i : \mathsf{L}_i \to Obs_i$ maps each term of the language L_i to its interpretation Obs_i(which is sometimes called the *observational outcome*); i ranges over the calculi of our interest. For example the interpretation defined for Turing machines may return the language accepted by the machine and the interpretation for sequential processes may return a relation between the input and the output of the processes. We denote the set of process expressions of $D\pi$ and $tD\pi$ by $\mathsf{L}_{D\pi}$, and respectively by $\mathsf{L}_{tD\pi}$.

We give first a *coding function* $\mathscr{C}o : \mathsf{L}_{D\pi} \to \mathsf{L}_{tD\pi}$ which is a mapping from $\mathsf{L}_{D\pi}$ to $\mathsf{L}_{tD\pi}$. We say that $\mathscr{C}o$ is a language embedding if it is homomorphic with respect to the operations defined over the two sets of process expressions (i.e. the parallel composition operators).

$$\forall P, Q \in \mathsf{L}_{D\pi} \text{ we have } \mathscr{C}o(P \,|\, Q) = \mathscr{C}o(P) \,|\,' \mathscr{C}o(Q).$$

In order to have a language embedding with interesting distinguishing power, $\mathscr{C}o$ must *preserve semantic distinction* and such an embedding is called a *sound* embedding. The soundness is formally defined as:

$$\forall P, Q \in \mathsf{L}_{D\pi} \text{ we have } P \not\simeq Q \Rightarrow \mathscr{C}o(P) \not\simeq' \mathscr{C}o(Q).$$

This is equivalent [147] to the existence of a mapping

$$\mathscr{D} : \mathscr{O}_{tD\pi}(\mathsf{L}_{tD\pi}) \to \mathscr{O}_{D\pi}(\mathsf{L}_{D\pi}),$$

called a *decoder* for which the following is true:

$$\forall P \in \mathsf{L}_{D\pi} \text{ we have } \mathscr{D}(\mathscr{O}_{tD\pi}(\mathscr{C}o(P))) = \mathscr{O}_{D\pi}(P).$$

For a clearer intuition we give the diagram of Figure 3.1 using general notations.

[1] The kernel of a function $f : X \to Y$ is defined as the equivalence relation $\simeq$ over the domain X of the function and satisfies: $\forall a, b \in X$, $a \simeq b$ *iff* $f(a) = f(b)$.

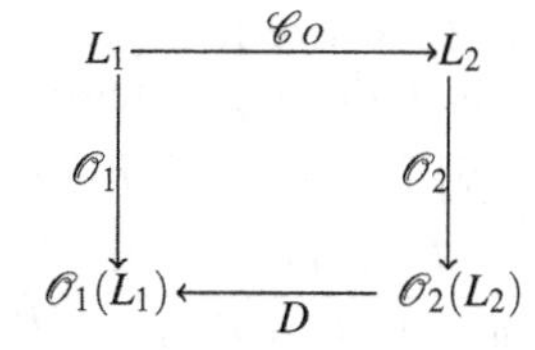

Fig. 3.1 A Sound Embedding of L_1 into L_2

Informally, to say that a language L_1 is embedded in a language L_2 we should map each process expression from L_1 into a corresponding process expression of L_2 through the coder $\mathscr{C}o$. Furthermore, the interpretation of the mapped expression of L_2 should behave the same as the interpretation of the process expression of L_1.

3.1.2 Barbed Bisimulations

The semantic interpretation for the family of π-like calculi is usually given through the operational semantics (i.e., the transition systems generated by the reduction rules or the LTS generated by the labelled transitions). Besides transitions, for every π-like calculus we can give a set of *barbs*. Thus, as observational equivalence we can use the barbed equivalence generated by the barbed bisimulation. The observational outcome of each process is the sets of barbs and the sequence of transitions. In an LTS framework we would have worked with *traces* of a process.

Two processes are equivalent if an observer cannot distinguish differences in their behaviours. Following the presentation of barbed bisimulation in [123, 143], we specify first which actions are observable, and which are considered as internal actions. To simplify the presentation we choose as observable only communication along the located channel names, without considering the transmitted messages. In $tD\pi$ we have synchronous communication on fixed located channels. In consequence, the observable actions can be both *input* and *output* communication. We consider as unobservable actions the *movement action go, application of the time-stepping function* ϕ_Δ, and *internal interaction* of processes. Intuitively an observer of process P is a process Q which runs in parallel with P.

For example, in order to observe if a process P communicates on the input channel $a?$ at location k, the observer waits for an output communication along the same channel $a!$ at the same location k. In $tD\pi$ there are mainly four observation coordinates: one involves the name of the communication channel (Milner and Sangiorgi's barbed bisimulation), another is given by locations, the third is given by type environment, and the fourth is given by timers. Here it suffices to give only the classical strong barbed bisimulation [123] for $tD\pi$. We define first the notion of barb. Barbs are sometimes called commitment predicates and define the possibility of a process to immediately communicate on a specific channel.

Definition 3.1. A barb predicate $\downarrow_\mu$ where $\mu \in \{a?, a!\}$ with a being any channel name, is defined inductively by the following system of rules.

$$\overline{a^{\Delta t}!\langle v\rangle.(P,R) \downarrow_{a!}} \quad \overline{a^{\Delta t}?(X:T).(P,R) \downarrow_{a?}}$$

$$\frac{P \downarrow_\mu}{P\,|\,Q \downarrow_\mu} \quad \frac{P \downarrow_\mu \; and \; a \neq \underline{\mu}}{(\nu a : A)P \downarrow_\mu} \quad \frac{P \downarrow_\mu}{{}^*P \downarrow_\mu}$$

We denote by $\underline{\mu}$ the names of the input or output channels (e.g. if $\mu = a?$ then $\underline{\mu} = a$).

Definition 3.2. A barbed bisimulation $\mathscr{S}$ is a symmetric binary relation over processes which for each $(P,Q) \in \mathscr{S}$ and for any barb $\downarrow_\mu$ implies

1. if $P \downarrow_\mu$, then $Q \downarrow_\mu$;
2. if $P \to P'$, then $Q \to Q'$, and $(P',Q') \in \mathscr{S}$

Two processes are barbed bisimilar, denoted $P \dot{\sim}_B Q$, if and only if $(P,Q) \in \mathscr{S}$ for some barbed bisimulation $\mathscr{S}$.

The barbs and the barbed bisimulation are naturally applied to located processes N. A general notion of barbs is given in [92] in terms of *acceptances* $s_1A_1 \ldots s_nA_n$, where s_i is a sequence of actions from the set of actions Act and $A_i \subset Act$. However we want to look at actions separately, not at sequences s_i of actions. In consequence we adopt a presentation based on the notion of barbs introduced by Milner and Sangiorgi [123]. Note that, despite the fact that in $tD\pi$ we do not have a summation operator, a process P may satisfy more than one barb because of the parallel composition operator. Thus, if $P \downarrow_\mu$ and $Q \downarrow_{\mu'}$ with $\mu \neq \mu'$, then both statements $P\,|\,Q \downarrow_\mu$ and $P\,|\,Q \downarrow'_\mu$ hold.

The barbed bisimulation by itself does not offer satisfactory properties. In order to obtain a barbed equivalence (barbed congruence), the bisimulation is closed under all static (respectively normal) contexts [143]. A context could be viewed as a process running in parallel with our equated processes. In the paper cited above it is shown that in the setting of the π-calculus, the barbed equivalence and barbed congruence coincide with the labelled early bisimilarity, respectively congruence relation on the class of *image-finite processes* [151]. We denote the barbed equivalence by $\simeq_B$. Thus, we have the semantic interpretations for $D\pi$ and $tD\pi$ given their corresponding reduction rules together with the observation predicates (the barbs); we denote them by $\simeq_B^{D\pi}$ respectively $\simeq_B^{tD\pi}$.

3.1.3 Coding Function

The definition of the coding function uses the syntax of the $D\pi$ calculus which we omit here (however it can be deduced from Definition 3.3). For a simpler presentation we choose a restricted form of $D\pi$ which does not take into account the matching process expression. The coding function is applied to each syntactic construct of $D\pi$, regardless of whether it is a type expression or a process expression.

Some entities (like locations) remain unchanged by $\mathscr{C}o$. Other entities (like channel names) are changed into timed entities of $tD\pi$. As expressed in [81], the coder must be defined in a compositional way, and the decoder definition should be element-wise. The compositionality of the coder is not hard to obtain, and it is rather natural in the definition below.

Definition 3.3. The coding function $\mathscr{C}o : \mathsf{L}_{D\pi} \rightarrow \mathsf{L}_{tD\pi}$ is defined in a compositional way for processes and types. We use α for any of the channel capabilities r/w of $D\pi$. For all expressions $E \in \mathsf{L}_{D\pi}$.

$$\mathscr{C}o(E) = \begin{cases} res:\{\alpha\langle \mathscr{C}o(T)\rangle\}\infty & \text{if } E = res:\{\alpha\langle T\rangle\}, \text{ a channel type} \\ gok.(\mathscr{C}o(P), ER) & \text{if } E = gok.P, \quad \text{movement process} \\ a^{\infty}!\langle v\rangle.(\mathscr{C}o(P), ER) & \text{if } E = a!\langle v\rangle.P, \quad \text{output process} \\ a^{\infty}?(X:\mathscr{C}o(T)).(\mathscr{C}o(P), ER) & \text{if } E = a?(X:T).P, \quad \text{input process} \\ \mathscr{C}o(P)\,|\,\mathscr{C}o(Q) & \text{if } E = P\,|\,Q, \quad \text{interaction} \\ (\nu a:\mathscr{C}o(A))\mathscr{C}o(P) & \text{if } E = (\nu a:A)P, \quad \text{restriction} \\ {}^{*}\mathscr{C}o(P) & \text{if } E = {}^{*}P\,, \quad \text{replication} \\ stop & \text{if } E = stop\,, \quad \text{termination} \end{cases}$$

where $ER \in \mathsf{L}_{tD\pi}$ always generates a runtime error in $tD\pi$, and it is defined as $l[[er^{\infty}!\langle v\rangle]]_{\Gamma}$ with $\Gamma(l,v) \not<: wobj(\Gamma(l,er))$.

We do not use located processes in the definition above. The extension of $\mathscr{C}o$ to tagged located processes requires changing the type environments by adding ∞ as timer value to channel types.

The element-wise property of the decoder is not so natural. The definition of a decoder applied to all the observational outcomes of the executions of a process is not so intuitive. The observable behaviour is defined via the operational semantics of the calculi. In our case we use a semantics based on a reduction relation and barbs (or commitment predicates). The set of observables of a process P is a powerset (i.e., the elements are sets of barbs). Moreover, the elements of the reduction relation are the corresponding reduction rules. Instead of defining a decoder, we test whether the reaction rules of $tD\pi$ can simulate the reaction rules of $D\pi$. This is equivalent to the constraint imposed on the decoder to be defined element-wise. Moreover, the barbs of one process must correspond to the barbs of the mapped process (the coded process). The barbs are equivalent to the observational outcome of a process expression.

3.1.4 Expressiveness and Faithfulness

Theorem 3.1. *$tD\pi$ can express any process described in $D\pi$.*

Proof. This means that our timed distributed π-calculus is more expressive than the underlying $D\pi$. We follow the steps of the method described above. We use the coding function $\mathscr{C}o$ to translate a process expression of $D\pi$ into a process expression of $tD\pi$. Then we prove that this coding function agrees with the dynamics given by the operational semantics (i.e., the processes have the same behaviour). Moreover, we prove that through the barbed bisimulation $\dot{\sim}_B$ we get the same set of barbs for the two processes (i.e., the processes have the same observational outcome).

By applying $\mathscr{C}o$, we get only timers with the value ∞. Consequently the application of the cleanup function ψ has no result. Precisely, ψ cannot decrease the value ∞ of the type timers, and thus the types are not removed anymore from the type environments of the processes.

We look at the behaviour of the time-stepping function ϕ_Δ by examining first the expressions of its definition which depend on the timer value t. Since the values of the timers are ∞, the timers are always greater than 1 ($\infty - 1 = \infty$), and we have no expired timers. The processes of form $a^\infty?(X:T).(R,Q)$ or $a^\infty!\langle v\rangle.(R,Q)$ remain unchanged after the application of the time-stepping function. The other cases in the definition of ϕ_Δ follow. If we have $P = gok.(R,Q)$ and $\Gamma(k) <: loc\{go\}$, then ϕ_Δ returns $k[[R]]_\Gamma$. This is the case when the location k has the go capability, and as a consequence, the rule R$_\Gamma$-IDLE of $tD\pi$ behaves as R$_\Gamma$-GO of $D\pi$. For the remaining expressions of the definition of ϕ_Δ, the rule R$_\Gamma$-IDLE should raise a runtime error as it happens in $D\pi$. This is achieved by defining an appropriate ER process instead of the second process Q of the pair (R,Q). An example of such a process is given when we have defined $\mathscr{C}o$. It is easy to see that all the other reduction rules of $D\pi$ have the same behaviour as the corresponding rules of $tD\pi$. The rule R$_\Gamma$-COM of $D\pi$ behaves as rule R$_\Gamma$-COM1 of $tD\pi$. Furthermore, rule R$_\Gamma$-COM2 of $tD\pi$ is not applicable anymore because the capability $ro\langle\rangle$ is not used (the coder $\mathscr{C}o$ does not use it in the translation of expressions from $D\pi$ into expressions in $tD\pi$).

A careful look should also be given to the corresponding error systems. For the error rules of $D\pi$ the reader should check [93]. We should get an error in $tD\pi$ whenever we have an error in $D\pi$, as well as the other way around. It is easy to see that the error rules E$_\Gamma$-GO$''$, E$_\Gamma$-SUBC$''$ and E$_\Gamma$-SND$''$ of $D\pi$ behave like the rules E$_\Gamma$-GO, E$_\Gamma$-SUBC and E$_\Gamma$-SND of $tD\pi$. Looking at the rules E$_\Gamma$-RCV and E$_\Gamma$-COM, the conditions involving the function $roobj$ are not valid anymore, and so they behave as the error rules E$_\Gamma$-RCV$''$ and E$_\Gamma$-COMM$''$ of $D\pi$.

We still have to prove that the process expression of $D\pi$ respects the same set of barbs as the process expression of $tD\pi$ resulting from the codification. In Definition 3.1 of a *simple barb* predicate we choose to observe only the name of the communication channels. The definition of the barbed bisimulation for the syntax of $D\pi$ is similar (we have the same choice of barbs since we do not have other actions that could be regarded as observable).

We consider the remaining process expressions from the definition of the coding function $\mathscr{C}o$. The process $gok.P$ and its codification $gok.(\mathscr{C}o(P),ER)$ are unobservable. The interaction between two processes running in parallel is unobservable in both formalisms. The output expression $a!\langle v\rangle.P$ of $D\pi$, as well as the output codification $a^\infty!\langle v\rangle.(\mathscr{C}o(P),ER)$ of $tD\pi$ have the same barb $\downarrow_{a!}$. Similarly, both the input expression of $D\pi$ and the codification of $tD\pi$ have the same barb. □

We have shown that the embedding of $D\pi$ into $tD\pi$ is *faithful* with respect to the observational equivalences generated by the simple barbed bisimulations of the calculi. Using a general notation, faithful means that, for all $a, b \in \mathsf{L}$ we have that $a \simeq b \Leftrightarrow \mathscr{C}o(a) \simeq' \mathscr{C}o(b)$. According to [147], faithfulness implies the soundness of the language embedding.

Corollary 3.1 (Faithfulness).

$$\forall P, Q \in \mathsf{L}_{D\pi},\ P \simeq_B^{D\pi} Q \Longleftrightarrow \mathscr{C}o(P) \simeq_B^{tD\pi} \mathscr{C}o(Q).$$

It is easy to see that $tD\pi$ cannot be embedded into $D\pi$ because we can give a counterexample of a $tD\pi$ process which cannot have a corresponding process in $D\pi$ with the same behaviour because $D\pi$ cannot manage the time aspects.

3.2 Pure Mobile Ambients into π-calculus

Although both the π-calculus and the calculus of mobile ambients are Turing-complete [42, 121] and they have almost the same field of application (mobile computations), it is widely believed (see [77]) that the π-calculus does not directly model phenomena such as the distribution of processes within different localities, their migrations, or their failures. At the same time the π-calculus provides a solid and useful foundation for concurrent programming languages [86, 135] and it is also supplied with a set of comprehensive techniques and tools for verification and analysis [84, 152]. Therefore, it is worthwhile to take those advantages of the π-calculus that could be useful for manipulation, implementation, verification, etc. of mobile ambients.

In a number of papers (see [42, 48, 111, 155]) it has been demonstrated that the calculus of mobile ambients can be used for simulating π-calculus computations. On the other hand, Fournet, Levy and Schmitt [88] have translated mobile ambients into the distributed join-calculus. The atomic steps of mobile ambient computation are decomposed into several elementary steps, each involving only local synchronization. By combining this translation with the encoding of the distributed join calculus into the join calculus [87] and then with the encoding of the join calculus into the asynchronous π-calculus [86] one could obtain the translation of mobile ambients into the asynchronous π-calculus. But to the best of our knowledge no efforts have been made to trace this chain of encodings from the beginning to the end. An attempt to build a straightforward translation from mobile ambients into a subset of the synchronous π calculus has been undertaken by Brodo, Degano and Priami [32]. In order to imitate the spatial structure of mobile ambients we impose some very rigid restrictions on the structural congruence rules of the π-calculus. It may be said that in [32] the encoding of mobile ambients into the π-calculus has been achieved on the purely syntactic level.

3.2.1 Main Idea

In what follows we also try to assess the capability of the π-calculus to encode mobile ambients. The topic is interesting because mobile ambients can be considered a higher-order calculus of fundamental character. Moreover, an encoding of mobile ambients into the π-calculus appears challenging, in particular because distributed conformance changes must be effectuated over varying numbers of encoded agents and capabilities (as encoded ambients migrate or open up). The main objective of our research is to build such a straightforward translation from the calculus of mobile ambients to the π-calculus which preserves the behavioural properties of the processes. This translation coupled with that of [42, 48, 155] may form a basis for the development of a uniform theory of mobile computations.

As the starting point we present a rather simple variant of the straightforward encoding of mobile ambients into the π-calculus to demonstrate the practicability of our expectancies. A key idea of the encoding is based on the separation of the spatial structure of mobile ambients from their operational semantics. The operational semantics of mobile ambients is given by a universal π-process *Ruler* which plays the role of an interpreter routine. Each mobile ambient A is encoded into a π-calculus term $Structure_A$ which simulates the spatial structure of A by means of channels. Each step of the encoding is explained in some detail. We also provide an operational correspondence between the two calculi [73].

To emphasize the key ideas of our encoding we confine ourselves to its most simple (sequential) variant which assumes the use of the unique *Ruler* for the whole system. As it can be readily seen from the description this encoding can be improved in such a way that it becomes both distributed and compositional. This can be achieved by supplying every π-process *Node* corresponding to an ambient with its own interpreter *Ruler*.

In what follows we define a relationship between MA-processes and π-processes. This relationship may be thought of as a non-deterministic encoding of pure ambients into π-processes: with every MA-process A it associates a set $[\![A]\!]$ of π-processes. In the next section we prove the operational correspondence of the encoding by demonstrating that each π-process from $[\![A]\!]$ corresponds to the behaviour of A. The only purpose of considering $[\![\cdot]\!]$ as a relation is to simplify the proofs. When using our translation in applications, one may take a single (minimum-size) π-process from $[\![A]\!]$ as a true image of A.

A specific feature of pure ambient calculus is that an MA-process A has a spatial tree-like structure which serves a dual function. On the one hand, mobile ambients control the run of processes in A by bounding the scope of actions. On the other hand, the mobile ambients of A are acted upon by the spatial structure of A. A similar idea of decomposing the ambient process into a tree and actions is used in [110] to define the normal semantics for mobile ambients. When translating A into a π-process we separate these functions of mobile ambients. An ambient A is translated into a π-process $Proc_A = Structure_A|Ruler|Environment$ which is a composition of three π-processes $Structure_A$, *Ruler* and *Environment*. The process $Structure_A$ is designed according to the following principles.

1. The mobile ambients and capabilities from A are represented by individual subprocesses Amb_i and Act_j; we will call these π-calculus terms *nodes*. The spatial structure of A is maintained by means of specific tree-wire subprocesses TW_k that are used for communication between nodes that represent ambients and actions. Thus, we have
$$Structure_A = Amb_1|\ldots|Amb_N|Act_1|\ldots|Act_M|TW_1|\ldots|TW_L$$
where Amb_i, Act_j, and TW_k represent generic notations for ambients, capabilities, and tree-wire structure.
2. Each subprocess Amb_i is associated with some mobile ambient $n[P]$ in A. It keeps the name n of the ambient and provides communication between the ambient and its upper and lower contexts.
3. Each subprocess Act_j is associated with some action of the form *in n.P*, *out n.P* or *open n.P* in A. It keeps the type of capability (*in*, *out* or *open*) and the name n and also provides communication between the action and its upper and lower context.
4. A subprocess TW_k is a set of wires arranged in a tree-like structure. TW_k delivers requests from its leaf nodes to the root and sends back replies from the root to the leaves. A tree-wire subprocess is intended to provide message exchange between the nodes and to accumulate consumed capabilities and dissolved ambients of A.
5. When a capability is consumed or an ambient is dissolved, a corresponding node becomes passive. A passive node becomes a wire and adds itself to some tree-wire. Thus, the wires of $Structure_A$ take account of computation steps generated by A. Since there are many different ways to derive the same mobile ambient term A, it may be encoded into a whole set of π-calculus terms which have the same nodes and differ only in the structure of their tree-wires.

The subprocess *Ruler* does not depend on A. It is a universal π-process intended for simulating the operational semantics of mobile ambients. Here the *Ruler* is presented as a central handler. It is also possible to have *Ruler* acting as a virtual machine at each location; in this way the encoding becomes distributed. An execution of *Ruler* conforms to the following scenario.

1. *Ruler* selects from $Structure_A$ an arbitrary triple of nodes Act, Amb_{i_1} and Amb_{i_2} and collects the information about their types, names and links.
2. If (1) the type of Act is *in*, (2) Act is linked with Amb_{i_1}, (3) Act stores the same name as Amb_{i_2}, and (4) Amb_{i_1} and Amb_{i_2} are linked with the same node in $Structure_A$, then the subprocess $Act|Amb_{i_1}|Amb_{i_2}$ corresponds to a mobile ambient pattern $n[in\ m.P|Q]|m[R]$. In this case *Ruler* simulates the implementation of an entry instruction by switching the link of Amb_{i_1} to Amb_{i_2} and converting the node Act into a wire. This changes the entry-pattern $n[in\ m.P|Q]|m[R]$ into $m[n[P|Q]|R]$.
3. If (1) the type of Act is *out*, (2) Act is linked with Amb_{i_1}, (3) Act keeps the same name as Amb_{i_2}, and (4) Amb_{i_1} is linked with Amb_{i_2}, then $Act|Amb_{i_1}|Amb_{i_2}$ corresponds to a pattern $m[n[out\ m.P|Q]|R]$. In this case *Ruler* simulates the implementation of an exit instruction by converting the node Act into a wire, and

directing the link of Amb_{i_1} to the same destination to which the link of Amb_{i_2} is directed. This changes the exit-pattern $m[n[out\ m.P|Q]|R$ into $m[R]|n[P|Q]$.

4. If (1) the type of Act is $open$, (2) Act keeps the same name as Amb_{i_1} and (3) both Act and Amb_{i_1} are linked with Amb_{i_2}, then $Act|Amb_{i_1}|Amb_{i_2}$ corresponds to a pattern $m[open\ n.P|n[Q]]$. In this case $Ruler$ simulates the implementation of an open instruction by converting both Act and Amb_{i_1} into wires. This changes the open-pattern $m[open\ n.P|n[Q]]$ into $m[P|Q]$.
5. If none of the above cases holds, then $Ruler$ tries another triple of nodes Act', Amb_{j_1} and Amb_{j_2} from $Structure_A$.

The subprocess $Environment$ plays the role of a virtual external environment of mobile ambients. It bounds every MA-process and can not be dissolved. When $Ruler$ simulates an open operation, it can select $Environment$ for Amb_{i_2}.

Now we are ready to present the formal description of the processes involved in $Structure_A$, $Ruler$ and $Environment$ and define the encoding relation between pure mobile ambients and π-processes.

3.2.2 Tree-Wire Processes

A spatial structure of mobile ambients is represented by specific π-processes which are called *tree-wires*. A tree-wire is a parallel composition of some basic processes which are called *wires*. A wire serves the message passing from one agent to another and back. By setting up an appropriate correspondence on the names of agents one can compose wires into any tree-like communication structure.

For $x \neq y$, we define a wire process by $W(x,y) = !\ (\nu u)\ x(v).\ \overline{y}\langle u\rangle.\ u(t).\ \overline{v}\langle t\rangle$. A wire has two parameters x and y. The parameter x is a name of a channel for communication with low-level components of the system, whereas y is a name of a channel for communication with its top-level component. The wire $W(x,y)$ receives a request $x(v)$ from one of the low-level components, re-addresses it to the top-level component $\overline{y}\langle u\rangle$, then receives a reply $u(t)$ via a private channel u, and finally re-addresses this reply $\overline{v}\langle t\rangle$ to the low-level component. It should be noticed that it may be the case that several low-level components try to communicate at the same time with a top-level component via $W(x,y)$. Then the top-level component can serve all the low-level components one by one. To avoid broadcasting a new private channel u is selected every time.

A tree-wire process $TW_{I,k}$, where I is a set of names and k is a name with $k \notin I$, is any process which is composed of basic wires $TW_{I,k} = W(x_1,y_1)|\ldots|W(x_m,y_m)$ in such a way that this parallel composition has a tree-like communication structure and provides message exchange between the set of leaf nodes I and the root k.

Definition 3.4. The set of tree-wires is the minimal set of π-processes satisfying the following requirements.

1. Every wire $W_{x,y}$ is a tree-wire $TW_{\{x\},y}$.

2. If $W_{x,y}$ is a wire, $TW_{I,x}$ is a tree-wire and $y \notin I$, then the term $(\nu x)\ (W_{x,y}|TW_{I,x})$ is a tree-wire $TW_{I,y}$.
3. If $TW_{I,y}$ and $TW_{J,y}$ are tree-wires such that $I \cap J = \emptyset$, then the term $TW_{I,y}|TW_{J,y}$ is a tree-wire $TW_{I \cup J,y}$.

When dealing with a tree-wire $TW_{I,y}$, where $I = \{x_1, x_2, \ldots, x_n\}$, we use the shorthand notation $(\nu I)(P|TW_{I,k})$ for $(\nu x_1)\ (\nu x_2) \ldots (\nu x_n)\ (P|TW_{I,k})$.

Proposition 3.1. *If $TW_{I,y}$ is a tree-wire, then $y \notin I$ and $fn(TW_{I,y}) = I \cup \{y\}$.*

When appending a tree-wire to a tree-wire we get again a tree-wire.

Proposition 3.2. *Let TW_{I_1,y_1} and TW_{I_2,y_2} be a pair of tree-wires such that $I_1 \cap I_2 = \emptyset$, $y_1 \notin I_2$, $y_2 \in I_1$. Then $(\nu y_2)\ (TW_{I_1,y_1}|TW_{I_2,y_2})$ is a tree-wire TW_{I_3,y_1}, where $I_3 = (I_1 - \{y_2\}) \cup I_2$.*

A tree-wire delivers requests from its leaf nodes to the root and replies from the root to leaf nodes.

Proposition 3.3. *Let $TW_{I,y}$ be a tree-wire and $x \in I$. Then the π-process*

$$(\nu v)\ \overline{x}\langle v\rangle.\ v(z).\ z|TW_{I,y}|y(u).\ \overline{u}\langle t\rangle$$

has a deterministic terminating run

$$(\nu v)\ \overline{x}\langle v\rangle.\ v(z).\ z|TW_{I,y}|y(u).\ \overline{u}\langle t\rangle \ \mapsto^* \ t|TW_{I,y}$$

3.2.3 Ambients and Actions

The main difficulty of encoding pure ambient processes into π-processes is that of simulating the consumption of capabilities, dissolving the boundaries of ambients, and changing the structure accordingly. In an MA-process, when an action *in n.P*, *out n.P* or *open n.P* is executed, the corresponding capability just disappears from the process (it is consumed). The same effect manifests itself when the boundary of an ambient named $m[Q]$ is dissolved. But when simulating actions and ambients as individual π-subprocesses of $Structure_A$ it is not possible just to reduce the consumed capabilities or dissolved ambients to inactive processes $\mathbf{0}$ since in this case we lose the links between the processes in $Structure_A$. The simplest way to make such processes inactive while preserving a tree-like structure of links between the remaining processes is to convert consumed capabilities and dissolved ambients into wires and use them merely to maintain links between the active processes. In this case the process $Structure_A$ corresponding to an MA-process A will depend not only on the spatial structure of A, but also on the way A is computed (the history of A). That is why instead of using a deterministic encoding which maps every MA-process into a single π-process we introduce an encoding relation $\models$ which associates every MA-process A with a set of π-processes $[\![A]\!]$. Each process $Structure_A$ from $[\![A]\!]$ keeps along with the spatial structure of A the possible history of A, i.e. the way the MA-process A can be computed from the other processes. This history is represented by wires which keep track of consumed capabilities and dissolved ambients. The history of A does not influence the functionality of $Structure_A$; its only purpose is to maintain the links between active nodes of $Structure_A$.

Definition 3.5. The formal description of a π-process $Node(a,n,u,d,s,l)$ is

$$\begin{array}{lcll}
Node(a,n,u,d,s,l) & = & Reply(d,s,l)|Main(a,n,u,d,s,l) & \\
Reply(d,s,l) & = & d(y).\ [y=l] & (1)\\
 & & (\ d(u).\ \overline{s}\langle l\rangle.\ W_{d,u}\ , & (2)\\
 & & \quad \overline{y}\langle l\rangle.\ Reply(d,s,l)\ ) & (3)\\
Main(a,n,u,d,s,l) & = & (\nu v)\ (\nu w)\ (\nu k) & (4)\\
 & & \overline{s}\langle v\rangle.\ v(x). & (5)\\
 & & \overline{u}\langle w\rangle.\ w(l_u). & (6)\\
 & & \overline{x}\langle a,n,l,u,d,l_u\rangle.\ v(y,z). & (7)\\
 & & [y=l] & (8)\\
 & & (\ \overline{d}\langle l\rangle.\ \overline{d}\langle z\rangle\ , & (9)\\
 & & \quad \overline{d}\langle k\rangle.\ k(w'). & (10)\\
 & & \quad \overline{s}\langle l\rangle.\ Main(a,n,z,d,s,l)\ ) & (11)
\end{array}$$

The π-processes *Act* and *Amb* associated with actions $cap\, n.P$, $cap \in \{in, out, open\}$, and ambients $n[P]$ in MA-process A have a similar arrangement. An action represented by a capability $cap\ n$ is encoded into $Node(cap,n,up,down,s_{act},label)$, where up and $down$ are names of channels used for communication with upper and lower contexts of the action, s_{act} is a channel name shared by all action-type nodes of $Structure_A$ for communications with *Ruler*, and *label* is an individual label of the action in A. An ambient n is encoded into the π-process $Node(amb,n,up,down,s_{amb},label)$, where up, $down$ and $label$ have the same meaning as above, amb is the key word for distinguishing ambients from capabilities, and s_{amb} is a channel name shared by all ambient-type nodes of $Structure_A$ for communications with *Ruler*.

The π-process $Node(a,n,u,d,s,l)$ is a recursive process composed of two subprocesses $Reply(d,s,l)$ and $Main(a,n,u,d,s,l)$. The subprocess *Reply* serves the dual function of providing communication with the lower context of a node (which is a set of nodes) and also of converting (if necessary) the node into a wire. The subprocess *Main* keeps the information about the node (its type, name and context) and communicates with *Ruler*.

3.2.4 Simulating the Operational Semantics of Pure Ambients

The π-process $Structure_A$ represents only the spatial structure of MA process A. The behaviour of A is simulated by a universal π-process *Ruler* which does not depend on A. This process has two parameters s_{act} and s_{amb} as channel names for receiving submissions from nodes corresponding to actions and ambients. The received submissions indicate the readiness of the nodes to participate in the simulation of some MA operation (entering, exiting or opening).

Definition 3.6. The formal description of a π-process *Ruler* is as follows.

$$
\begin{aligned}
Ruler(s_{act}, s_{amb}) \;=\;& (\nu x_0)\,(\nu x_1)\,(\nu x_2)\,(\nu y) && (12)\\
& s_{act}(v_0).\; s_{amb}(v_1).\; s_{amb}(v_2). && (13)\\
& \overline{v}_0\langle x_0\rangle.x_0(t_c, n_c, l_c, u_c, d_c, ul_c). && (14)\\
& \overline{v}_1\langle x_1\rangle.x_1(t_{a,1}, n_{a,1}, l_{a,1}, u_{a,1}, d_{a,1}, ul_{a,1}). && (15)\\
& \overline{v}_2\langle x_2\rangle.x_2(t_{a,2}, n_{a,2}, l_{a,2}, u_{a,2}, d_{a,2}, ul_{a,2}). &&\\
& [t_c = in \wedge n_c = n_{a,2} \wedge ul_c = l_{a,1} \wedge ul_{a,1} = ul_{a,2}] && (16)\\
& \quad (\\
& \quad\quad \overline{v}_0\langle l_c, u_c\rangle.\overline{v}_1\langle y, d_{a,2}\rangle.\overline{v}_2\langle y, u_{a,2}\rangle\;, && (17)\\
& \quad\quad [t_c = out \wedge n_c = n_{a,2} \wedge ul_c = l_{a,1} \wedge ul_{a,1} = l_{a,2}] && (18)\\
& \quad\quad (\\
& \quad\quad\quad \overline{v}_0\langle l_c, u_c\rangle.\overline{v}_1\langle y, u_{a,2}\rangle.\overline{v}_2\langle y, u_{a,2}\rangle\;, && (19)\\
& \quad\quad\quad [t_c = open \wedge n_c = n_{a,1} \wedge ul_c = l_{a,2} \wedge ul_{a,1} = l_{a,2}] && (20)\\
& \quad\quad\quad\quad (\; \overline{v}_0\langle l_c, u_c\rangle.\overline{v}_1\langle l_{a,1}, u_{a,1}\rangle.\overline{v}_2\langle y, u_{a,2}\rangle\;, && (21)\\
& \quad\quad\quad\quad\quad \overline{v}_0\langle y, u_c\rangle.\overline{v}_1\langle y, u_{a,1}\rangle.\overline{v}_2\langle y, u_{a,2}\rangle\;\;) && (22)\\
& \quad\quad)\\
& \quad).s_{act}(z_0).\; s_{amb}(z_1).\; s_{amb}(z_2).\; Ruler(s_{act}, s_{amb}) && (23)
\end{aligned}
$$

Environment:

The environment is considered as a top-most ambient encompassing MA-process A. But unlike conventional ambients it cannot be dissolved and no ambient can exit from it. The environment plays the role of an upper context for unguarded actions and ambients; it can also participate in the simulation of open operations as Amb_2. Moreover, for the sake of uniformity it is convenient to compose an environment out of two ambient-type processes. Only one of these processes can actually participate in simulation of some MA-operation. The other is just a dummy which gives *Ruler* a possibility to operate while at least one active action-type node remains in $Structure_A$.

Definition 3.7. The formal description of a π-process *Environment* is

$$
\begin{aligned}
Environment(env, \top, d, s_{amb}) \;=\;& Top(env, \top, d, s_{amb})\;|\\
& (\nu\; d')\; Top(env, \top, d', s_{amb}) && (24)\\
Top(env, \top, d, s) \;=\;& (\nu v)\,(\nu l)\,(\nu u)\,(\nu w) && (25)\\
& \overline{s}\langle v\rangle.\; v(x). && (26)\\
& \overline{x}\langle env, \top, l, u, d, w\rangle. && (27)\\
& v(y,z).\; \overline{s}\langle l\rangle.\; Top(env, \top, d, s) && (28)
\end{aligned}
$$

3.2.5 The Intended Meaning of the Encoding Constructions

We present here some details about our encoding. First we briefly explain the intended meaning of some fragments of *Reply* and *Main* from the formal description of $Node(a,n,u,d,s,l)$:

(1) $Reply(d,s,l)$ can receive via the channel d either a request from the lower context of the node, or an instruction from *Main* which changes the whole node into a wire. In the former case y is evaluated as a private channel name for emitting at y the label l of the node. In the latter case the main process evaluates y as the label l (see line (9)) which does not match any private name. Therefore, after receiving the label l from *Main* the process *Reply* is reduced to the line (2), whereas after receiving a request from the lower context of the node it is reduced to the line (3).

(2) After receiving the label l of the node from the main subprocess of the node, *Reply* receives an updated channel name for communication with the upper context of the node (see line (9)), sends a synchronization message to *Ruler* indicating thus the completion of the instruction processing, and evolves into a wire which connects the lower and the upper contexts of the node. This may happen when the node becomes passive since the corresponding capability is consumed or the ambient boundary is dissolved.

(3) If *Reply* receives a request from the lower context of the node it considers the value of y as a private name of a channel and sends via this channel the label l of the node. As a consequence, the label of the node becomes available to its lower context. Afterwards *Reply* reverts to the original state.

(4) The subprocess $Main(a,n,u,d,s,l)$ uses the following private names:

- v as a name of a channel for receiving acknowledgments and instructions from *Ruler*,
- w as a name of a channel for receiving the label of a node that precedes our node in $Structure_A$,
- k as an arbitrary fresh name different from the label of the node to switch *Reply* to the line (3).

(5) *Main* begins with sending a message to *Ruler* to indicate the readiness of the node to participate in the simulation of some MA operation. *Ruler* will consider this message as a private name of a channel for communication with the node. If *Ruler* selects the node for simulating MA operation it sends via v another private name (see lines (14),(15)). This name will be used as a channel for sending to *Ruler* additional information: the name, the type and the environment of the node.

(6) The node sends a request to its upper context to know the label of the preceding node in $Structure_A$. The upper context replies via w and evaluates l_u to the label of the predecessor (see lines (1), (3) and Proposition 3).

(7) The node sends to *Ruler* its type, name and label, the channel names for communication with the upper and lower context, and the label of its predecessor in $Structure_A$. After processing this information *Ruler* replies via v to the node and informs it about its new status (active or passive) and its new upper context

(see lines (17), (19), (21) and (22)). If the node does not match a pattern for simulating an MA operation or corresponds to a component of MA which is not consumed or dissolved along the operation, then *Ruler* emits at v some private name which does not match l. The node considers this private name as an instruction to remain active. Otherwise *Ruler* evaluates y as the label l of the node and this is considered as an instruction to alter the node into a wire. In both cases *Ruler* evaluates z as a name of a new channel for communication with the upper context of the node (since the context of the node may also be changed as a result of simulation of MA operation).

(8)*Main* checks the instruction.

(9)If the node becomes passive due to the consumption of a capability or dissolution of an ambient, then *Reply* is instructed to become a wire and receives an updated channel name for communication with its upper context. In this case the main subprocess of the node is reduced to **0**.

(10) Otherwise the subprocess *Reply* is informed that the node remains active. The input action $k(w')$ is used just for the sake of uniformity.

(11) Afterwards the main subprocess sends to *Ruler* a synchronization message which indicates completion of the instruction processing, updates its upper context and reverts to the original state.

The intended meaning of the lines in the definition of $Ruler(s_{act}, s_{amb})$ is as follows.

(12) The subprocess *Ruler* uses the following private names:

- x_0, x_1, x_2 for communication with the nodes selected for simulating MA operations, and
- y for instructing the selected nodes to remain active.

(13) *Ruler* non-deterministically selects three nodes representing an action and a pair of ambients in an MA-process that can participate in the simulation of MA operation. Selection is put into effect by receiving requests via public channels s_{act} and s_{amb} (see line (5)). The first selected node is an action-type node since the request from this node is received via the channel s_{act} which is shared by action-type nodes only. Two others are ambient-type nodes. The requests include private channel names for communication with the selected nodes.

(14) Using this private channel, *Ruler* sends a fresh channel name x_0 to the action-type node. The node considers this message as an inquiry about its characteristics (type t_c, name n_c, label l_c, channel names u_c, d_c for communication with upper and lower contexts, label ul_c of the preceding node). It delivers the required names to *Ruler* via the private channel x_0 (see line (7)).

(15) As in the case of actions (see line (14)), *Ruler* asks the selected ambient-type nodes to provide the information on the names $n_{a,1}$ and $n_{a,2}$, labels $l_{a,1}$ and $l_{a,2}$, channel names for communication with lower contexts $d_{a,1}$ and $d_{a,2}$, and labels of the preceding nodes $ul_{a,1}$ and $ul_{a,2}$ of these nodes.

(16) From this point *Ruler* begins to check which operation on MA can be executed by means of the capability represented by the selected action-type node on the ambients represented by the selected ambient-type nodes. There are three

conditions to ensure that *Ruler* simulates the application of the entering reduction step to the MA process. First, the selected action-type node labelled with l_c should represent an action which has the capability to enter ($t_c = in$) into the ambient named $n_{a,2}$ ($n_c = n_{a,2}$); secondly, it lies on the top level in the ambient named $n_{a,1}$ ($ul_c = l_{a,1}$); and finally, the ambients named $n_{a,1}$ and $n_{a,2}$ are siblings ($ul_{a,1} = ul_{a,2}$).

(17) The entering reduction step is simulated by changing the communication net in the π-process $Structure_A$ which represents the spatial structure of MA-process A. Since the capability represented by the selected action-type node is consumed, *Ruler* sends to this node its label l_c via the private channel. After receiving this message the node evolves into a wire (see lines (7), (8), (9), (1), (2)). Since the ambient named $n_{a,1}$ enters the sibling ambient named $n_{a,2}$, the corresponding ambient-type node has to change its upper context. *Ruler* sends to this node the channel name $d_{a,2}$ for communication with its new upper context which is the node corresponding to the ambient named $n_{a,2}$. The upper context of the node corresponding to the ambient named $n_{a,2}$ remains the same, and *Ruler* acknowledges this by communicating back the value $u_{a,2}$. A private name y which does not match the labels $l_{a,1}$ and $l_{a,2}$ is sent to the ambient-type nodes as an instruction to remain active (see lines (7), (8), (10), (11), (1), (3)).

(18) If the selected nodes do not match an entry-pattern, then *Ruler* checks for an exit-pattern. There are three conditions to ensure that *Ruler* simulates the application of an exiting reduction step to the MA process. First, the selected action-type node labelled with l_c should represent an action which has the capability to exit ($t_c = out$) from the ambient named $n_{a,2}$ ($n_c = n_{a,2}$); secondly, it lies on the top level in the ambient named $n_{a,1}$ ($ul_c = l_{a,1}$); and finally, the ambient named $n_{a,1}$ lies on the top level of the ambient named $n_{a,2}$ ($ul_{a,1} = l_{a,2}$).

(19) Since the capability represented by the selected action-type node is consumed, *Ruler* sends to this node its label l_c to evolve the node into a wire (see lines (7), (8), (9), (1), (2)). Since the ambient named $n_{a,1}$ is transformed into a sibling of the ambient named $n_{a,2}$, it changes the upper context from $u_{a,1}$ to $u_{a,2}$. The upper context of the node corresponding to the ambient named $n_{a,2}$ remains the same. *Ruler* sends a private name y which does not match the labels $l_{a,1}$ and $l_{a,2}$ to instruct the ambient-type nodes to remain active (see lines (7), (8), (10), (11), (1), (3)).

(20) If the selected nodes do not match an exit-pattern, then *Ruler* checks for an open-pattern. If the selected action-type node labelled with l_c represents an action which has the capability to dissolve the boundary ($t_c = open$) of the ambient named $n_{a,1}$ ($n_c = n_{a,1}$), lies on the top level in the ambient named $n_{a,2}$ ($ul_c = l_{a,2}$), and the ambient named $n_{a,1}$ also lies on the top level of the ambient $n_{a,2}$ ($ul_{a,1} = l_{a,2}$), then *Ruler* simulates the application of an opening reduction step to the MA process.

(21) To simulate an opening MA-operation, *Ruler* sends l_c to the action-type node and $l_{a,1}$ to the ambient-type node named $n_{a,1}$ to evolve these nodes into wires (see lines (7), (8), (9), (1), (2)) since the capability is consumed and the boundary of the ambient is dissolved. *Ruler* sends a private name y which does not match the

label $l_{a,2}$ to instruct the ambient-type node named $n_{a,2}$ to remain active (see lines (7), (8), (10), (11), (1), (3)).

(22) If the selected nodes do not match any pattern, then *Ruler* informs them via private channels to remain active and to keep their channels for communication with upper contexts unchanged.

(23) After this *Ruler* waits till the nodes which participated in this round of simulation send their synchronization messages (the labels) to indicate that instructions sent to them (see lines (17), (19), (21), or (22)) are performed (see lines (2) and (11)), reverts to the initial state and tries another triple of nodes.

Finally, we comment briefly on the intended meaning of *Environment*.

(24) The environment is composed of two processes *Top*. They have the same functionality, but only the first one has a global name for communication with its lower context (top-level actions and ambients encompassed by the environment). Nevertheless both processes can communicate with *Ruler* via the channel s_{amb} shared by ambient-type nodes. The name $\top$ is an arbitrary name which is different from any free name in an ambient encompassed by the environment.

(25) A process *Top* uses the following private names:

- v as a name of a channel for receiving acknowledgments and instructions from *Ruler*,
- l, u, w as dummy names that are used only for the sake of uniformity in communications with *Ruler* (they stand for the label, upper context and label of the predecessor of the node that are of no importance for the environment).

(26) The environment processes begin with sending to *Ruler* their requests for participation in the simulation of MA operations. When participation is granted *Top* receives a private channel name for communication with *Ruler* (see line (15)).

(27) Using this channel *Top* sends to *Ruler* the information about its type (a keyword *env*), name (it should be different from any name in $Structure_A$), channel names for communication with its upper context (since it does not exist any private name is possible) and lower context (d), the label of the preceding node (it does not exist also and *Top* uses any private name for this purpose).

(28) Like any other node participating in the simulation of MA operation as seen in line (7), *Top* receives a pair of names (y and z) which are interpreted as instructions for changing its status and updating its context. But since the environment cannot be dissolved and it has no upper context, these names do not affect its functionality. This input is used only for the sake of uniformity which gives *Ruler* the possibility not to distinguish *Top* as a specific node. Therefore, *Top* just acknowledges the receipt of these names by sending a synchronization message to *Ruler* and reverts to the original state.

3.2.6 Encoding of Pure Ambients into π-processes

The encoding of pure ambients in π-processes is defined in terms of two relations $\models_0$ and $\models$. We use $\models_0$ for constructing $Structure_A$ corresponding to MA-process A and $\models$ for constructing the ultimate π-process out of $Structure_A$, *Ruler* and *Environment*.

Definition 3.8. The encoding relation $\models_0$ is defined inductively by the following axioms and rules. In every pair $[P,k]$ to the right of $\models_0$ the second component k stands for the free channel name used in the π-process P for communication with its upper context.

Axioms:

Ax1 (Simple Inactivity) $\mathbf{0} \models_0 [\mathbf{0},k]$, *where* $k \in \mathcal{N}$;

Ax2 (Tree-wire) $\mathbf{0} \models_0 [(\nu I)\ TW_{I,k},k]$, *where* $I \subset \mathcal{N}$, $k \in \mathcal{N}$;

Rules:

R1 (Add tree-wire) $\dfrac{A \models_0 [P,k]}{A \models_0 [(\nu\, I)\ (W_{I,m}|P),m]}$,

where $k \in I, fn(P) \cap I \subseteq \{k\},\ m \notin fn(P) \cup I$;

R2 (Composition) $\dfrac{A_1 \models_0 [P_1,k]\ ,\ A_2 \models_0 [P_2,k]}{A_1|A_2 \models_0 [P_1|P_2,k]}$;

R3 (Restriction) $\dfrac{A \models_0 [P,k]}{(\nu n)\ A \models_0 [(\nu\, n)\ P,k]}$; *R4 (Replication)* $\dfrac{A \models_0 [P,k]}{!A \models_0 [!P,k]}$;

R5 (Action) $\dfrac{A \models_0 [P,k]}{cap\ n.\ A \models_0 [(\nu\, k)\ (\nu\, l)(Node(cap,n,m,k,s_{act},l)|P),m]}$,

where $cap \in \{in,out,open\}$, $l,m \notin fn(P) \cup \{n\}$;

R6 (Ambient) $\dfrac{A \models_0 [P,k]}{n[A] \models_0 [(\nu\, k)\ (\nu\, l)\ (Node(amb,n,m,k,s_{amb},l)|P),m]}$,

where $l,m \notin fn(P) \cup \{n\}$.

The encoding relation $\models$ is defined by the single rule

R0 $(MA - to - \pi)$

$$\frac{A \models_0 [Structure_A,k]}{A \models (\nu\, \Sigma)\ (Structure_A|Ruler(s_{act},s_{amb})|Environment(env,\top,k,s_{amb}))}$$

where $\nu\Sigma$ stands for the prefix

$$(\nu\ in)\ (\nu\ out)\ (\nu\ open)\ (\nu\ amb)\ (\nu\ env)\ (\nu\ s_{act})\ (\nu\ s_{amb})\ (\nu\ \top)\ (\nu\ k)\ ,$$

and $\top$ is any name from $\mathcal{N} - fn(A)$.

Proposition 3.4.

1. *Let* A,B *be two MA-processes such that* $A \equiv_a B$, *and* $A \models P$. *Then there exists a derivation* $B \models Q$ *such that* $P \equiv_\pi Q$.
2. *If* $A \models P$, *then* $fn(A) = fn(P)$.

3.2.7 Operational Correspondence

In this section we will demonstrate that the encoding of pure ambients into the π-calculus is complete and sound. By completeness we mean that any π-process P associated with an MA-process A through the encoding relation $A \models P$ admits only those π-calculus reductions $P \to_\pi^* P'$ that can be interpreted in terms of pure ambient reductions $A \to_a A'$ such that $A' \models P'$. Soundness means that any reduction $A \to_a A'$ corresponds to some chain of π-calculus reductions $P \to_\pi^* P'$ of P such that $A' \models P'$. Thus, we may speak of a homomorphic embedding of pure mobile ambients into the π-calculus.

Theorem 3.2 (Completeness). *Let A_0, A_1 be MA-processes and P_0 be a π-process such that $A_0 \to_a A_1$ and $A_0 \models P_0$. Then there exists a chain of π-calculus reduction steps*

$$P_0 \hookrightarrow_\pi^3 P_1' \mapsto_\pi^* P_1$$

such that $A_1 \models P_1$.

The proof follows straightforwardly from the description of processes *Main*, *Reply*, *Ruler*, and *Top* and Proposition 3.3. The only non-deterministic steps in the reduction $P_0 \to_\pi^* P_1$ are three communications steps when *Ruler* selects nodes representing a capability and a pair of ambients for simulating a reduction step $A_0 \to_a A_1$. Afterwards the reduction of P_0 is completely deterministic until all communication actions in the bodies of subprocesses *Ruler*, *Main* and *Reply* are executed to an end.

Theorem 3.3 (Soundness). *Let A_0 be an MA-process and P_0 be a π-process such that $A_0 \models P_0$. Let $P_0 \to_\pi^* P$ be a chain of π-calculus reduction steps. Then there exist an integer N, $0 \le N \le 2$, a sequence of π-calculus terms $P_1', P_1, P_2', P_2, \dots, P_n', P_n$ and a sequence of pure ambient terms $A_1, A_2, \dots, A_n$ such that the following conditions hold*

1. $P \hookrightarrow_\pi^N P_n' \mapsto_\pi^* P_n$;
2. *The chain of π-calculus reductions $P_0 \to_\pi^* P \to_\pi^* P_n$ can be partitioned as follows:*

$$P_0 \hookrightarrow_\pi^3 P_1' \mapsto_\pi^* P_1 \hookrightarrow_\pi^3 P_2' \mapsto_\pi^* P_2 \hookrightarrow_\pi^3 \cdots \hookrightarrow_\pi^3 P_{n-1}' \mapsto_\pi^* P_{n-1} \hookrightarrow_\pi^{3-N} P \hookrightarrow_\pi^N P_n' \mapsto_\pi^* P_n$$

such that

a. $A_i \models P_i$ for every i, $0 \le i \le n$;
b. for every i, $0 \le i < n$, either $A_i \equiv_a A_{i+1}$ or $A_i \to_a A_{i+1}$.

The intended meaning of this theorem is as follows. Suppose that a π-process P_0 encodes an MA-process A_0, and it can be reduced to a π-process P. Then either P encodes an MA-process A_n, or P is in an "intermediate" form and it can be further reduced to a π-process P_n which encodes A_n. In the latter case, the reduction of P to P_n is a composition of

- N non-deterministic reduction steps $P \hookrightarrow_\pi^N P_n'$, where $0 \le N \le 2$; these steps complete, if necessary, a non-deterministic selection of nodes representing an action and ambients in an MA-process (see Subsection 3.2.5.(13)), and

- a finite number of deterministic reduction steps $P'_n \mapsto^*_\pi P_n$ corresponding to the interaction between the *Ruler* and the selected process *Node* (see Subsection 3.2.5.(14)-(23)).

When reduction of P to P_n is completed, the whole chain of π-calculus reductions $P_0 \rightarrow^*_\pi P \rightarrow^*_\pi P_n$ becomes a step-by-step simulation of some mobile ambient computation $A_0 \rightarrow^*_a A_n$.

The proof of this theorem is by induction on the number of non-deterministic steps $\hookrightarrow_\pi$ in a reduction of P_0. Each triple of non-deterministic steps in such a reduction is followed by a chain of deterministic reduction steps that either simulate the execution of some MA-reduction step if the selected nodes in a π-process P_i comply with one of the MA-reduction rules, or restore P_i otherwise.

We may note that our translation has diverging reductions whenever the selected nodes do not conform to any MA-reduction rule. In this case we may obtain an infinite chain $P \hookrightarrow^3_\pi P' \mapsto^*_\pi P \hookrightarrow^3_\pi P' \mapsto^*_\pi P \hookrightarrow^3_\pi \ldots$

Proposition 3.5. *If A is an MA-process and P is a π-process such that $A \models P$, then we can see that P is a localized π-calculus term.*

3.2.8 Further Extensions

The encoding makes it possible to analyze some properties of mobile ambients by means of static analysis and the congruence-checking machinery developed for the π-calculus. The fact that we restrict ourselves to localized and sum-free π-terms substantially alleviates the analysis. Moreover, our encoding does not involve any sophisticated structures that can affect the precision of such analysis.

The encoding can be extended to a full ambient calculus, adding a communication channel per ambient. This implies a "merging" of channels when an ambient is opened; we may use the same mechanism: the *Ruler* randomly selects an input and an output, checks if they belong to the same ambient, and performs communication. On the other hand, it is worth noticing that our encoding is slightly more general in the sense that the target language is even simpler. We can use the asynchronous π-calculus, which is even simpler than the synchronous π-calculus, by using the standard encoding [95].

It can readily be imagined that the encoding could be made more advanced by combining every π-process *Node* corresponding to an ambient with its own interpreter *Ruler*. In this case a distributed interaction between the components of a system can be achieved, and the encoding becomes compositional. To study the behavioural properties of mobile ambients that are preserved by such an advanced encoding would be our next step. The ultimate aim is to build a fully abstract translation which preserves behavioural equalities between processes, such as reduction barbed congruence [117, 144]. If such a translation should be obtained, it provides a sound basis for a uniform framework of the theory of mobile computations.

3.3 Safe Mobile Ambients into Mutual Mobile Membranes

Membrane systems [127, 128] and mobile ambients [42] have similar structures and common concepts. Both have a hierarchical structure representing locations, and are used to model various aspects of biological systems. Mobile ambients are suitable to represent the movement of ambients through ambients and the communication which takes place inside the boundaries of ambients. Membrane systems are suitable to represent the movement of objects and membranes through membranes. We consider these new computing models used in describing various biological phenomena [40, 65], and encode the ambients into membrane systems [12, 19]. We present such an encoding, and use it to describe the sodium-potassium exchange pump [14]. We provide an operational correspondence between safe ambients and their encodings, as well as various related properties of membrane systems [14].

3.3.1 Main Idea

In what follows we describe a relationship between ambients and membrane systems. This relationship is mainly provided by an encoding of safe mobile ambients into systems of mutual mobile membranes. We use the following translation steps:

- every safe process $\mathbf{0}$ is replaced by the empty multiset λ;
- every ambient n is translated into a membrane labelled by n;
- every capability $cap\ n$ is translated both into an object "$cap\ n$" and into a membrane labelled by "$cap\ n$", both placed in the same region;
- every path of capabilities is translated into a nested structure of membranes (e.g., $in\ m.\ \overline{out}\ n$ is translated into $in\ m\ [\ \overline{out}\ n\ [\ ]_{\overline{out}\ n}\ ]_{in\ m}$);

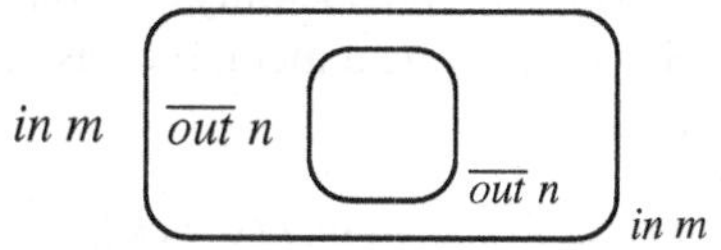

- an object $dlock$ is placed near the membrane structure after all the translation is done; the additional object $dlock$ prevents the consumption of capability objects in a membrane system which corresponds to a mobile ambient from the set $\mathscr{D}_{amb}$.

A feature of pure safe ambients is that they have a spatial tree-like structure. The nodes in this structure are represented by ambients and capabilities. When translating a pure safe ambient into a membrane system we obtain the same tree structure by means of membranes: every node is a membrane having the same label as the corresponding ambient or capability. Let us consider the following mobile ambient

$$n[\ in\ m \mid t[\ ]\ ] \mid m[\ \overline{in}\ m\ ].$$

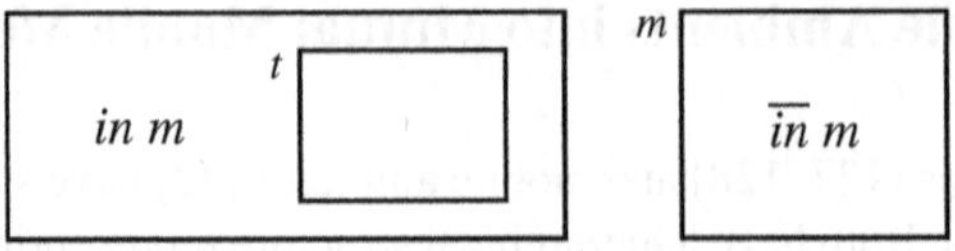

Translating it into a mobile membrane system, we obtain

$$dlock\ [\ in\ m\ [\ ]_{in\ m}\ [\ ]_t\ ]_n\ [\ \overline{in}\ m\ [\ ]_{\overline{in}\ m}]_m.$$

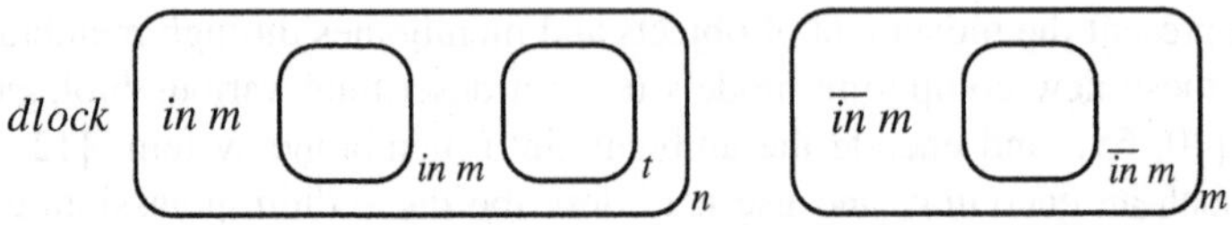

Remark 3.1. Whenever we encode a path of capabilities, we wish to preserve the order in which the capabilities are consumed. This order is preserved by the translation given above, even it requires lots of resources. Another solution is to encode every capability only into an object, and to preserve the order of the objects by adding extra objects and rules into the system. This can be done by introducing objects able to enchain a certain sequence of rules: for instance, if we have *in n. in m*... then in the corresponding membrane system we have the rules:

$$in\ n \to in\ n\ x, \quad in\ n\ x \to in\ m\ y, \quad \ldots$$

Cardelli and Gordon in [42] use the following structure

$$p[\ succ[\ open\ op\ ]\ ]\ \mid\ open\ q.open\ p.P\ \mid\ op[\ in\ succ.in\ p.in\ succ.$$
$$(q[\ out\ succ.out\ succ.out\ p\ ]\ \mid\ open\ op)\ ]$$

Starting from such a structure, we look for a translation such that the capabilities' order is preserved. For every consumed pair of capabilities in safe ambients, there is a change in the ambient structure. We simulate this with the help of some special developmental rules in membrane systems. An object *one* is used to ensure that no more than one pair (capability, co-capability) is consumed at every tick of the universal clock. Using rules for moving a membrane as in [104, 108], we define the following developmental rules:

a) $[\ in\ m\ dlock\ one\ ]_n\ [\ \overline{in}\ m\ ]_m \to [\ in^*m\ in_*m\ dlock\ ]_n\ [\ \overline{in}_*m\ ]_m$
If a membrane n (containing the objects *in m*, *dlock*, *one*) and a membrane m (containing an object $\overline{in}\ m$) are placed in the same region, then the objects *in m* and *one* are replaced by the objects in_*m and in^*m; the object $\overline{in}\ m$ is replaced by the object $\overline{in}_*m$. The object in^*m is used to control the process of introducing membrane n into membrane m, and the objects in_*m, $\overline{in}_*m$ are used to dissolve the membranes *in m* and $\overline{in}\ m$.

b) $cap_*m\ [\ ]_{cap\ m} \to [\ \delta\ ]_{cap\ m}$
If an object cap_*m and the membrane labelled by *cap m* are placed in the same region, then the object cap_*m is consumed and the membrane labelled by *cap m* is dissolved (this is denoted by the symbol δ). This rule simulates the consumption of a capability *cap m* in ambients.

c) $[\,in^{*}m\,]_n\,[\,]_m \rightarrow [\,[\,]_n\,]_m \;|_{[\,\neg cap^{*}\,]_m}$
If an elementary membrane n (containing an object $in^{*}m$) and a membrane labelled by m (which does not contain star objects – this is denoted by $|_{[\,\neg cap^{*}\,]_m}$) are placed in the same membrane, then the membrane n enters the membrane labelled by m and the object $in^{*}m$ is consumed in this process. The | operator is used to denote the fact that the rule can be applied only if the conditions on the right hand side are initially satisfied.

d) $[\,\overline{out}\;m\,[\,out\;m\;dlock\;one\,]_n\,]_m \rightarrow [\,\overline{out}_{*}m\,[\,out^{*}m\;out_{*}m\;dlock\,]_n\,]_m$
If a membrane m contains both an object $\overline{out}\;m$ and a membrane n (having the objects $out\;m$, $dlock$, one), then the objects $out\;m$ and one are replaced by $out_{*}m$, $out^{*}m$; moreover, $\overline{out}\;m$ is replaced by the object $\overline{out}_{*}m$. The object $out^{*}m$ is used to control the process of extracting membrane n from membrane m, and the objects $out_{*}m$, $\overline{out}_{*}m$ are used to dissolve the membranes $out\;m$ and $\overline{out}\;m$, respectively.

e) $[\,[\,out^{*}m\,]_n\,]_m \rightarrow [\,]_n\,[\,]_m$
If a membrane m contains an elementary membrane n which has an object $out^{*}m$, then membrane n is extracted from the membrane labelled by m, and object $out^{*}m$ is consumed in this process.

f) $[\overline{open}\;m]_m\;open\;m\;dlock\;one \rightarrow [\,\delta\,]_m open_{*}m\;\overline{open}_{*}m\;dlock$
If a membrane m and the objects $open\;m$, $dlock$, one are placed inside the same region, then membrane m is dissolved, the object $\overline{open}\;m$ is consumed, and the objects $open\;m$ and one are replaced by the objects $open_{*}m$ and $\overline{open}_{*}m$.

g) $[\,U^{*}\,[\,]_t\,]_n \rightarrow [\,U^{*}\,[out^{*}n\;in^{*}n\;U^{*}]_t\,]_n \;|_{[\,\neg cap^{*}\,]_t}$
We denote by U^{*} an arbitrary non-empty multiset of star objects placed in membrane n. If a membrane n contains a multiset of star objects U^{*} and a membrane t which does not contain star objects (this is denoted by $|_{[\,\neg cap^{*}\,]_t}$), then a copy of set U^{*} and two new objects $in^{*}n$ and $out^{*}n$ are created inside membrane t. The existence of a multiset U^{*} of star objects indicates that membrane n can be used by rules c), e) to enter/exit into/from another membrane. In order to move, membrane n must be elementary; to accomplish this, the objects $out^{*}n$, $in^{*}n$ and a copy of the multiset U^{*} are created inside membrane t such that membrane t can be extracted. After membrane n completes its movement (this is denoted by the fact that the membrane labelled by n does not contain star objects), membrane t is introduced back into membrane n.

h) $dlock\,[\,]_n \rightarrow dlock\,[\,dlock\,]_n \mid_{\neg n\,[\,\neg dlock\,]_n}$
If an object $dlock$ and a membrane n (which does not already contain an object $dlock$, i.e., $[\,\neg dlock\,]_n$) are placed inside the same region, and there is no object n placed in that region (denoted by $\neg n$), then a new object $dlock$ is placed inside membrane n. This rule specifies the fact that object $dlock$ can only pass through membranes corresponding to translated ambients; this makes impossible the consumption of capability objects from the translated structures from $\mathscr{D}_{amb}$.

i) $[\,dlock\,]_n \rightarrow [\,]_n$
The object $dlock$ created by an application of rule h and located inside membrane n is removed.

j) $[\,dlock\,]_n \rightarrow [\,dlock\;one\,]_n$

If a membrane n contains an object *dlock*, then an additional object *one* is created in membrane n.

k) $one \to [\,\delta\,]$
An object *one* is consumed; the last two rules ensure that at most one object *one* exists in the membrane system at any moment.

Remark 3.2. Whenever we get the membrane system

$$dlock\ [\ in\ m\ [\ ]_{in\ m}\ \overline{out}\ n\ [\ ]_{\overline{out}\ n}\ [\ out\ n\ [\ ]_{out\ n}\ ]_t\ ]_n\ [\ \overline{in}\ m\ [\ ]_{\overline{in}\ m}\ ]_m$$

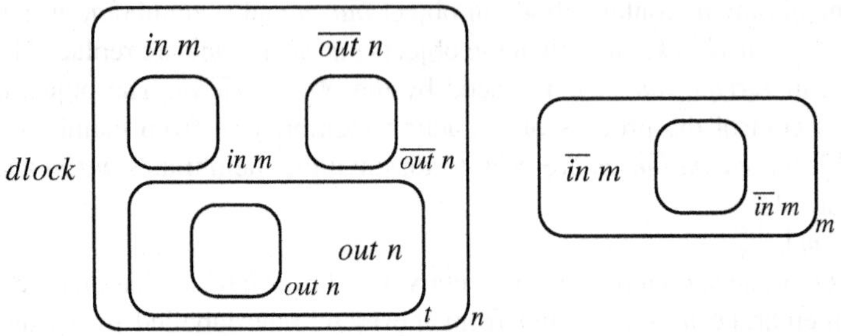

after applying in a maximally parallel manner the developmental rules from the set defined above, we obtain either the membrane system $[\ [\ ]_n\ [\ ]_t\ ]_m$ or $[\ ]_t\ [\ [\ ]_n\ ]_m$. The order in which the pairs of objects corresponding to translated capabilities are consumed in the membrane encoding should be the same as the order in which the pairs of capabilities are consumed in the encoded ambient. However in the example above this order cannot be established; the non-star objects can be consumed by two rules applied in parallel. For this reason we have imposed the following priorities between the developmental rules defined above:

b) > c), e), g) > a), d), f) > k) > h), i), j)

According to these priorities, the membrane system

$$dlock\ [\ in\ m\ [\ ]_{in\ m}\ \overline{out}\ n\ [\ ]_{\overline{out}\ n}\ [\ out\ n\ [\ ]_{out\ n}\ ]_t\ ]_n\ [\ \overline{in}\ m\ [\ ]_{\overline{in}\ m}\ ]_m$$

evolves only to the membrane system $[\ [\ ]_n\ [\ ]_t\ ]_m$ if the objects *in m* and $\overline{in}\ m$ are consumed before the objects *out n* and $\overline{out}\ n$ by a rule from the set given above. The applied rules are the ones defined in what follows. After each evolution step we represent graphically the membrane system obtained.

r_1 : $dlock\ [\ ]_n \to dlock\ [\ dlock\ ]_n$ — type k) — a copy of the object *dlock* is created inside the membrane n, which does not contain a *dlock* object;

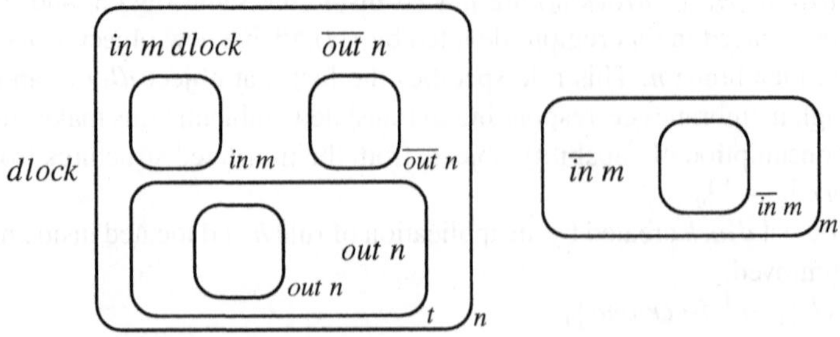

r_2 : $dlock\,[\ ]_t \rightarrow dlock\,[\,dlock\,]_t$ — type k) — a copy of the object *dlock* is created inside the membrane *t*, which does not contain a *dlock* object;

r_3 : $[\,dlock\,]_n \rightarrow [\,dlock\ one\,]_n$ — type j) — an object *one* is created in the membrane *n* which contains a *dlock* object;

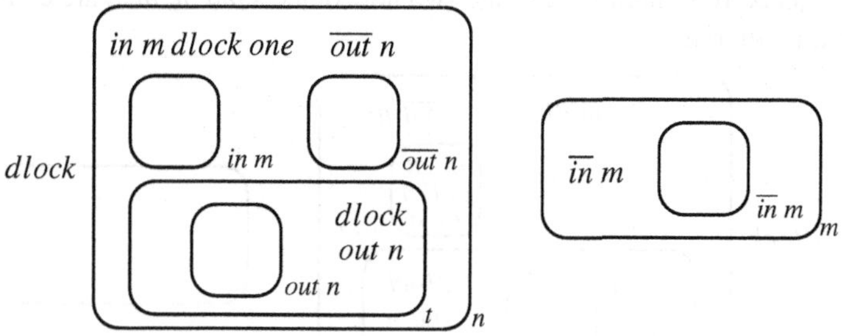

r_4 : $[\,in\ m\ dlock\ one\,]_n\,[\,\overline{in}\ m\,]_m \rightarrow [\,in^{*}m\ in_{*}m\ dlock\,]_n\,[\,\overline{in}_{*}m\,]_m$ — type a) — the objects *in m*, *one* in membrane *n* are replaced by the objects $in_{*}m$, $in^{*}m$, and the object $\overline{in}\ m$ in membrane *m* is replaced by the object $\overline{in}_{*}m$;

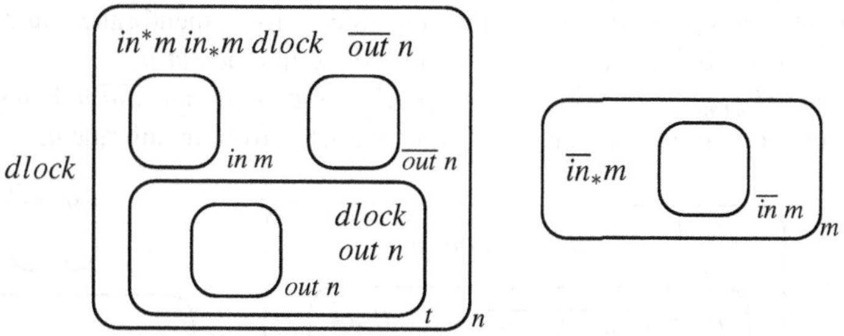

r_5 : $in_{*}m\,[\ ]_{in\ m} \rightarrow [\,\delta\,]_{in\ m}$ — type b) — in the presence of an object $in_{*}m$ the membrane labelled by *in m* is dissolved; the object $in_{*}m$ signals the fact that the object *in m* has been consumed;

r_6 : $\overline{in}_{*}m\,[\ ]_{\overline{in}\ m} \rightarrow [\,\delta\,]_{\overline{in}\ m}$ — type b) — in the presence of an object $\overline{in}_{*}m$ the membrane labelled by $\overline{in}\ m$ is dissolved; the object $\overline{in}_{*}m$ signals the fact that the object $\overline{in}\ m$ has been consumed;

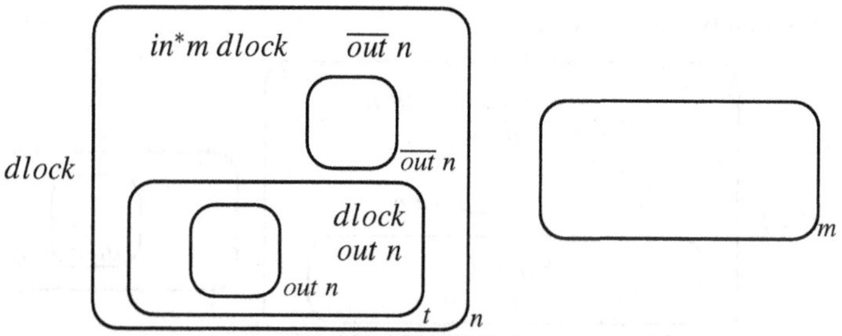

r_7 : $[\,in^{*}m\,[\ ]_t\,]_n \rightarrow [\,in^{*}m\,[\,in^{*}m\ out^{*}n\ in^{*}n\,]_t\,]_n$ — type g) — in the presence of star objects in membrane *n* (which is not an elementary one), a copy of all the star objects from membrane n and two new objects out^*n, in^*n are created in the nested membrane t;

r_8 : $[\, in^*m\, [\,]_{\overline{out}\ n}\,]_n \rightarrow [\, in^*m\, [\, in^*m\ out^*n\ in^*n\,]_{\overline{out}\ n}\,]_n$ — type g) — in the presence of star objects in membrane n (which is not an elementary one), a copy of all the star objects from membrane n and two new objects out^*n, in^*n are created in the nested membrane $\overline{out}\ n$;

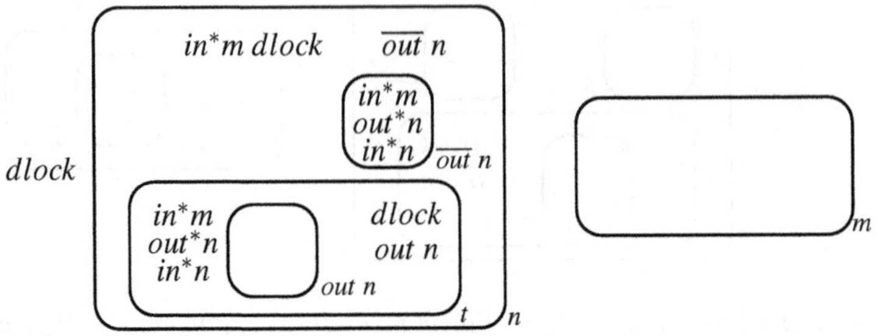

r_9 : $[\, in^*m\ out^*n\ in^*n\, [\,]_{out\ n}\,]_t \rightarrow [\, in^*m\ out^*n\ in^*n\, [\, in^*m\ out^*n\ in^*n\ in^*t\ out^*t\,]_{out\ n}\,]_t$ — type g) — in the presence of star objects in membrane t (which is not an elementary one), a copy of all the star objects from membrane t and two new objects out^*t, in^*t are created in the nested membrane $out\ n$;

r_{10} : $[\,[\, out^*n\,]_{\overline{out}\ n}\,]_n \rightarrow [\,]_n\, [\,]_{\overline{out}\ n}$ — type e) — the membrane $\overline{out}\ n$, being elementary and containing the object out^*n, is extracted from membrane n;

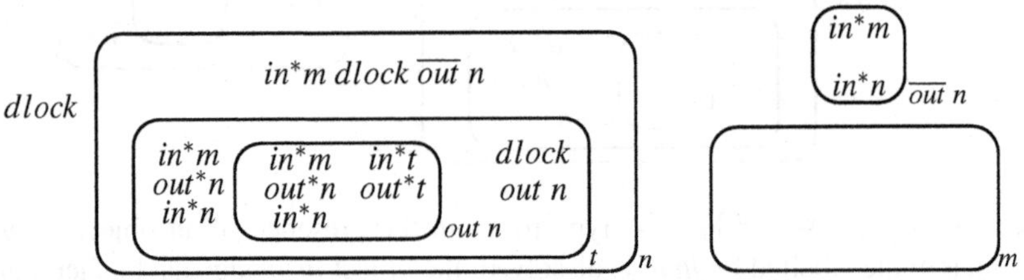

r_{11} : $[\,[\, out^*t\,]_{out\ n}\,]_t \rightarrow [\,]_t\, [\,]_{out\ n}$ — type e) — the membrane $out\ n$, being elementary and containing the object out^*t, is extracted from membrane t;

r_{12} : $[\, in^*m\,]_{\overline{out}\ n}\, [\,]_m \rightarrow [\,[\,]_{\overline{out}\ n}\,]_m$ — type c) — the membrane $\overline{out}\ n$, being elementary and containing an object in^*m, is introduced into membrane m which does not contain any star objects;

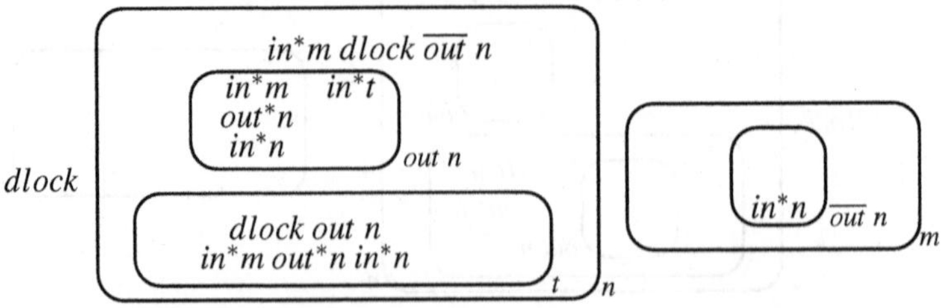

r_{13} : $[\,[\, out^*n\,]_{out\ n}\,]_n \rightarrow [\,]_n\, [\,]_{out\ n}$ — type e) — the membrane $out\ n$, being elementary and containing an object out^*n, is extracted from membrane n;

r_{14} : $[\,[\,out^*n\,]_t\,]_n \rightarrow [\,]_n\,[\,]_t$ — type e) — the membrane t, being elementary and containing an object out^*n, is extracted from membrane n;

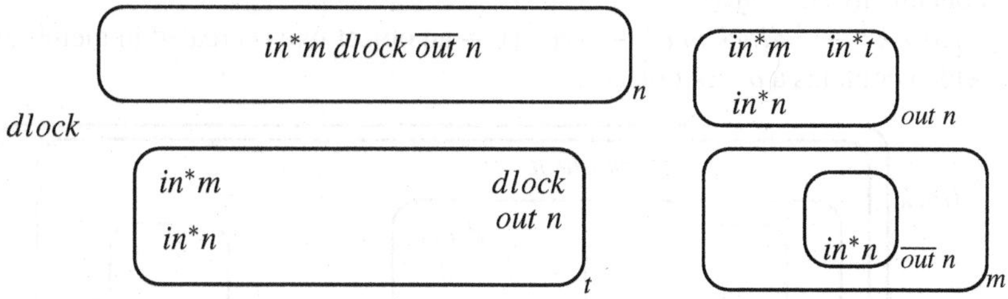

r_{15} : $[\,in^*m\,]_n[\,]_m \rightarrow [\,[\,]_n\,]_m$ — type c) — the membrane n, being elementary and containing an object in^*m, is introduced into membrane m which does not contain any star objects;

r_{16} : $[\,in^*m\,]_t\,[\,]_m \rightarrow [\,[\,]_t\,]_m$ — type c) — the membrane t, being elementary and containing an object in^*m, is introduced into membrane m which does not contain any star objects;

r_{17} : $[\,in^*m\,]_{out\ n}\,[\,]_m \rightarrow [\,[\,]_{out\ n}\,]_m$ — type c) — the membrane *out n*, being elementary and containing an object in^*m, is introduced into membrane m which does not contain any star objects;

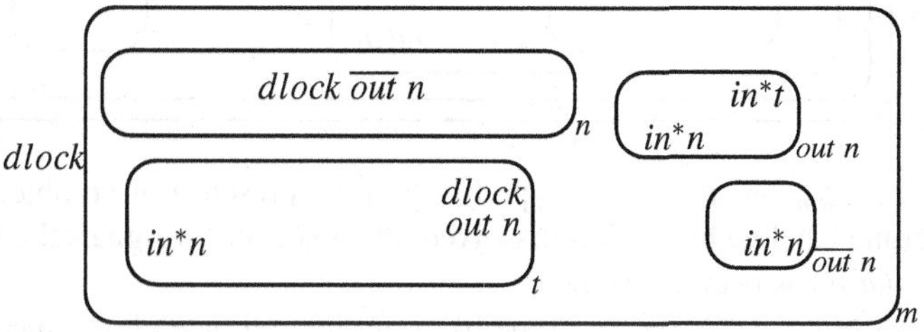

r_{18} : $[\,in^*n\,]_t\,[\,]_n \rightarrow [\,[\,]_t\,]_n$ — type c) — the membrane t, being elementary and containing an object in^*n, is introduced into membrane n which does not contain any star objects;

r_{19} : $[\,in^*n\,]_{out\ n}\,[\,]_n \rightarrow [\,[\,]_{out\ n}\,]_n$ — type c) — the membrane *out n*, being elementary and containing an object in^*n, is introduced into membrane n which does not contain any star objects;

r_{20} : $[\,in^*n\,]_{\overline{out}\ n}\,[\,]_n \rightarrow [\,[\,]_{\overline{out}\ n}\,]_n$ — type c) — the membrane $\overline{out}\ n$, being elementary and containing an object in^*n, is introduced into membrane n which does not contain any star objects;

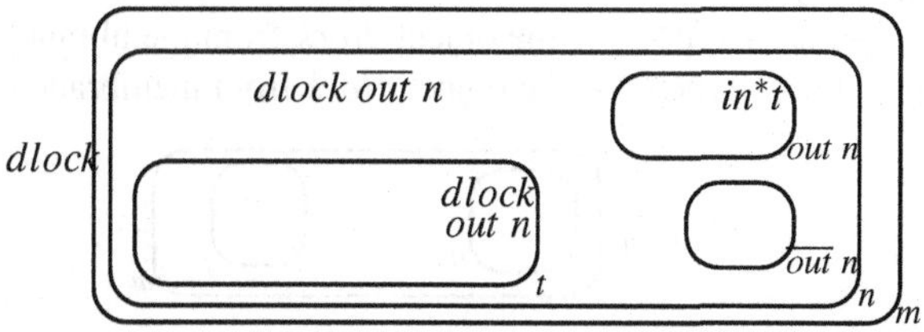

$r_{21} : [\, in^{*}t \,]_{out\ n} [\,]_t \rightarrow [\, [\,]_{out\ n} \,]_t$ — type c) — the membrane *out n*, being elementary and containing an object $in^{*}t$, is introduced into membrane t which does not contain any star objects;

$r_{22} : [\, dlock \,]_t \rightarrow [\, dlock\ one \,]_t$ — type j) — an object *one* is created in membrane t which contains a *dlock* object;

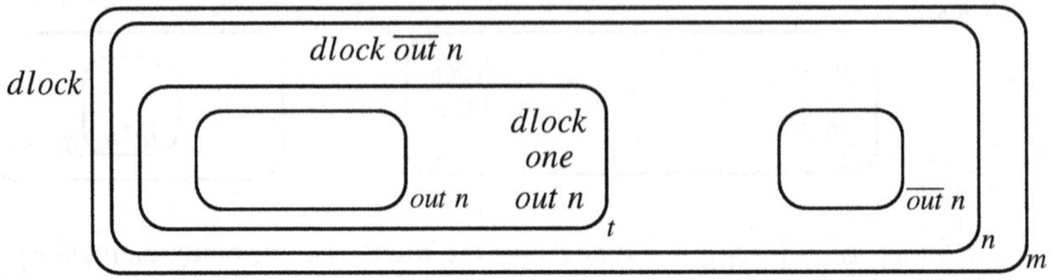

$r_{23} : [\, \overline{out}\ n\ [\, out\ n\ dlock\ one \,]_t \,]_n \rightarrow [\, \overline{out}_{*}n\ [\, out^{*}n\ out_{*}n\ dlock \,]_t \,]_n$ — type d) — objects *one*, *out n* are replaced in membrane t by objects $out^{*}n$, $out_{*}n$, and object $\overline{out}\ n$ in membrane n is replaced by object $\overline{out}_{*}n$;

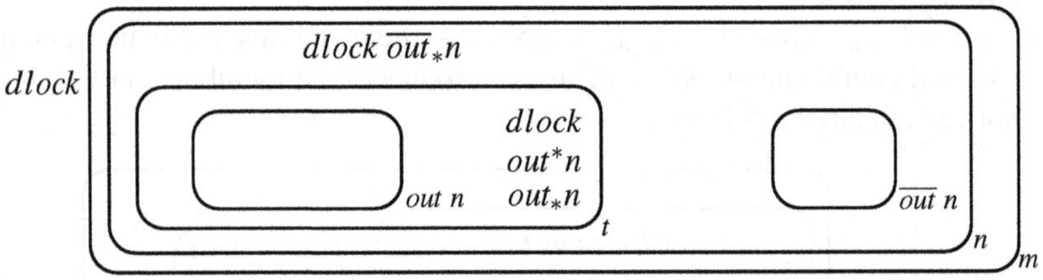

$r_{24} : out_{*}n\ [\,]_{out\ n} \rightarrow [\, \delta \,]_{out\ n}$ — type b) — in the presence of an object $out_{*}n$ the membrane labelled by *out n* is dissolved; the object $out_{*}n$ signals the fact that the object *out n* has been consumed;

$r_{25} : \overline{out}_{*}n\ [\,]_{\overline{out}\ n} \rightarrow [\, \delta \,]_{\overline{out}\ n}$ — type b) — in the presence of an object $\overline{out}_{*}n$ the membrane labelled by $\overline{out}\ n$ is dissolved; the object $\overline{out}_{*}n$ signals the fact that the object $\overline{out}\ n$ has been consumed;

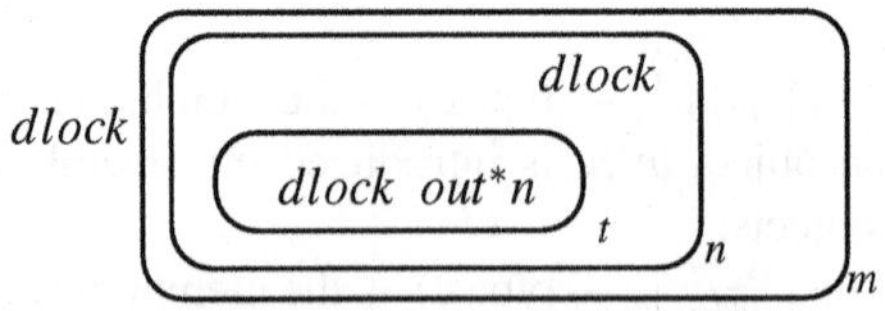

$r_{26} : [\, [\, out^{*}n \,]_t \,]_n \rightarrow [\,]_t [\,]_n$ — type e) — the membrane t, being elementary and containing an object $out^{*}n$, is extracted from membrane n;

$r_{27} : [\, dlock \,]_n \rightarrow [\,]_n$ — type i) — the object *dlock* from membrane n is consumed;

$r_{28} : [\, dlock \,]_t \rightarrow [\,]_t$ — type i) — the object *dlock* from membrane t is consumed;

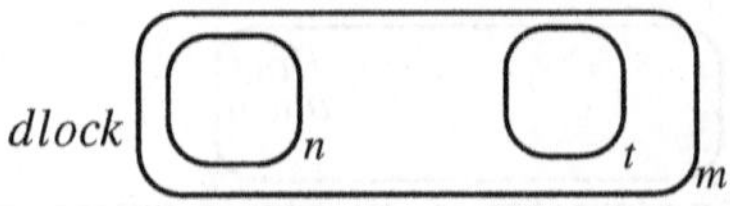

The rules are applied in the order presented above, with the additional remark that the tuples of rules (r_5,r_6), (r_7,r_8), (r_9,r_{10}), (r_{11},r_{12}), (r_{13},r_{14}), (r_{15},r_{16},r_{17}), (r_{18},r_{19},r_{20}), (r_{21},r_{22}), (r_{24},r_{25}), (r_{26},r_{27},r_{28}) can be applied in any order or in parallel. The computation stops when, after introducing all the possible objects $dlock$ by applying rules of form $h)$, none of the sequences of rules $j),a)$ or $j),d)$ or $j),f)$ can be applied.

Remark 3.3. At a certain moment, the membrane $out\ n$ of the example above contains the objects in^*m, out^*n, in^*n, in^*t, out^*t. In order to avoid an unexpected return of membrane $out\ n$ into membrane t, we restrict the use of the object in^*t by imposing the lack of star objects in the target membrane. This is enough to distinguish membrane t (having star objects) from membrane m which has no such star objects. It is worth noting that the membrane system

$$dlock\ [\, in\ m\ [\,]_{in\ m}\ \overline{out}\ n\ [\,]_{\overline{out}\ n}\ [\, out\ n\ [\,]_{out\ n}\ ]_t\ ]_n\ [\, \overline{in}\ m\ [\,]_{\overline{in}\ m}\ ]_m$$

evolves to the membrane system $[\,]_t\ [\,[\,]_n\ ]_m$ if the objects $out\ n$ and $\overline{out}\ n$ are consumed before the objects $in\ m$ and $\overline{in}\ m$.

3.3.2 Translation

We denote membrane systems by M, M', M_i, N, N', and the labels of the membranes by $n, m, \ldots$.

Definition 3.9. The set $\mathscr{M}$ of membrane configurations M is defined by

$$M ::= \lambda \mid O \mid [\, M\,]_n \mid (\nu n)M \mid M_1\ M_2$$

where by O we denote a finite multiset of objects.

We can write O or $M_1\ M_2$ omitting the surrounding membrane, because all the membrane structures are placed inside a *skin* membrane. Similar to the restriction operator presented in [44], we consider in [13] an operator $(\nu n)M$ for the restriction of a name n to a membrane configuration M.

In order to give a formal encoding of pure safe ambients into simple mobile membranes, we define the following function:

Definition 3.10. A translation $\mathscr{T} : \mathscr{A} \to \mathscr{M}$ is given by $\mathscr{T}(A) = dlock\ \mathscr{T}_1(A)$, where $\mathscr{T}_1 : \mathscr{A} \to \mathscr{M}$ is

$$\mathscr{T}_1(A) = \begin{cases} cap\ n[\,]_{cap\ n} & \text{if } A = cap\ n \\ cap\ n[\ \mathscr{T}_1(A_1)\ ]_{cap\ n} & \text{if } A = cap\ n.A_1 \\ [\ \mathscr{T}_1(A_1)\ ]_n & \text{if } A = n[\ A_1\ ] \\ [\,]_n & \text{if } A = n[\,] \\ (\nu n)\mathscr{T}_1(A_1) & \text{if } A = (\nu n)A_1 \\ \mathscr{T}_1(A_1)\ \mathscr{T}_1(A_2) & \text{if } A = A_1 \mid A_2 \\ \lambda & \text{if } A = 0 \end{cases}$$

A membrane structure can be represented in a natural way as a Venn diagram. This makes clear the fact that the order of membrane structures and objects placed in the same region in a large membrane structure is irrelevant; what matters is the relationships between membranes and objects. A rule of the form $a\,[\,b\,]_n \to b\,[\,a\,]_n$ has the same meaning as any of the rules $[\,b\,]_n\, a \to [\,a\,]_n\, b,\ \ a\,[\,b\,]_n \to [\,a\,]_n\, b$, and $[\,b\,]_n\, a \to b\,[\,a\,]_n$. Inspired by [40], we formally define the notion of structural congruence which clarifies this aspect, and reduces the number of rules written for a membrane system.

3.3.3 Properties Preserved Through Translation

Definition 3.11. The structural congruence $\equiv_{mem}$ over $\mathscr{M}$ is the smallest congruence relation satisfying:

$M\,N \equiv_{mem} N\,M \quad M\,(N\,M') \equiv_{mem} (M\,N)\,M'$;
$M\,\lambda \equiv_{mem} M \quad (\nu n)(\nu m)M \equiv_{mem} (\nu m)(\nu n)M$;
$(\nu m)M \equiv_{mem} (\nu n)M\{n/m\}$, where n is not a membrane name in M;
$(\nu n)(N\,M) \equiv_{mem} M\,(\nu n)N$ where n is not a membrane name in M;
$n \neq m$ implies $(\nu n)[\,M\,]_m \equiv_{mem} [\,(\nu n)M\,]_m$.

The restriction operator can float outward to extend the scope of a membrane name, and can float inward to restrict the scope of a membrane name.

We deal with multisets of objects, and multisets of membranes. For example, we have $[\,]_n\,[\,]_m \equiv_{mem} [\,]_m\,[\,]_n$, $in\ m\,[\,]_n \equiv_{mem} [\,]_n\ in\ m$ and $in^*n\ out^*m \equiv_{mem} out^*m\ in^*n$.

Proposition 3.6. *The structural congruence has the following properties:*

$M \equiv_{mem} M$;
$M \equiv_{mem} N$ *implies* $N \equiv_{mem} M$;
$M \equiv_{mem} N$ *and* $N \equiv_{mem} M'$ *implies* $M \equiv_{mem} M'$;
$M \equiv_{mem} N$ *implies* $M\,M' \equiv_{mem} N\,M'$;
$M \equiv_{mem} N$ *implies* $M'\,M \equiv_{mem} M'\,N$;
$M \equiv_{mem} N$ *implies* $[\,M\,]_n \equiv_{mem} [\,N\,]_n$;
$M \equiv_{mem} N$ *implies* $(\nu n)M \equiv_{mem} (\nu n)N$.

Proposition 3.7. *Structurally congruent ambients are translated into structurally congruent membrane systems; moreover, structurally congruent translated membrane systems correspond to structurally congruent ambients:*

$$A \equiv_{amb} B \text{ iff } \mathscr{T}(A) \equiv_{mem} \mathscr{T}(B).$$

$$\begin{array}{ccc} A & \cdots \equiv_{amb} \cdots & B \\ \downarrow \mathscr{T} & & \downarrow \mathscr{T} \\ \mathscr{T}(A) & \equiv_{mem} & \mathscr{T}(B) \end{array} \quad \text{and} \quad \begin{array}{ccc} A & \equiv_{amb} & B \\ \downarrow \mathscr{T} & & \downarrow \mathscr{T} \\ \mathscr{T}(A) & \cdots \equiv_{mem} \cdots & \mathscr{T}(B) \end{array}$$

Proof. We prove that if $A \equiv_{amb} B$ then $\mathscr{T}(A) \equiv_{mem} \mathscr{T}(B)$.

If $A = A_1 \mid A_2$ and $A \equiv_{amb} B$, then $B = A_2 \mid A_1$. Through $\mathscr{T}$ are obtained $\mathscr{T}(A) = dlock\ \mathscr{T}_1(A_1)\ \mathscr{T}_1(A_2)$ and $\mathscr{T}(B) = dlock\ \mathscr{T}(A_2)\ \mathscr{T}(A_1)$. From the definition of $\equiv_{mem}$ it results that $\mathscr{T}(A) \equiv_{mem} \mathscr{T}(B)$.

$$\begin{array}{ccc} A_1 \mid A_2 & \equiv_{amb} & A_2 \mid A_1 \\ \downarrow \mathscr{T} & & \downarrow \mathscr{T} \\ dlock\ \mathscr{T}_1(A_1)\ \mathscr{T}_1(A_2) & \cdot\cdot \equiv_{mem} \cdot\cdot & dlock\ \mathscr{T}_1(A_2)\ \mathscr{T}_1(A_1) \end{array}$$

If $A = A_1 \mid (A_2 \mid A_3)$ and $A \equiv_{amb} B$, then $B = (A_1 \mid A_2) \mid A_3$. Through $\mathscr{T}$ are obtained $\mathscr{T}(A) = dlock\ \mathscr{T}_1(A_1)\ (\mathscr{T}_1(A_2)\ \mathscr{T}_1(A_3))$ and $\mathscr{T}(B) = dlock(\mathscr{T}_1(A_1)\ \mathscr{T}_1(A_2))$ $\mathscr{T}_1(A_3)$. From the definition of $\equiv_{mem}$ it results that $\mathscr{T}(A) \equiv_{mem} \mathscr{T}(B)$.

$$\begin{array}{ccc} A_1 \mid (A_2 \mid A_3) & \equiv_{amb} & (A_1 \mid A_2) \mid A_3 \\ \downarrow \mathscr{T} & & \downarrow \mathscr{T} \\ dlock\ \mathscr{T}_1(A_1)\ (\mathscr{T}_1(A_2)\ \mathscr{T}_1(A_3)) & \cdot\cdot \equiv_{mem} \cdot\cdot & dlock(\mathscr{T}_1(A_1)\ \mathscr{T}_1(A_2))\ \mathscr{T}_1(A_3) \end{array}$$

If $A = A_1 \mid \mathbf{0}$ and $A \equiv_{amb} B$, then $B = A_1$. Through $\mathscr{T}$ are obtained $\mathscr{T}(A) = dlock\ \mathscr{T}_1(A_1)\ \lambda$ and $\mathscr{T}(B) = dlock\ \mathscr{T}_1(A_1)$. From the definition of $\equiv_{mem}$ it results that $\mathscr{T}(A) \equiv_{mem} \mathscr{T}(B)$.

$$\begin{array}{ccc} A_1 \mid \mathbf{0} & \equiv_{amb} & A_1 \\ \downarrow \mathscr{T} & & \downarrow \mathscr{T} \\ dlock\ \mathscr{T}_1(A_1)\ \lambda & \cdot\cdot \equiv_{mem} \cdot\cdot & dlock\ \mathscr{T}_1(A_1) \end{array}$$

If $A = (\nu n)(\nu m)A_1$ and $A \equiv_{amb} B$, then $B = (\nu m)(\nu n)A_1$. Through $\mathscr{T}$ are obtained $\mathscr{T}(A) = dlock(\nu n)(\nu m)\mathscr{T}_1(A_1)$ and $\mathscr{T}(B) = dlock(\nu m)(\nu n)\ \mathscr{T}_1(A_1)$. From the definition of $\equiv_{mem}$ it results that $\mathscr{T}(A) \equiv_{mem} \mathscr{T}(B)$.

$$\begin{array}{ccc} (\nu n)(\nu m)A_1 & \equiv_{amb} & (\nu m)(\nu n)A_1 \\ \downarrow \mathscr{T} & & \downarrow \mathscr{T} \\ dlock(\nu n)(\nu m)\mathscr{T}_1(A_1) & \cdot\cdot \equiv_{mem} \cdot\cdot & dlock(\nu m)(\nu n)\mathscr{T}_1(A_1) \end{array}$$

If $A = (\nu n)A_1$, $n \notin fn(A_1)$ and $A \equiv_{amb} B$, then $B = (\nu m)A_1\{m/n\}$. Through $\mathscr{T}$ are obtained $\mathscr{T}(A) = dlock(\nu n)\mathscr{T}_1(A_1)$ and $\mathscr{T}(B) = dlock(\nu m)\mathscr{T}_1(A_1)\{m/n\}$. From the definition of $\equiv_{mem}$ it results that $\mathscr{T}(A) \equiv_{mem} \mathscr{T}(B)$.

$$\begin{array}{ccc}
(\nu n)A_1 & \equiv_{amb} & (\nu m)A_1\{m/n\} \\
\Big\downarrow \mathscr{T} & & \Big\downarrow \mathscr{T} \\
dlock(\nu n)\mathscr{T}_1(A_1) & \cdot\cdot \equiv_{mem} \cdot\cdot & dlock(\nu m)\mathscr{T}_1(A_1)\{m/n\}
\end{array}$$

If $A = (\nu n)(A_1 \mid A_2)$, $n \notin fn(A_1)$ and $A \equiv_{amb} B$, then $B = A_1 \mid (\nu n)A_2$. Through $\mathscr{T}$ are obtained $\mathscr{T}(A) = dlock(\nu n)\mathscr{T}_1(A_1)\mathscr{T}_1(A_2)$ and $\mathscr{T}(B) = dlock\mathscr{T}_1(A_1)(\nu n)\mathscr{T}_1(A_2)$. From the definition of $\equiv_{mem}$ it results that $\mathscr{T}(A) \equiv_{mem} \mathscr{T}(B)$.

$$\begin{array}{ccc}
(\nu n)(A_1 \mid A_2) & \equiv_{amb} & A_1 \mid (\nu n)A_2 \\
\Big\downarrow \mathscr{T} & & \Big\downarrow \mathscr{T} \\
dlock(\nu n)\ \mathscr{T}_1(A_1) & \cdot\cdot \equiv_{mem} \cdot\cdot & dlock\mathscr{T}_1(A_1)(\nu n)\mathscr{T}_1(A_2)
\end{array}$$

If $A = (\nu n)m[A_1]$, $n \neq m$ and $A \equiv_{amb} B$, then $B = m[(\nu n)A_1]$. Through $\mathscr{T}$ are obtained $\mathscr{T}(A) = dlock(\nu n)[\mathscr{T}_1(A_1)]_m$ and $\mathscr{T}(B) = dlock[(\nu n)\mathscr{T}_1(A_1)]_m$. From the definition of $\equiv_{mem}$ it results that $\mathscr{T}(A) \equiv_{mem} \mathscr{T}(B)$.

$$\begin{array}{ccc}
(\nu n)m[A_1]) & \equiv_{amb} & m[(\nu n)A_1] \\
\Big\downarrow \mathscr{T} & & \Big\downarrow \mathscr{T} \\
dlock(\nu n)[\mathscr{T}_1(A_1)] & \cdot\cdot \equiv_{mem} \cdot\cdot & dlock[(\nu n)\mathscr{T}_1(A_1)]_m
\end{array}$$

□

Exercise 3.1. Prove the other implication in Proposition 3.7.

In [46] the authors put together the concept of "behaviour" of a biological system and the concept of "observer". "Biological System" represents a mathematical model of a biological system; such a system evolves by passing from one configuration to another, and producing in this way a "behaviour". An "Observer" is placed outside the biological system, and watches its behaviour.

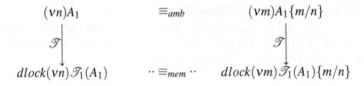

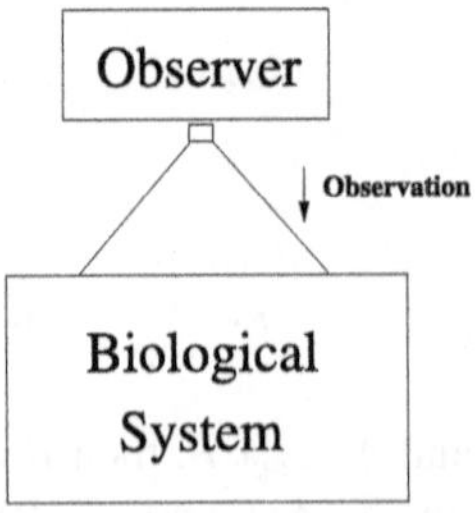

Fig. 3.2 Observer

Similar to protein observation defined in [79], we introduce a relation called barbed bisimulation which equates systems if they are indistinguishable under certain observations. In membrane systems the observer has the possibility of watching only the top-level membranes at any step of the computation, where the set of top-level membranes TL is defined as follows:

if $M = O$, then $TL(M) = \emptyset$
if $M = [\,N\,]_n$, then $TL(M) = \{n\}$;
if $M = (\nu n)N$, then $TL(M) = TL(N) \backslash \{n\}$;
if $M = M_1\, M_2$, then $TL(M) = TL(M_1) \cup TL(M_2)$.

For the case $M = (\nu n)N$ we have that $TL(M) = TL(N) \backslash \{n\}$ because an observer does not have the power to observe the membranes with restricted names.

From now on, we work with a subclass of $\mathcal{M}$, namely the systems obtained from the translation of safe ambients. Representing by r one of the rules $a), \ldots, k)$ from our particular set of developmental rules, we use $M \xrightarrow{r} N$ to denote the transformation of a membrane system M into a membrane system N by applying a rule r. Similar to [22] where a structural operational semantics for a particular class of P systems was defined, we can define the corresponding relation $\Rightarrow_{mem}$. Considering two membrane systems M and N with only one object *dlock*, we say that $M \Rightarrow_{mem} N$ if there is a sequence of rules $r_1, \ldots, r_i$ such that $M \xrightarrow{r_1} \ldots \xrightarrow{r_i} N$. The *operational semantics* of the membrane systems is defined in terms of the transformation relation $\xrightarrow{r}$ by the following rules:

$$\textit{(DRule)}\ M \xrightarrow{r} N \text{ for each developmental rule } a), \ldots, k)$$

$$\textit{(Res)}\ \frac{M \xrightarrow{r} M'}{(\nu n)M \xrightarrow{r} (\nu n)M'}\ ;\ \textit{(Comp)}\ \frac{M \xrightarrow{r} M'}{M\,N \xrightarrow{r} M'\,N};$$

$$\textit{(Amb)}\ \frac{M \xrightarrow{r} M'}{[\,M\,]_n \xrightarrow{r} [\,M'\,]_n}\ ;\ \textit{(Struc)}\ \frac{M \equiv_{mem} M',\ M' \xrightarrow{r} N'\ N' \equiv_{mem} N}{M \xrightarrow{r} N}.$$

The key ingredient of barbed bisimulation is the notion of barb. A barb is a predicate which describes the observed elements of a certain structure.

Definition 3.12. A barb $\downarrow^{mem}$ is defined inductively by the following rules:

$M \downarrow^{mem} n$	if $n \in TL(M)$
$M_1 \cdots M_k \downarrow^{mem} n_1 \cdots n_k$	if $n_i \in TL(M_j)$ $1 \le i,j \le k$
$(\nu k)M \downarrow^{mem} n$	if $k \neq n$

We write $M \Downarrow^{mem} n$ if either $M \downarrow^{mem} n$ or $M \Rightarrow^{+}_{mem} M'$ and $M' \downarrow^{mem} n$. Formally, we have:

$$M \downarrow^{mem} n \stackrel{def}{=} M \equiv_{mem} (\nu m_1)\ldots(\nu m_i)[\,M_1\,]_n\, M_2, \text{ and } n \notin \{m_1, \ldots, m_i\}.$$

$$M \Downarrow^{mem} n \stackrel{def}{=} \text{either } M \downarrow^{mem} n \text{ or } M \Rightarrow^{+}_{mem} M' \text{ and } M' \downarrow^{mem} n.$$

The following result reflects a relationship between structural congruence and barb predicates.

Proposition 3.8. *Structurally congruent membrane systems have the same top level membranes. If $M \equiv_{mem} N$, then $M \downarrow^{mem} n$ iff $N \downarrow^{mem} n$, for all $n \in TL(M,N)$.*

Proof. $M \downarrow^{mem} n$ means that $M \equiv_{mem} (\nu m_1)\dots(\nu m_i)[\,M_1\,]_n\,M_2$, where n is a label different from $m_1,\dots,m_i$. From $M \equiv_{mem} (\nu m_1)\dots(\nu m_i)[\,M_1\,]_n\,M_2$ and $M \equiv_{mem} N$, we get $N \equiv_{mem} (\nu m_1)\dots(\nu m_i)\,[\,M_1\,]_n\,M_2$, which means that $N \downarrow^{mem} n$. □

The set of membrane labels ML is defined as follows:
if $M = O$, then $ML(M) = \emptyset$;
if $M = [\,N\,]_n$, then $ML(M) = ML(N) \cup \{n\}$;
if $M = (\nu n)N$, then $ML(M) = ML(N)$;
if $M = M_1\,M_2$, then $ML(M) = ML(M_1) \cup ML(M_2)$.
If a system contains a top level membrane after applying a number of computation steps, then a structurally congruent membrane system contains the same top level membrane after applying the same number of computation steps.

Proposition 3.9. *If $M \equiv_{mem} N$, then $M \Downarrow^{mem} n$ iff $N \Downarrow^{mem} n$ for all $n \in ML(M,N)$.*

Proof. We prove only the first implication, the other being treated similarly by switching M and N.

If $M \Downarrow^{mem} n$, then either $M \downarrow^{mem} n$ or $M \Rightarrow^{+}_{mem} M'$ and $M' \downarrow^{mem} n$. The first case was studied in the previous proposition, so only the second case is presented. If $M \Rightarrow^{+}_{mem} M'$ and $M \equiv_{mem} N$, then exists N' with $N \Rightarrow^{+}_{mem} N'$ and $M' \equiv_{mem} N'$. From $M' \equiv_{mem} N'$ and $M' \downarrow^{mem} n$ we have that $N' \downarrow^{mem} n$, which together with $N \Rightarrow^{+}_{mem} N'$ implies that $N \Downarrow^{mem} n$. □

Exercise 3.2. Prove the other implication from Proposition 3.9.

Proposition 3.10. *An ambient contains a top ambient labelled by n if and only if the translated membrane system contains a top level membrane labelled by n.*

Formally, $A \downarrow^{amb} n$ iff $\mathcal{T}(A) \downarrow^{mem} n$ for all $n \in TL(\mathcal{T}(A))$.

Proof. We prove only the first implication, the other being treated similarly.

If $A \downarrow^{amb} n$, then we have A_1, A_2 such that $A = (\nu m_1)\dots(\nu m_i)n[\,A_1\,] \mid A_2$, where $n \notin \{m_1,\dots,m_i\}$. From $A = (\nu m_1)\dots(\nu m_i)n[\,A_1\,] \mid A_2$ and the definition of the translation function we have that $\mathcal{T}(A) = (\nu m_1)\dots(\nu m_i)dlock\ [\,\mathcal{T}_1(A_1)\,]_n\,\mathcal{T}_1(A_2)$, which means that $\mathcal{T}(A) \downarrow^{mem} n$. □

Exercise 3.3. Prove the other implication from Proposition 3.10.

Proposition 3.11. *An ambient eventually contains a top ambient n if and only if the translated membrane system, after applying the same number of steps, eventually contains a top level membrane n.*

Formally, $A \Downarrow^{amb} n$ iff $\mathcal{T}(A) \Downarrow^{mem} n$ for all $n \in ML(\mathcal{T}(A))$.

Proof. We prove only the first implication, the other being treated similarly.

If $A \Downarrow^{amb} n$, then either $A \downarrow^{amb} n$ or $A \Rightarrow^{+}_{amb} B$ and $B \downarrow^{amb} n$. The first case was studied in the previous proposition, so only the second case is presented. If $A \Rightarrow^{+}_{amb} B$ then $\mathcal{T}(A) \Rightarrow^{+}_{mem} \mathcal{T}(B)$. From $B \downarrow^{amb} n$ then according to the previous proposition we have that $\mathcal{T}(B) \downarrow^{mem} n$. From $\mathcal{T}(B) \downarrow^{mem} n$ and $\mathcal{T}(A) \Rightarrow^{+}_{mem} \mathcal{T}(B)$, we get that $\mathcal{T}(A) \Downarrow^{mem} n$. □

We consider that two membrane configurations are similar if they behave in the same way when they are placed in the same (arbitrary) context. We define a contextual bisimulation as in [109].

Considering a pair $(M;N)$ of membrane systems, we construct all the possible context pairs $(\mathscr{C}_{mem}(M);\mathscr{C}_{mem}(N))$ using the following recursive definition:

$$(\mathscr{C}_{mem}(M);\mathscr{C}_{mem}(N)) = (M;N) \;\mid\; ([\,\mathscr{C}_{mem}(M)\,]_n;[\,\mathscr{C}_{mem}(N)\,]_n) \;\mid$$
$$((\nu n)\mathscr{C}_{mem}(M);(\nu n)\mathscr{C}_{mem}(N)) \;\mid\; (\mathscr{C}_{mem}(M)\,M';\mathscr{C}_{mem}(N)\,M') \;\mid$$
$$(M'\,\mathscr{C}_{mem}(M);M'\,\mathscr{C}_{mem}(N)).$$

We define a contextual equivalence $\simeq_{mem}$ over membrane systems by

$$M \simeq_{mem} N \stackrel{def}{=} \text{for all } n \text{ and for all the pairs } (\mathscr{C}_{mem}(M);\mathscr{C}_{mem}(N)),$$
$$\mathscr{C}_{mem}(M) \Downarrow^{mem} n \text{ iff } \mathscr{C}_{mem}(N) \Downarrow^{mem} n.$$

Proposition 3.12. *If $M \equiv_{mem} N$ then $M \simeq_{mem} N$.*

Proof. Consider an arbitrary pair $(\mathscr{C}_{mem}(M);\mathscr{C}_{mem}(N))$, and a name n such that $\mathscr{C}_{mem}(M) \Downarrow^{mem} n$. We show that $\mathscr{C}_{mem}(N) \Downarrow^{mem} n$. $\mathscr{C}_{mem}(M) \equiv_{mem} \mathscr{C}_{mem}(N)$ is proved by induction on the size of $\mathscr{C}_{mem}(M)$ and $\mathscr{C}_{mem}(N)$. According to Proposition 3.9 we have that $\mathscr{C}_{mem}(M) \Downarrow^{mem} n$ implies $\mathscr{C}_{mem}(N) \Downarrow^{mem} n$; the other implication is proved by switching M and N. □

Proposition 3.13. *If $\mathscr{T}_1(A) \simeq_{mem} \mathscr{T}_1(B)$ then $A \simeq_{amb} B$.*

$$\begin{array}{ccc} A & \cdots\simeq_{amb}\cdots & B \\ \Big\downarrow{\scriptstyle \mathscr{T}_1} & & \Big\downarrow{\scriptstyle \mathscr{T}_1} \\ \mathscr{T}_1(A) & \simeq_{mem} & \mathscr{T}_1(B) \end{array}$$

Proof. $\mathscr{T}_1(A) \simeq_{mem} \mathscr{T}_1(B)$ means that for all n and for all pairs $(\mathscr{C}_{mem}(\mathscr{T}_1(A));$ $\mathscr{C}_{mem}(\mathscr{T}_1(B)))$ we have that $\mathscr{C}_{mem}(\mathscr{T}_1(A)) \Downarrow^{mem} n$ iff $\mathscr{C}_{mem}(\mathscr{T}_1(B)) \Downarrow^{mem} n$. For all the contexts, the pair $(\mathscr{C}_{amb}(A);\mathscr{C}_{amb}(B))$ is translated into a pair $(\mathscr{T}(\mathscr{C}_{amb}(A));$ $\mathscr{T}(\mathscr{C}_{amb}(B)))$. By applying the translation function in the second pair, we obtain the pair $(\mathscr{C}_{mem}(\mathscr{T}_1(A));\mathscr{C}_{mem}(\mathscr{T}_1(B)))$, where $\mathscr{C}_{mem}$ corresponds to $\mathscr{C}_{amb}$ by translation.

We have $\mathscr{C}_{mem}(\mathscr{T}_1(A)) \Downarrow^{mem} n \Leftrightarrow \mathscr{T}(\mathscr{C}_{amb}(A)) \Downarrow^{mem} n \stackrel{Prop.3.11}{\Leftrightarrow} \mathscr{C}_{amb}(A) \Downarrow^{amb} n.$

Similarly $\mathscr{C}_{mem}(\mathscr{T}_1(B)) \Downarrow^{mem} n \Leftrightarrow \mathscr{T}(\mathscr{C}_{amb}(B)) \Downarrow^{mem} n \stackrel{Prop.3.11}{\Leftrightarrow} \mathscr{C}_{amb}(B) \Downarrow^{amb} n.$
It follows that $\mathscr{C}_{amb}(A) \Downarrow^{amb} n \Leftrightarrow \mathscr{C}_{amb}(B) \Downarrow^{amb} n$ which implies $A \simeq_{amb} B$. □

Remark 3.4. $A \simeq_{amb} B$ does not necessarily imply that $\mathscr{T}_1(A) \simeq_{mem} \mathscr{T}_1(B)$, because the translated contexts from mobile ambients do not represent all the contexts from membrane systems (the set of contexts in membrane systems is larger than the set of translated contexts). $A \simeq_{amb} B$ implies that for all $n \in ML(\mathscr{T}(A \mid B))$ and for all $\mathscr{C}_{amb}$ we have $\mathscr{C}_{amb}(A) \Downarrow^{amb} n \Leftrightarrow \mathscr{C}_{amb}(B) \Downarrow^{amb} n$. According to Proposition 3.11, we have $\mathscr{T}(\mathscr{C}_{amb}(A)) \Downarrow^{mem} n \Leftrightarrow \mathscr{T}(\mathscr{C}_{amb}(B)) \Downarrow^{mem} n$. We should check that $\mathscr{C}_{mem}(\mathscr{T}_1(A)) \Downarrow^{mem} n \Leftrightarrow \mathscr{C}_{mem}(\mathscr{T}_1(B)) \Downarrow^{mem} n$ for all the contexts (not only for a particular set) in order to have $\mathscr{T}_1(A) \simeq_{mem} \mathscr{T}_1(B)$, and in general this cannot be done.

We define deadlock for a membrane system. Deadlock is a predicate

$$deadlock_{mem}(M) = \begin{cases} \text{true} & \text{if } M \in \mathcal{D}_{mem} \\ \text{false} & \text{if } M \notin \mathcal{D}_{mem}, \end{cases}$$

where $\mathcal{D}_{mem}$ is a set of membrane systems in which, after introducing all the possible objects $dlock$ by applying rules of form h), none of the sequences of rules $j),a)$ or $j),d)$ or $j),f)$ can be applied.

The next result relates the two notions of deadlock (in ambients and in membrane systems) through the defined translation function.

Proposition 3.14. *$A \in \mathcal{D}_{amb}$ iff $\mathcal{T}(A) \in \mathcal{D}_{mem}$.*
Furthermore $deadlock_{amb}(A) = deadlock_{mem}(\mathcal{T}(A))$.

Proof. $\mathcal{T}(A) \in \mathcal{D}_{mem}$ means that after applying all the possible rules of type h), none of the rules of type a), d) or f) can be applied. This is equivalent to the fact that no object corresponding to a translated capability is consumed, and from the definition of the object $dlock$ this means that the translated ambient is a deadlock, so $A \in \mathcal{D}_{amb}$.

We demonstrate the other implication by checking all the rules from the definition of $\mathcal{D}_{amb}$, and we proceed using structural induction.

If $A = \mathbf{0}$ then $\mathcal{T}(A) = dlock$. No rule of the type h), a), d) or f) can be applied which means that $\mathcal{T}(A) \in \mathcal{D}_{mem}$.

If $A = cap\ n.A_1$ then $\mathcal{T}(A) = dlock\ cap\ n\ [\ \mathcal{T}_1(A_1)\ ]_{cap\ n}$. Note that no rule of the type h), a), d) or f) can be applied which means that $\mathcal{T}(A) \in \mathcal{D}_{mem}$.

If $A = n[\ A_1\ ]$, $A_1 \in \mathcal{D}_{amb}$, and $TA(A_1) = \emptyset$ then we have $A_1 = A'_1 \mid \ldots \mid A'_k$ where each $A'_i = cap\ n_i.A''_i,\ for\ all\ i \in \{1,\ldots,k\}$. This means that $\mathcal{T}(A) = dlock\ [\ cap\ n_1[\ \mathcal{T}_1(A''_1)\ ]_{cap\ n_1} \ldots cap\ n_k[\ \mathcal{T}_1(A''_k)\ ]_{cap\ n_k}\ ]_n$. Only one rule of type h) can be applied; in the resulting configuration we cannot apply a rule of type a), d) or f), which means that $\mathcal{T}(A) \in \mathcal{D}_{mem}$.

If $A = n[\ A_1\ ]$, $A_1 \in \mathcal{D}_{amb}$, $TA(A_1) \neq \emptyset$, $open\ m \mid A_1 \Rightarrow_{amb} A''_m$ together with $out\ n \notin TC(A''_m)$, for all $m \in TA(A_1)$ then we have that $A_1 = A'_1 \mid \ldots \mid A'_k$, where each $A'_i = cap\ n_i.A''_i$ or $A'_i = n_i[\ A''_i\ ]$ for all $i \in \{1,\ldots,k\}$, and it results that $\mathcal{T}(A) = dlock\ [cap\ n_{i_1}[\mathcal{T}_1(A''_{i_1})]_{cap\ n_{i_1}} \ldots cap\ n_{i_s}[\mathcal{T}_1(A''_{i_s})]_{cap\ n_{i_s}}[\mathcal{T}_1(A''_{j_1})]_{n_{j_1}} \ldots [\mathcal{T}_1(A''_{j_t})]_{n_{j_t}}]_n$. After applying all the possible rules of type h), and using the fact that the membrane systems $\mathcal{T}_1(A''_{j_1}) \ldots \mathcal{T}_1(A''_{j_t})$ do not contain the object $out\ n$ and $\mathcal{T}_1(A_1) \in \mathcal{D}_{mem}$, in the resulting configuration we cannot apply a rule of type a), d) or f), which means that $\mathcal{T}(A) \in \mathcal{D}_{mem}$.

If $A = A_1 \mid A_2$, $A_1, A_2 \in \mathcal{D}_{amb}$, $open\ m \notin TC(A_i)$, $open\ k \mid A_i \Rightarrow_{amb} A''_i$, $in\ m \notin TC(A''_i)$ for all $k \in TA(A_i)$, for all $m \in TA(A_j)$, $i \neq j$, $i,j = 1,2$ then we have that $A = A'_1 \mid \ldots \mid A'_k$, where each $A'_i = cap\ n_i.A''_i$ or $A'_i = n_i[\ A''_i\ ]$ for all $i \in \{1,\ldots,k\}$, and it results that $\mathcal{T}(A) = dlock\ \ cap\ n_{i_1}[\mathcal{T}_1(A''_{i_1})]_{cap\ n_{i_1}} \cdots cap\ n_{i_s}[\mathcal{T}_1(A''_{i_s})]_{cap\ n_{i_s}}[\mathcal{T}_1(A''_{j_1})]_{n_{j_1}} \ldots [\mathcal{T}_1(A''_{j_t})]_{n_{j_t}}$. After applying all possible rules of type h), using the fact that the membrane system $\mathcal{T}_a(A''_{j_d})$, $d = \{1,\ldots,t\}$, does not contain objects in n_{j_c}, $c = \{1,\ldots,t\}$, $c \neq d$, and if $cap\ n_{i_c} = open\ n_{i_c}$ then $n_{i_c} \neq n_{j_d}$ for all $d \in \{1\ldots t\}$, and $\mathcal{T}_i(A_i) \in \mathcal{D}_{mem}$, $i = 1,2$, in the resulting configuration we cannot apply a rule of type a), d) or f), which means that $\mathcal{T}(A) \in \mathcal{D}_{mem}$. □

3.3.4 Operational Correspondence

Proposition 3.15. *If A and B are two ambients and M is a membrane system such that $A \Rightarrow_{amb} B$ and $M = \mathscr{T}(A)$, then there exists a chain of transitions $M \xrightarrow{r_1} \dots \xrightarrow{r_k} N$ such that $r_1, \dots, r_k$ are developmental rules, and $N = \mathscr{T}(B)$.*

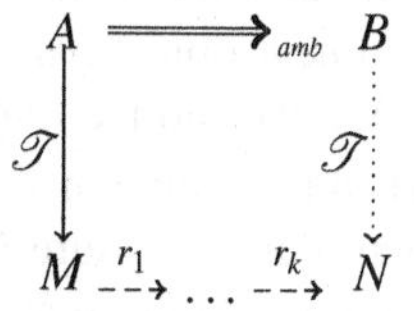

Proof. Since $A \Rightarrow_{amb} B$, then one of the requirements In, Out or $Open$ is fulfilled for ambients A' and B' which are included in A and B respectively. We treat all the possible cases:

1. $A' = n[\ in\ m\ ] \ \mid\ m[\ \overline{in}\ m\ ]$ and $B' = m[\ n[\ ]\ ]$, where n is an ambient which contains only the capability *in m*. Then according to the definition of the translation function $\mathscr{T}$, M contains the membrane structure

$$[\ in\ m\ [\ ]_{in\ m}\ ]_n\ [\ \overline{in}\ m\ [\ ]_{\overline{in}\ m}\ ]_m,$$

and applying some rules of form $h)$ we obtain the following structure:

$$[\ in\ m\ dlock\ [\ ]_{in\ m}\ ]_n\ [\ \overline{in}\ m\ [\ ]_{\overline{in}\ m}\ ]_m.$$

Using the rules

$r_1 : [\ dlock\ ]_n \rightarrow [\ dlock\ one\ ]_n$

$r_2 : [\ in\ m\ dlock\ one\ ]_n\ [\ \overline{in}\ m\ ]_m \rightarrow [\ in^*m\ in_*m\ dlock\ ]_n\ [\overline{in}_*m]_m$

$r_3 : in_*m\ [\ ]_{in\ m} \rightarrow [\ \delta\ ]_{in\ m}$

$r_4 : \overline{in}_*m\ [\ ]_{\overline{in}\ m} \rightarrow [\ \delta\ ]_{\overline{in}\ m}$

$r_5 : [in^*m]_n[\ ]_m \rightarrow [[\ ]_n]_m,$

and some rules of the form $i)$ there exists the following sequence of transitions $M \overset{k)}{\rightarrow^*} M_1 \xrightarrow{r_1} \dots M_5 \xrightarrow{r_5} M_6 \overset{i)}{\rightarrow^*} N$, where $M_1, \dots, M_6$ are intermediary configurations, and the membrane structure N contains the membrane structure $[\ [\ ]_n\ ]_m$. Once the objects *dlock* and *one* are created near object *in m*, these transitions are the only deterministic steps which can be performed. We can notice that $\mathscr{T}_1(B') = [\ [\ ]_n\ ]_m$. Hence, according to the definition of translation function $\mathscr{T}$ and transition relation $\xrightarrow{r}$, we reach the conclusion that the membrane structure M admits the required sequence of transitions leading to the membrane structure N, and $\mathscr{T}(B) = N$.

2. $A' = m[\ \overline{out}\ m\ n[\ out\ m\ ]\ ]$ and $B' = m[\ ]\ n[\ ]$, where n is an ambient which contains only the capability *out m*. Then according to the definition of the translation function $\mathscr{T}$, M contains the membrane structure

$$[\ \overline{out}\ m\ [\ ]_{\overline{out}\ m}[\ out\ m\ [\ ]_{out\ m}]_n\ ]_m,$$

and applying some rules of form $h)$ we obtain the following structure:

$$[\ \overline{out}\ m\ [\ ]_{\overline{out}\ m}[\ dlock\ out\ m\ [\ ]_{out\ m}]_n\ ]_m.$$

Using the rules

$r_1 : [\ dlock\ ]_n \rightarrow [\ dlock\ one\ ]_n$

$r_2 : [\ \overline{out}\ m\ [\ out\ m\ dlock\ one]_n\ ]_m \rightarrow [\ \overline{out}_*m\ [\ out^*m\ out_*m]_n]_m$

$r_3 : out_* m\,[\,]_{out\ m} \rightarrow [\,\delta\,]_{out\ m}$
$r_4 : \overline{out}_* m\,[\,]_{\overline{out}\ m} \rightarrow [\,\delta\,]_{\overline{out}\ m}$
$r_5 : [\,[\,out^* m\,]_n\,]_m \rightarrow [\,]_m\,[\,]_n,$

and some rules of the form *i*) there exists the following sequence of transitions $M \stackrel{k)}{\rightarrow^*} M_1 \stackrel{r_1}{\rightarrow} \ldots M_5 \stackrel{r_5}{\rightarrow} M_6 \stackrel{i)}{\rightarrow^*} N$, where $M_1, \ldots, M_6$ are intermediary configurations, and the membrane structure N contains the membrane structure $[\,]_m\,[\,]_n$. Once the objects *dlock* and *one* are created near object *out m*, these transitions are the only steps which can be performed. We can notice that $\mathscr{T}_1(B') = [\,]_m[\,]_n$. Hence, according to the definition of translation function $\mathscr{T}$ and transition relation $\stackrel{r}{\rightarrow}$, we reach the conclusion that the membrane structure M admits the required sequence of transitions leading to the membrane structure N, and $\mathscr{T}(B) = N$.

3. $A' = m[\,n[\,A_1\ \overline{open}\ n\,]\ open\ n\,]$ and $B' = m[\,A_1\,]$, where n is an ambient containing the ambient structure A_1. Then according to the definition of the translation function $\mathscr{T}$, M contains the membrane structure

$$[\,[\,\mathscr{T}_1(A_1)\ \overline{open}\ n\,[\,]_{\overline{open}\ n}\,]_n\ open\ n\,[\,]_{open\ n}\,]_m,$$

and applying some rules of form *h*) we obtain the following structure:

$$[\,[\,\mathscr{T}_1(A_1)\ \overline{open}\ n\,[\,]_{\overline{open}\ n}\,]_n\ dlock\ open\ n\,[\,]_{open\ n}\,]_m$$

Using the rules

$r_1 : [\,dlock\,]_m \rightarrow [\,dlock\ one\,]_m$
$r_2 : [\overline{open}\ n]_n\ open\ n\ dlock\ one \rightarrow [\,\delta\,]_n\ \overline{open}_* n\ open_* n\ dlock$
$r_3 : open_* n\,[\,]_{open\ n} \rightarrow [\,\delta\,]_{open\ n}$
$r_4 : \overline{open}_* n\,[\,]_{\overline{open}\ n} \rightarrow [\,\delta\,]_{\overline{open}\ n}$

and some rules of the form *i*) there exists the following sequence of transitions $M \stackrel{k)}{\rightarrow^*} M_1 \stackrel{r_1}{\rightarrow} \ldots M_4 \stackrel{r_4}{\rightarrow} M_5 \stackrel{i)}{\rightarrow^*} N$, where $M_1, \ldots, M_5$ are intermediary configurations, and the membrane structure N contains the membrane structure $[\,\mathscr{T}_1(A_1)\,]_m$. Once the objects *dlock* and *one* are created near object *open m*, these transitions are the only steps which can be performed. We can notice that $\mathscr{T}_1(B') = [\,\mathscr{T}_1(A_1)\,]_m$. Hence, according to the definition of translation function $\mathscr{T}$ and transition relation $\stackrel{r}{\rightarrow}$, we reach the conclusion that the membrane structure M admits the required sequence of transitions leading to the membrane structure N, and $\mathscr{T}(B) = N$.

4. $A' = n[\,in\ m \ldots] \mid m[\,\overline{in}\ m\,]$ and $B' = m[\,n[\,]\,]$, where n is an ambient which contains the capability *in m* and another capabilities or ambients. We treat only the case with $A' = n[\,in\ m\ t[\,]\,] \mid m[\,\overline{in}\ m\,]$ and $B' = m[\,n[\,t[\,]\,]\,]$, where t is an empty ambient and give some ideas on how to treat the cases with a more nested structure for n. Then according to the definition of the translation function $\mathscr{T}$, M contains the membrane structure

$$[\,in\ m\,[\,]_{in\ m}\,[\,]_t\,]_n\,[\,\overline{in}\ m\,[\,]_{\overline{in}\ m}]_m,$$

and applying some rules of form *h*) we obtain the following structure:

$$[\,in\ m\ dlock\,[\,]_{in\ m}\,[\,]_t\,]_n\,[\,\overline{in}\ m\,[\,]_{\overline{in}\ m}]_m.$$

Using the rules

$r_1 : [\,dlock\,]_n \rightarrow [\,dlock\ one\,]_n$
$r_2 : [\,in\ m\ dlock\ one\,]_n\,[\,\overline{in}\ m\,]_m \rightarrow [\,in^* m\ in_* m\ dlock\,]_n\,[\,\overline{in}_* m\,]_m$
$r_3 : in_* m\,[\,]_{in\ m} \rightarrow [\,\delta\,]_{in\ m}$

$r_4 : \overline{in}_* m\, [\,]_{\overline{in}\ m} \to [\,\delta\,]_{\overline{in}\ m}$
$r_5 : [\, in^*m\, [\,]_t\,]_n \to [\, in^*m\, [\, in^*m\ out^*n\ in^*n\,]_t\,]_n$
$r_6 : [\,[\, out^*n\,]_t\,]_n \to [\,]_n\, [\,]_t$
$r_7 : [\, in^*m\,]_n\, [\,]_m \to [\,[\,]_n\,]_m$
$r_8 : [\, in^*m\,]_t\, [\,]_m \to [\,[\,]_t\,]_m$
$r_9 : [\, in^*n\,]_t\, [\,]_n \to [\,[\,]_t\,]_n,$

and some rules of the form $i)$ there exists the following sequence of transitions $M \overset{k)}{\to^*} M_1 \overset{r_1}{\to} \ldots M_9 \overset{r_9}{\to} M_{10} \overset{i)}{\to^*} N$, where $M_1, \ldots, M_{10}$ are intermediary configurations, and the membrane structure N contains the membrane structure $[\,[\,[\,]_t\,]_n\,]_m$. Once the objects *dlock* and *one* are created near object *in m*, these transitions are the only steps which can be performed. We can notice that $\mathscr{T}_1(B') = [\,[\,[\,]_t\,]_n\,]_m$. Hence, according to the definition of translation function $\mathscr{T}$ and transition relation $\overset{r}{\to}$, we reach the conclusion that the membrane structure M admits the required sequence of transitions leading to the membrane structure N, and $\mathscr{T}(B) = N$.

If the number of membranes nested in the ambient n is more than one or we have more capabilities, the number of rules applied increases, but the result is the same: the membrane n is transformed into an elementary membrane, it is introduced in m, where it regains the same nested structure, all this process being controlled by the objects created by the sequence of rules applied. The process stops when all the star objects are consumed and there is only one *dlock* object in the membrane system.

5. $A' = m[\, n[\, out\ m \ldots]\ \overline{out}\ m\,]$ and $B' = m[\,]\ n[\,]$, where n is an ambient which contains the capability *out m* and other capabilities or ambients. We treat only the case with $A' = m[\, n[\, out\ m\ t[\,]\ \overline{out}\ m\,]] \mid m[\,]$ and $B' = n[\, t[\,]\,]\ m[\,]$, where t is an empty ambient and give some ideas on how to treat the cases with a more nested structure for n. Then according to the definition of the translation function $\mathscr{T}$, M contains the membrane structure

$$[\,[\, out\ m\, [\,]_{out\ m}\, [\,]_t\,]_n\ \overline{out}\ m\, [\,]_{\overline{out}\ m}\,]_m,$$

and applying some rules of form $h)$ we obtain the following structure:

$$[\,[\, dlock\ out\ m\, [\,]_{out\ m}\, [\,]_t\,]_n\ \overline{out}\ m\, [\,]_{\overline{out}\ m}\,]_m.$$

Using the rules

$r_1 : [\, dlock\,]_m \to [\, dlock\ one\,]_m$
$r_2 : [\,[\, out\ m\ dlock\ one\,]_n\ \overline{out}\ m\,]_m \to [\,[\, out^*m\ out_*m\ dlock\,]_n\ \overline{out}_*m\,]_m$
$r_3 : out_*m\, [\,]_{out\ m} \to [\,\delta\,]_{out\ m}$
$r_4 : \overline{out}_*m\, [\,]_{\overline{out}\ m} \to [\,\delta\,]_{\overline{out}\ m}$
$r_5 : [\, out^*m\, [\,]_t\,]_n \to [\, out^*m\, [\, out^*m\ out^*n\ in^*n\,]_t\,]_n$
$r_6 : [\,[\, out^*n\,]_t\,]_n \to [\,]_n\, [\,]_t$
$r_7 : [\,[\, out^*m\,]_n\,]_m \to [\,]_n\, [\,]_m$
$r_8 : [\,[\, out^*m\,]_t\,]_m \to [\,]_t\, [\,]_m$
$r_9 : [\, in^*n\,]_t\, [\,]_n \to [\,[\,]_t\,]_n,$

and some rules of the form $i)$ there exists the following sequence of transitions $M \overset{k)}{\to^*} M_1 \overset{r_1}{\to} \ldots M_9 \overset{r_9}{\to} M_{10} \overset{i)}{\to^*} N$, where $M_1, \ldots, M_{10}$ are intermediary configurations, and the membrane structure N contains the membrane structure $[\,]_m\, [\,[\,]_t\,]_n$. Once the objects *dlock* and *one* are created near object *out m*, these transitions are the only steps which can be performed. We can notice that $\mathcal{T}_1(B') = [\,]_m\,[\,[\,]_t\,]_n$. Hence, according to the definition of translation function $\mathcal{T}$ and transition relation $\xrightarrow{r}$, we reach the conclusion that the membrane structure M admits the required sequence of transitions leading to the membrane structure N, and $\mathcal{T}(B) = N$.

If the number of membranes nested in the ambient n is more than one or we have more capabilities, the number of rules applied increases, but the result is the same: the membrane n is transformed into an elementary membrane, it is extracted from m, then it regains the same nested structure, all this process being controlled by the objects created by the sequence of rules applied. The process stops when all the star objects are consumed. □

Proposition 3.16. *Let M and N be two membrane systems with only one dlock object, and an ambient A such that $M = \mathcal{T}(A)$. If there is a sequence of transitions $M \xrightarrow{r_1} \ldots \xrightarrow{r_k} N$, then there exists an ambient B with $A \Rightarrow^*_{amb} B$ and $N = \mathcal{T}(B)$. The number of pairs of non-star objects consumed in membrane systems is equal to the number of pairs of capabilities consumed in ambients.*

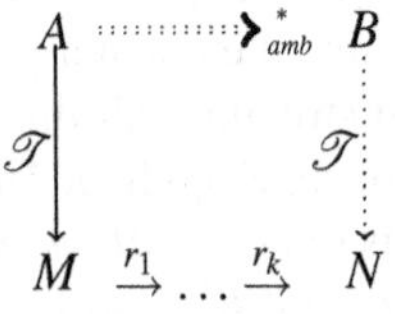

Proof. We proceed by structural induction. Since M does not contain any star object, the first rule which consumes a translated capability object has one of the following forms:

- $[\, in\ m\ dlock\ one\,]_n\ [\overline{in}\ m]_m \to [\, in^* m\ in_* m\ dlock\,]_n\ [\,\overline{in}_* m\,]_m$,
- $[\,[\, out\ m\ dlock\ one\,]_n\ \overline{out}\ m\,]_m \to [\,[\, out^* m\ out_* m\ dlock\,]_n\ \overline{out}_* m\,]_m$,
- $[\,\overline{open}\ m\,]_m open\ m\ dlock\ one \to [\,\delta\,]_m\ dlock\ open_* m\ \overline{open}_* m$.

We treat all the possible cases:

1. If the first rule applied is

$$[\, in\ m\ dlock\ one\,]_n\ [\overline{in}\ m]_m \to [\, in^* m\ in_* m\ dlock\,]_n\ [\,\overline{in}_* m\,]_m$$

where the membrane n contains only the capability object *in m* and the corresponding membrane labelled *in m*, then M contains the membrane structure

$$[\, in\ m\, [\,]_{in\ m}\,]_n\, [\,\overline{in}\ m\, [\,]_{\overline{in}\ m}\,]_m.$$

According to the definition of $\mathcal{T}$, M can be written as $M_1\ M'$ or $M_2[\,M'\,]$, where $M' = [\, in\ m\, [\,]_{in\ m}\,]_n\, [\,\overline{in}\ m\, [\,]_{\overline{in}\ m}\,]_m$ and M_2 represents a membrane structure in which M' is placed inside a nested structure of translated ambients. If A is a mobile ambient encoded by $M = M_1\ M'$, then according to the definition of $\mathcal{T}$ it contains two ambients $A' = n[\, in\ m\,] \mid m[\,\overline{in}\ m\,]$ and A_1 such that $A = A_1 \mid A'$, $\mathcal{T}_1(A') = M'$, and $\mathcal{T}_1(A_1) = M_1$. If A is a mobile ambient encoded by $M = M_2[\,M'\,]$, then according to the definition of $\mathcal{T}$ it contains two ambients $A' = n[\, in\ m\,] \mid m[\,\overline{in}\ m\,]$ and A_2 such that $A = A_2[\,A'\,]$, $\mathcal{T}_1(A') = M'$, and $\mathcal{T}_1(A_2) = M_2$.

The application of the rule defined above to the membrane system M changes only the membrane system M'. The newly created objects $in^* m$, $in_* m$ and $\overline{in}_* m$

control the moving of membrane n into membrane m, and are consumed by the following rules:

- $in_*m\,[\,]_{in\ m} \rightarrow [\,\delta\,]_{in\ m}$
- $\overline{in}_*m\,[\,]_{\overline{in}\ m} \rightarrow [\,\delta\,]_{\overline{in}\ m}$
- $[\,in^*m\,]_n\,[\,]_m \rightarrow [\,[\,]_n\,]_m$.

After the application of these rules, M' evolves to $N' = [\,[\,]_n\,]_m$. The inductive hypothesis expresses that N' encodes an ambient B'. After obtaining N', N has the structure $N = M_1\ N'$ if $M = M_1\ M'$ and it encodes the mobile ambient $B = A_1 \mid B'$, or N has the structure $N = M_2[\,N'\,]$ if $M = M_2[\,M'\,]$ and it encodes the mobile ambient $B = A_2[B']$. The transition from M' to N' represents also the transition from M to N.

It should be noticed that by consuming the capability $in\ m$ we have $A' \Rightarrow_{amb} B'$. So the transition from M to N with the consumption of only one non-star object is simulated by the transition of A to B.

2. If the first rule applied is

$$[\,in\ m\ dlock\ one\,]_n\,[\overline{in}\ m]_m \rightarrow [\,in^*m\ in_*m\ dlock\,]_n\,[\,\overline{in}_*m\,]_m$$

where the membrane n contains the object $in\ m$, the corresponding membrane labelled $in\ m$ and another nested membrane $[\,]_t$, then M contains the membrane structure

$$[\,in\ m\,[\,]_{in\ m}\,[\,]_t\,]_n\,[\,\overline{in}\ m\,[\,]_{\overline{in}\ m}\,]_m.$$

The cases in which the ambient n contains more capabilities or/and more nested ambients are treated using structural induction on the membrane structure. According to the definition of $\mathscr{T}$, M can be written as $M_1\ M'$ or $M_2[M']$, where $M' = [\,in\ m\,[\,]_{in\ m}\,[\,]_t\,]_n\,[\,\overline{in}\ m\,[\,]_{\overline{in}\ m}\,]_m$ and M_2 represents a membrane structure in which M' is placed inside a nested structure of translated ambients. If A is a mobile ambient encoded by $M = M_1\ M'$, then according to the definition of $\mathscr{T}$ it contains two ambients $A' = n[\,in\ m\,t[\,]\,] \mid m[\,\overline{in}\ m\,]$ and A_1 such that $A = A_1 \mid A'$, $\mathscr{T}_1(A') = M'$, and $\mathscr{T}_1(A_1) = M_1$. If A is a mobile ambient encoded by $M = M_2[\,M'\,]$, then according to the definition of $\mathscr{T}$ it contains two ambients $A' = n[\,in\ m\,t[\,]\,] \mid m[\,\overline{in}\ m\,]$ and A_2 such that $A = A_2[\,A'\,]$, $\mathscr{T}_1(A') = M'$, and $\mathscr{T}_1(A_2) = M_2$.

The application of the rule defined above to the membrane system M changes only the membrane system M'. The newly created objects in^*m, in_*m and $\overline{in}_*m$ control the moving of membrane n into membrane m, and are consumed by the following rules:

- $in_*m\,[\,]_{in\ m} \rightarrow [\,\delta\,]_{in\ m}$
- $\overline{in}_*m\,[\,]_{\overline{in}\ m} \rightarrow [\,\delta\,]_{\overline{in}\ m}$
- $[\,in^*m\,[\,]_t\,]_n\,[\,]_m \rightarrow [\,in^*m\,[\,in^*m\ in^*n\ out^*n\,]_t\,]_n\,[\,]_m$
- $[\,[\,out^*n\,]_t\,]_n \rightarrow [\,]_n\,[\,]_t$
- $[\,in^*m\,]_n\,[\,]_m \rightarrow [\,[\,]_n\,]_m$
- $[\,in^*m\,]_t\,[\,]_m \rightarrow [\,[\,]_t\,]_m$
- $[\,in^*n\,]_t\,[\,]_n \rightarrow [\,[\,]_t\,]_n$

After the application of these rules, M' evolves to $N' = [\,[\,[\,]_t\,]_n\,]_m$. The inductive hypothesis expresses that N' encodes an ambient B'. After obtaining N', N has the structure $N = M_1\ N'$ if $M = M_1\ M'$ and it encodes the mobile ambient $B = A_1 \mid B'$, or N has the structure $N = M_2[\,N'\,]$ if $M = M_2[\,M'\,]$ and it encodes the mobile ambient $B = A_2[B']$. The transition from M' to N' represents also the transition from M to N.

It should be noticed that by consuming the capability *in m* we have $A' \Rightarrow_{amb} B'$. So the transition from M to N with the consumption of only one non-star object is simulated by the transition of A to B.

3. If the first rule applied is

$[\,[\,out\ m\ dlock\ one\,]_n\ \overline{out}\ m\,]_m \rightarrow [\,[\,out^{*}m\ out_{*}m\ dlock\,]_n\ \overline{out}_{*}m\,]_m$

where the membrane n contains only the object *out m* and the corresponding membrane labelled *out m*, then M contains the membrane structure

$$[\,[\,out\ m\,[\,]_{out\ m}]_n\ \overline{out}\ m\,[\,]_{\overline{out}\ m}\,]_m.$$

According to the definition of $\mathscr{T}$, M can be written as $M_1\ M'$ or $M_2[\,M'\,]$, where $M' = [\,[\,out\ m\,[\,]_{out\ m}]_n\ \overline{out}\ m\,[\,]_{\overline{out}\ m}\,]_m$ and M_2 represents a membrane structure in which M' is placed inside a nested structure of translated ambients. If A is a mobile ambient encoded by $M = M_1\ M'$, then according to the definition of $\mathscr{T}$ it contains two ambients $A' = m[\,n[\,out\ m\,]\ \overline{out}\ m\,]$ and A_1 such that $A = A_1 \mid A'$, $\mathscr{T}_1(A') = M'$, and $\mathscr{T}_1(A_1) = M_1$. If A is a mobile ambient encoded by $M = M_2[\,M'\,]$, then according to the definition of $\mathscr{T}$ it contains two ambients $A' = m[\,n[\,out\ m\,]\ \overline{out}\ m\,]$ and A_2 such that $A = A_2[\,A'\,]$, $\mathscr{T}_1(A') = M'$, and $\mathscr{T}_1(A_2) = M_2$.

The application of the rule defined above to the membrane system M changes only the membrane system M'. The newly created objects $in^{*}m$, $in_{*}m$ and $\overline{in}_{*}m$ are consumed by the following rules:

- $out_{*}m\,[\,]_{out\ m} \rightarrow [\,\delta\,]_{out\ m}$
- $\overline{out}_{*}m\,[\,]_{\overline{out}\ m} \rightarrow [\,\delta\,]_{\overline{out}\ m}$
- $[\,[\,out^{*}m\,]_n\,]_m \rightarrow [\,]_n\,[\,]_m.$

After the application of these rules, M' evolves to $N' = [\,]_n\,[\,]_m$. The inductive hypothesis expresses that N' encodes an ambient B'. After obtaining N', N has the structure $N = M_1\ N'$ if $M = M_1\ M'$ and it encodes the mobile ambient $B = A_1 \mid B'$, or N has the structure $N = M_2[\,N'\,]$ if $M = M_2[\,M'\,]$ and it encodes the mobile ambient $B = A_2[B']$. The transition from M' to N' represents also the transition from M to N.

It should be noticed that by consuming the capability *out m* we have $A' \Rightarrow_{amb} B'$. So the transition from M to N with the consumption of only one non-star object is simulated by the transition of A to B.

4. If the first rule applied is

$[\,[\,out\ m\ dlock\ one\,]_n\ \overline{out}\ m\,]_m \rightarrow [\,[\,out^{*}m\ out_{*}m\ dlock\,]_n\ \overline{out}_{*}m\,]_m,$

where the membrane n contains only the object *out m*, the corresponding membrane labelled *out m* and another nested membrane $[\,]_t$, then M contains the membrane structure

$$[\,[\,out\ m\,[\,]_{out\ m}\,]_n\ \overline{out}\ m\,[\,]_{\overline{out}\ m}]_m.$$

The cases in which the ambient n contains more capabilities or/and more nested ambients are treated using structural induction on the membrane structure. According to the definition of $\mathscr{T}$, M can be written as $M_1\ M'$ or $M_2[\,M'\,]$, where

$M' = [\,[\,out\ m\,[\,]_{out\ m}\,]_n\,\overline{out}\ m\,[\,]_{\overline{out}\ m}]_m$ and M_2 represents a membrane structure in which M' is placed inside a nested structure of translated ambients. If A is a mobile ambient encoded by $M = M_1\ M'$, then according to the definition of $\mathcal{T}$ it contains two ambients $A' = m[\,n[\,out\ m\,]\,\overline{out}\ m\,]$ and A_1 such that $A = A_1 \mid A'$, $\mathcal{T}_1(A') = M'$, and $\mathcal{T}_1(A_1) = M_1$. If A is a mobile ambient encoded by $M = M_2[\,M'\,]$, then according to the definition of $\mathcal{T}$ it contains two ambients $A' = m[\,n[\,out\ m\,]\,\overline{out}\ m\,]$ and A_2 such that $A = A_2[\,A'\,]$, $\mathcal{T}_1(A') = M'$, and $\mathcal{T}_1(A_2) = M_2$.

The application of the rule defined above to the membrane system M changes only the membrane system M'. The newly created objects in^*m, in_*m and $\overline{in}_*m$ determine the application of the following rules:

- $out_*m\,[\,]_{out\ m} \to [\,\delta\,]_{out\ m}$
- $\overline{out}_*m\,[\,]_{\overline{out}\ m} \to [\,\delta\,]_{\overline{out}\ m}$
- $[\,out^*m\,[\,]_t\,]_n \to [\,out^*m\,[\,out^*m\ out^*n\ in^*n\,]_t\,]_n$
- $[\,[\,out^*n\,]_t\,]_n \to [\,]_n\,[\,]_t$
- $[\,[\,out^*m\,]_n\,]_m \to [\,]_n\,[\,]_m$
- $[\,[\,out^*m\,]_t\,]_m \to [\,]_t\,[\,]_m$
- $[\,in^*n\,]_t\,[\,]_n \to [\,[\,]_t\,]_n$

After the application of these rules, M' evolves to $N' = [\,[\,]_t\,]_n\,[\,]_m$. The inductive hypothesis expresses that N' encodes an ambient B'. After obtaining N', N has the structure $N = M_1\ N'$ if $M = M_1\ M'$ and it encodes the mobile ambient $B = A_1 \mid B'$, or N has the structure $N = M_2[\,N'\,]$ if $M = M_2[\,M'\,]$ and it encodes the mobile ambient $B = A_2[B']$. The transition from M' to N' represents also the transition from M to N.

It should be noticed that by consuming the capability *out m* we have $A' \Rightarrow_{amb} B'$. So the transition from M to N with the consumption of only one non-star object is simulated by the transition of A to B.

5. If the first rule applied is

$$[\,\overline{open}\ m\,]_m open\ m\ dlock\ one \to [\,\delta\,]_m\ dlock\ open_*m\ \overline{open}_*m,$$

where the membrane m contains the object $\overline{open}\ m$, the corresponding membrane labelled $\overline{open}\ m$ and the membrane structure M_1, whereas membrane $open\ m$ contains the membrane structure M_2, then M contains the membrane structure

$$[\,\overline{open}\ m\,[\,]_{\overline{open}\ m}\ M_1]_m\ open\ m\,[\,M_2\,]_{open\ m}.$$

According to the definition of $\mathcal{T}$, M can be written as $M_3\ M'$ or $M_4[\,M'\,]$, where $M' = [\,\overline{open}\ m\,[\,]_{\overline{open}\ m}\ M_1]_m\ open\ m\,[\,M_2\,]_{open\ m}$ and M_4 represents a membrane structure in which M' is placed inside a nested structure of translated ambients. We denote with the word *path* a path of capabilities encoded in the membrane structure M_2 with length greater or equal to zero and with the word *nest* the mobile structure contained in membrane m; we have $\mathcal{T}_1(nest) = M_3$ and $\mathcal{T}_1(path) = M_4$. If A is a mobile ambient encoded by $M = M_3\ M'$, then according to the definition of $\mathcal{T}$ it contains two ambients $A' = m[\,nest\ \overline{open}\ m\,] \mid open\ m.path$ and A_3 such that $A = A_3 \mid A'$, $\mathcal{T}_1(A') = M'$, and $\mathcal{T}_1(A_3) = M_3$. If A is a mobile ambient encoded by $M = M_4[\,M'\,]$, then according to the definition of $\mathcal{T}$ it contains two ambients $A' = m[\,nest\ \overline{open}\ m\,] \mid open\ m.path$ and A_4 such that $A = A_4[\,A'\,]$, $\mathcal{T}_1(A') = M'$, and $\mathcal{T}_1(A_4) = M_4$.

The application of the rule defined above to the membrane system M changes only the membrane system M'. The newly created objects $open_* m$ and $\overline{open}_* m$ are consumed by the following rules:

- $open_* m\, [\,]_{open\ m} \rightarrow [\,\delta\,]_{open\ m}$
- $\overline{open}_* m\, [\,]_{\overline{open}\ m} \rightarrow [\,\delta\,]_{\overline{open}\ m}$

After the application of these rules, M' evolves to $N' = M_1\, M_2$. The inductive hypothesis expresses that N' encodes an ambient B'. After obtaining N', N has the structure $N = M_3\, N'$ if $M = M_3\, M'$ and it encodes the mobile ambient $B = A_1 \mid B'$, or N has the structure $N = M_4[\, N'\,]$ if $M = M_4[\, M'\,]$ and it encodes the mobile ambient $B = A_4[B']$. The transition from M' to N' represents also the transition from M to N.

It should be noticed that by consuming the capability $open\ m$ we have $A' \Rightarrow_{amb} B'$. So the transition from M to N with the consumption of only one non-star object is simulated by the transition of A to B.

6. If the first rule applied is

$$[\, in\ m\ dlock\ one\,]_n\ [\overline{in}\ m]_m \rightarrow [\, in^* m\ in_* m\ dlock\,]_n\ [\,\overline{in}_* m\,]_m$$

where the membrane n contains the object $in\ m$, the corresponding membrane labelled $in\ m$ and another nested membrane structure $[\ldots[\,]_{t_{n+1}}\cdots]_{t_1}$, then M contains the membrane structure

$$[\, in\ m\, [\,]_{in\ m}\, [\ldots[\,]_{t_{n+1}}\cdots]_{t_1}\,]_n\, [\,\overline{in}\ m\, [\,]_{\overline{in}\ m}\,]_m.$$

We suppose that for membrane n with depth less than or equal to s the proposition is true. We prove the result for depth $s+1$. According to the definition of $\mathcal{T}$, M can be written as $M_1\, M'$ or $M_2[\, M'\,]$, where $M' = [\, in\ m\, [\,]_{in\ m}\, [\ldots[\,]_{t_{n+1}}\cdots]_{t_1}\,]_n$ $[\,\overline{in}\ m\, [\,]_{\overline{in}\ m}\,]_m$ and M_2 represents a membrane structure in which M' is placed inside a nested structure of translated ambients. If A is a mobile ambient encoded by $M = M_1\, M'$, then according to the definition of $\mathcal{T}$ it contains two ambients $A' = n[\, in\ m\ t_1[\ldots t_{s+1}[\,]\,]\,]\,] \mid m[\,]$ and A_1 such that $A = A_1 \mid A'$, $\mathcal{T}_1(A') = M'$, and $\mathcal{T}_1(A_1) = M_1$. If A is a mobile ambient encoded by $M = M_2[\, M'\,]$, then according to the definition of $\mathcal{T}$ it contains two ambients $A' = n[\, in\ m\ t_1[\ldots t_{s+1}[\,]\,]\,]\,] \mid m[\,]$ and A_2 such that $A = A_2[\, A'\,]$, $\mathcal{T}_1(A') = M'$, and $\mathcal{T}_1(A_2) = M_2$.

The application of the rule defined above to the membrane system M changes only the membrane system M'. The newly created objects $in^* m$, $in_* m$ and $\overline{in}_* m$ determine the application of other rules. After applying all the rules such that the object $in\ m$ is consumed and no star objects exist in the membrane system, the only membrane structures modified by the application of these rules is M' which evolves to $N' = [\,[\,[\,\cdots[\,]_{t_{s+1}}\,]_{t_1}\,]_n\,]_m$ and M_1 or M_2 remains the same. We know, from the inductive hypothesis, that an ambient $B'' = m[\, n[\, t_1[\ldots t_s[\,]\,]\,]\,]\,]$ is encoded into the structure $N'' = [\,[\,[\ldots[\,]_{t_s}\,]_{t_1}\,]_n\,]_m$, and that $B''' = t_{s+1}[\,]$ is encoded into the structure $N''' = [\,]_{t_{s+1}}$. According to the definition of $\mathcal{T}$ there exists an ambient B' with the structure $m[\, n[\, t_1[\ldots t_{s+1}[\,]\,]\,]\,]\,]$ such that $\mathcal{T}_1(B') = N'$. After obtaining N', N has the structure $N = M_1\, N'$ if $M = M_1\, M'$ and it encodes the mobile ambient $B = A_1 \mid B'$, or N has the structure $N = M_2[\, N'\,]$ if $M = M_2[\, M'\,]$ and it encodes the mobile ambient $B = A_2[\, B'\,]$. The transition from M' to N' represents also the transition from M to N.

It should be noticed that by consuming the capability *in m* we have $A' \Rightarrow_{amb} B'$. So the transition from M to N with the consumption of only one non-star object is simulated by the transition of A to B.

7. If the first rule applied is

$$[\,[\,out\ m\ dlock\ one\,]_n\ \overline{out}\ m\,]_m \rightarrow [\,[\,out^*m\ out_*m\ dlock\,]_n\ \overline{out}_*m\,]_m,$$

where the membrane n contains only the object *out m*, the corresponding membrane labelled *out m* and another nested membrane $[\ldots[\,]_{t_{s+1}}\cdots]_{t_1}$, then M contains the membrane structure

$$[\,[\,out\ m\,[\,]_{out\ m}\,[\cdots[\,]_{t_{s+1}}\cdots]_{t_1}]_n\ \overline{out}\ m\,[\,]_{\overline{out}\ m}]_m.$$

We suppose that for membrane n with the depth less than or equal to s the proposition is true. We prove the result for depth $s+1$. According to the definition of $\mathscr{T}$, M can be written as $M_1\ M'$ or $M_2[\,M'\,]$, where $M' = [\,[\,out\ m\,[\,]_{out\ m}\,[\cdots[\,]_{t_{s+1}}\cdots]_{t_1}]_n\ \overline{out}\ m\,[\,]_{\overline{out}\ m}]_m$ and M_2 represents a membrane structure in which M' is placed inside a nested structure of translated ambients. If A is a mobile ambient encoded by $M = M_1\ M'$, then according to the definition of $\mathscr{T}$ it contains two ambients $A' = m[\,n[\,out\ m\ t_1[\ldots t_{s+1}[\,]\,]\,]\,]$ and A_1 such that $A = A_1 \mid A'$, $\mathscr{T}_1(A') = M'$, and $\mathscr{T}_1(A_1) = M_1$. If A is a mobile ambient encoded by $M = M_2[M']$, then according to the definition of $\mathscr{T}$ it contains two ambients $A' = m[\,n[\,out\ m\ t_1[\ldots t_{s+1}[\,]\,]\,]\,]$ and A_2 such that $A = A_2[\,A'\,]$, $\mathscr{T}_1(A') = M'$, and $\mathscr{T}_1(A_2) = M_2$.

The application of the rule defined above to the membrane system M changes only the membrane system M'. The newly created objects out^*m, out_*m and $\overline{out}_*m$ determine the application of other rules. After applying all the rules such that the object *out m* is consumed and no star objects exist in the membrane system, the only membrane structures modified by the application of these rules is M' which evolves to $N' = [\,]_m\,[\,[\ldots[\,]_{t_{s+1}}\,]_{t_1}\,]_n$ and M_1 or M_2 remains the same. We know, from the inductive hypothesis, that an ambient $B'' = m[\,]\ n[\,t[\ldots t_s[\,]\,]\,]$ is encoded into the structure $N'' = [\,]_m\,[\,[\ldots[\,]_{t_s}\,]_{t_1}\,]_n$, and that $B''' = t_{s+1}[\,]$ is encoded into the structure $N''' = [\,]_{t_{s+1}}$. According to the definition of $\mathscr{T}$ there exists an ambient B' with the structure $m[\,]n[\,t[\ldots t_{s+1}[\,]\,]\,]$ such that $\mathscr{T}_1(B') = N'$. After obtaining N', N has the structure $N = M_1\ N'$ if $M = M_1\ M'$ and it encodes the mobile ambient $B = A_1 \mid B'$, or N has the structure $N = M_2[\,N'\,]$ if $M = M_2[\,M'\,]$ and it encodes the mobile ambient $B = A_2[\,B'\,]$. The transition from M' to N' represents also the transition from M to N.

It should be noticed that by consuming the capability *out m* we have $A' \Rightarrow_{amb} B'$. So the transition from M to N with the consumption of only one non-star object is simulated by the transition of A to B.

All the sequences of rules from all the cases above are determined uniquely by the star objects and by the priorities imposed. □

Remark 3.5. If $M \xrightarrow{r_1} \ldots \xrightarrow{r_k} N$, and both M and N contain only one *dlock* object, then the number of steps which transform ambient A into ambient B is the number of pairs of non-star objects consumed during the computation in the membrane evolution. The order in which the reductions take place in ambients is the order in which the pairs of non-star objects are consumed in the membrane systems.

Considering together the previous two propositions, we have in [12] the following result.

Theorem 3.4 (Operational correspondence).

1. If $A \Rightarrow_{amb} B$, then $\mathscr{T}(A) \Rightarrow_{mem} \mathscr{T}(B)$.

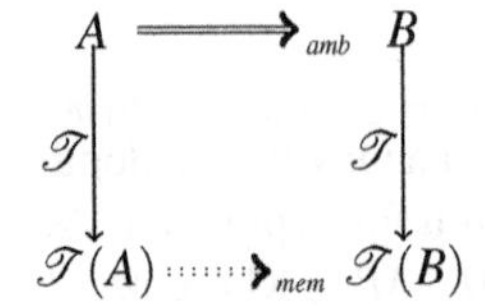

2. If $\mathscr{T}(A) \Rightarrow_{mem} M$, then there exists B such that $A \Rightarrow_{amb} B$ and $M = \mathscr{T}(B)$.

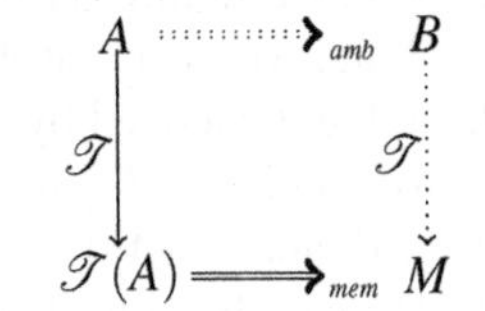

3.4 Branes into Mobile Membranes with Objects on Surface

Some work has been done trying to relate membrane systems and brane calculi [35, 37, 47, 105, 106]. Inspired by brane calculi, a model of membrane systems having objects attached to the membranes has been introduced in [45]. In [31], a class of membrane systems containing both free floating objects and objects attached to membranes have been proposed. We are continuing this research line, and simulate a fragment of brane calculi by using systems of mutual membranes with objects on surface.

"At the first sight, the role of objects placed on membranes is different in membrane and brane systems: in membrane computing, the focus is on the evolution of objects themselves, while in brane calculi the objects ("proteins") mainly control the evolution of membranes" [129]. By defining an encoding of the PEP fragment of brane calculi into systems of mutual membranes with objects on surface, we show that the difference between the two models is not significant. Another difference regarding the semantics of the two formalisms is expressed in [35]: "whereas brane calculi are usually equipped with an interleaving, sequential semantics (each computational step consists of the execution of a single instruction), the usual semantics in membrane computing is based on maximal parallelism (a computational step is composed of a maximal set of independent interactions)."

3.4.1 Translation

Definition 3.13. A translation $\mathcal{T}_2 : \mathcal{P} \to \mathcal{M}$ is given by

$$\mathcal{T}_2(P) = \begin{cases} [\,]_{\mathcal{S}(\sigma)} & \text{if } P = \sigma(\,) \\ [\mathcal{T}_2(R)]_{\mathcal{S}(\sigma)} & \text{if } P = \sigma(R) \\ \mathcal{T}_2(Q)\ \mathcal{T}_2(R) & \text{if } P = Q \mid R \end{cases}$$

where $\mathcal{S} : \mathcal{P} \to A$ is defined as:

$$\mathcal{S}(\sigma) = \begin{cases} \sigma & \text{if } \sigma = phago_n \text{ or } \sigma = exo_n \text{ or } \sigma = \overline{exo_n} \\ \overline{phago_n}\ S(\rho) & \text{if } \sigma = \overline{phago_n}(\rho) \\ pino\ S(\rho) & \text{if } \sigma = pino(\rho) \\ \mathcal{S}(a)\ \mathcal{S}(\rho) & \text{if } \sigma = a.\rho \\ \mathcal{S}(\tau)\ \mathcal{S}(\rho) & \text{if } \sigma = \tau \mid \rho \\ \lambda & \text{if } \sigma = 0 \end{cases}$$

The rules of systems of mutual membranes with objects on surface are in Table 3.1.

Table 3.1 Pino/Exo/Phago Rules of M^2OS

$[\,]_{S(phago_n.\sigma\mid\sigma_0)}[\,]_{S(\overline{phago_n}(\rho).\tau\mid\tau_0)} \to_m [[[\,]_{S(\sigma\mid\sigma_0)}]_{S(\rho)}]_{S(\tau\mid\tau_0)}$
$[[\,]_{S(exo_n.\sigma\mid\sigma_0)}]_{S(\overline{exo_n}.\tau\mid\tau_0)} \to_m [\,]_{S(\sigma\mid\sigma_0\mid\tau\mid\tau_0)}$
$[\,]_{S(pino(\rho).\sigma\mid\sigma_0)} \to_m [[\,]_{S(\rho)}]_{S(\sigma\mid\sigma_0)}$

3.4.2 Preservation of Properties Through Translation

The next proposition states that two PEP systems which are structurally equivalent are translated into two systems of mutual membranes with objects on surface which are structurally equivalent.

Proposition 3.17. *If $P \equiv_b Q$ then $\mathcal{T}_2(P) \equiv_m \mathcal{T}_2(Q)$.*

$$\begin{array}{ccc} P & \equiv_b & Q \\ \downarrow{\scriptstyle \mathcal{T}_2} & & \downarrow{\scriptstyle \mathcal{T}_2} \\ \mathcal{T}_2(P) & \equiv_m & \mathcal{T}_2(Q) \end{array}$$

Proof. We proceed by structural induction.

If $P = P_1 \circ P_2$ where P_1 and P_2 are two brane systems, then from the definition of $\equiv_b$ we have that $Q = P_2 \circ P_1$. Using the definition of $\mathcal{T}_2$ and $P = P_1 \circ P_2$ we have that $\mathcal{T}_2(P) = \mathcal{T}_2(P_1)\mathcal{T}_2(P_2)$. From $Q = P_2 \circ P_1$ and the definition of $\mathcal{T}_2$ we have that $\mathcal{T}_2(Q) = \mathcal{T}_2(P_2)\mathcal{T}_2(P_1)$. From the definition of $\equiv_m$ we get $\mathcal{T}_2(P) \equiv_m \mathcal{T}_2(Q)$.

If $P = P_1 \circ (P_2 \circ P_3)$ where P_1, P_2 and P_3 are three brane systems, then from the definition of $\equiv_b$ we have that $Q = (P_1 \circ P_2) \circ P_3$. Using the definition of $\mathcal{T}_2$ and $P = P_1 \circ (P_2 \circ P_3)$ we have that $\mathcal{T}_2(P) = \mathcal{T}_2(P_1)(\mathcal{T}_2(P_2)\mathcal{T}_2(P_3))$. From $Q = (P_1 \circ P_2) \circ P_3$ and the definition of $\mathcal{T}_2$ we have that $\mathcal{T}_2(Q) = (\mathcal{T}_2(P_1)\mathcal{T}_2(P_2))\mathcal{T}_2(P_3)$. From the definition of $\equiv_m$ we get $\mathcal{T}_2(P) \equiv_m \mathcal{T}_2(Q)$.

Let $P = P_1 \circ P_3$ and $Q = P_2 \circ P_3$ such that $P_1 \equiv_b P_2$. From the definition of $\equiv_b$ it results that $P \equiv_b Q$. Using the definition of $\mathcal{T}_2$ we have that $\mathcal{T}_2(P) = \mathcal{T}_2(P_1)\mathcal{T}_2(P_3)$ and $\mathcal{T}_2(Q) = \mathcal{T}_2(P_2)\mathcal{T}_2(P_3)$. Using the structural induction, from $P_1 \equiv_b P_2$ it results that $\mathcal{T}_2(P_1) \equiv_m \mathcal{T}_2(P_2)$. From the definition of $\equiv_m$ we get $\mathcal{T}_2(P) \equiv_m \mathcal{T}_2(Q)$.

Let $P = \rho(P_1)$ and $Q = \tau(P_2)$ such that $P_1 \equiv_b P_2$ and $\rho \equiv_b \tau$. From the definition of $\equiv_b$ it results that $P \equiv_b Q$. Using the definition of $\mathcal{T}_2$ we have that $\mathcal{T}_2(P) = [\mathcal{T}_2(P_1)]_{\mathcal{S}(\rho)}$ and $\mathcal{T}_2(Q) = [\mathcal{T}_2(P_2)]_{\mathcal{S}(\tau)}$. Using the structural induction, from $P_1 \equiv_b P_2$ it results that $\mathcal{T}_2(P_1) \equiv_m \mathcal{T}_2(P_2)$ and from $\rho \equiv_b \tau$ it results that $S(\rho) \equiv_m S(\tau)$. From the definition of $\equiv_m$ we get $\mathcal{T}_2(P) \equiv_m \mathcal{T}_2(Q)$.

In what follows we prove that indeed from $\rho \equiv_b \tau$ it results that $S(\rho) \equiv_m S(\tau)$. We proceed also by structural induction.

If $\rho = \rho_1 \mid \rho_2$ where ρ_1 and ρ_2 are two combinations of brane actions, then from the definition of $\equiv_b$ we have that $\tau = \rho_2 \mid \rho_1$. Using the definition of $\mathcal{S}$ and $\rho = \rho_1 \mid \rho_2$ we have that $\mathcal{S}(\rho) = \mathcal{S}(\rho_1)\mathcal{S}(\rho_2)$. From $\tau = \rho_2 \mid \rho_1$ and the definition of $\mathcal{S}$ we have that there exist v such that $\mathcal{S}(\tau) = \mathcal{S}(\tau_2)\mathcal{S}(\tau_1)$. From the definition of $\equiv_m$ we get $S(\rho) \equiv_m S(\tau)$.

If $\rho = \rho_1 \mid (\rho_2 \mid \rho_3)$ where ρ_1, ρ_2 and ρ_3 are three combinations of brane actions, then from the definition of $\equiv_b$ we have that $\tau = (\rho_1 \mid \rho_2) \mid \rho_3$. Using the definition of $\mathcal{S}$ and $\rho = \rho_1 \mid (\rho_2 \mid \rho_3)$ we have that $\mathcal{S}(\rho) = \mathcal{S}(\rho_1)(\mathcal{S}(\rho_2)\mathcal{S}(\rho_3))$. From $\tau = (\rho_1 \mid \rho_2) \mid \rho_3$ and the definition of $\mathcal{S}$ we have that $\mathcal{S}(\tau) = (\mathcal{S}(\rho_1)\mathcal{S}(\rho_2))\mathcal{S}(\rho_3)$. From the definition of $\equiv_m$ we get $S(\rho) \equiv_m S(\tau)$.

If $\rho = \rho_1 \mid 0$ where ρ_1 is a combination of brane actions, then from the definition of $\equiv_b$ we have that $\tau = \rho_1$. Using the definition of $\mathcal{S}$ and $\rho = \rho_1 \mid 0$ we have that $\mathcal{S}(\rho) = \mathcal{S}(\rho_1)\lambda$. From $\tau = \rho_1$ and the definition of $\mathcal{S}$ we have that $\mathcal{S}(\tau) = \mathcal{S}(\tau_1)$. From the definition of $\equiv_m$ we get $S(\rho) \equiv_m S(\tau)$.

Let $\rho = \rho_1 \mid \rho_3$ and $\tau = \rho_2 \mid \rho_3$ such that $\rho_1 \equiv_b \rho_2$. From the definition of $\equiv_b$ it results that $\rho \equiv_b \tau$. Using the definition of $\mathcal{S}$ we have that $\mathcal{S}(\rho) = \mathcal{S}(\rho_1)\mathcal{S}(\rho_3)$ and $\mathcal{S}(\tau) = \mathcal{S}(\rho_2)\mathcal{S}(\rho_3)$. Using the structural induction, from $\rho_1 \equiv_b \rho_2$ it results that $\mathcal{S}(\rho_1) \equiv_m \mathcal{S}(\rho_2)$. From the definition of $\equiv_m$ we get $S(\rho) \equiv_m S(\tau)$.

Let $\rho = a.\rho_1$ and $\tau = a.\rho_2$ such that $\rho_1 \equiv_b \rho_2$. From the definition of $\equiv_b$ it results that $\rho \equiv_b \tau$. Using the definition of $\mathcal{S}$ we have that $\mathcal{S}(\rho) = \mathcal{S}(a)\mathcal{S}(\rho_1)$ and $\mathcal{S}(\tau) = \mathcal{S}(a)\mathcal{S}(\rho_2)$. Using the structural induction, from $\rho_1 \equiv_b \rho_2$ it results that $\mathcal{S}(\rho_1) \equiv_m \mathcal{S}(\rho_2)$. From the definition of $\equiv_m$ we get $S(\rho) \equiv_m S(\tau)$.

Proposition 3.18. *If $\mathcal{T}_2(P) \equiv_m M$ then there exists Q such that $M = \mathcal{T}_2(Q)$.*

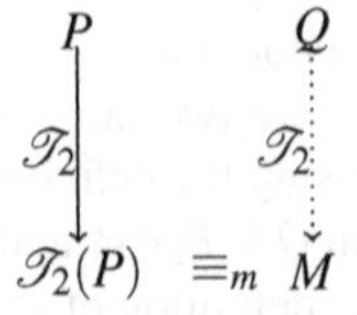

Proof. If $P = P_1 \circ P_2$ where P_1 and P_2 are two brane systems, then from the definition of $\mathcal{T}_2$ we have that $\mathcal{T}_2(P) = \mathcal{T}_2(P_1)\mathcal{T}_2(P_2)$. From the definition of $\equiv_m$ we get $\mathcal{T}_2(P) \equiv_m M$, where $M = \mathcal{T}_2(P_2)\mathcal{T}_2(P_1)$. For this M there exists Q with $Q = P_2 \circ P_1$ such that $M = \mathcal{T}_2(Q)$.

If $P = P_1 \circ (P_2 \circ P_3)$ where P_1, P_2 and P_3 are three brane systems, then from the definition of $\mathcal{T}_2$ we have that $\mathcal{T}_2(P) = \mathcal{T}_2(P_1)(\mathcal{T}_2(P_2)\mathcal{T}_2(P_3))$. From the definition of $\equiv_m$ we get $\mathcal{T}_2(P) \equiv_m M$, where $M = (\mathcal{T}_2(P_1)\mathcal{T}_2(P_2))\mathcal{T}_2(P_3)$. For this M there exists Q with $Q = (P_1 \circ P_2) \circ P_3$ such that $M = \mathcal{T}_2(Q)$.

If $P = P_1 \circ P_3$, where P_1 and P_3 are two brane systems, then from the definition of $\mathcal{T}_2$ we have that $\mathcal{T}_2(P) = \mathcal{T}_2(P_1)\mathcal{T}_2(P_2)$. From the definition of $\equiv_m$ we get $\mathcal{T}_2(P) \equiv_m M$, where $M = \mathcal{T}_2(P_2)\mathcal{T}_2(P_3)$ for $\mathcal{T}_2(P_1) \equiv_m \mathcal{T}_2(P_2)$. For this M there exists Q with $Q = P_2 \circ P_3$ such that $M = \mathcal{T}_2(Q)$.

If $P = \rho(P_1)$, where P_1 is a brane system and ρ is a combination of brane actions, then from the definition of $\mathcal{T}_2$ we have that $\mathcal{T}_2(P) = [\mathcal{T}_2(P_1)]_{\mathcal{S}(\rho)}$. From the definition of $\equiv_m$ we get $\mathcal{T}_2(P) \equiv_m M$, where $M = [\mathcal{T}_2(P_2)]_{\mathcal{S}(\tau)}$ for $\mathcal{T}_2(P_1) \equiv_b \mathcal{T}_2(P_2)$ and $\mathcal{S}(\rho) \equiv_m \mathcal{S}(\tau)$. For this M there exists Q with $Q = \sigma(P_2)$ such that $M = \mathcal{T}_2(Q)$.

Remark 3.6. In Proposition 3.18 it is possible that $P \not\equiv_b Q$. If $P = phago_n.\ exo_n(\ )$, then by translation $M = \mathcal{T}_2 = [\ ]_{phago_n exo_n} \equiv_m [\ ]_{exo_n phago_n}$=N. It is possible to have $Q = exo_n.phago_n(\ )$ or $Q = exo_n \mid phago_n(\ )$ such that $N = \mathcal{T}_2(Q)$, but $P \not\equiv_b Q$.

Proposition 3.19. *If $P \to_b Q$ then $\mathcal{T}_2(P) \to_m \mathcal{T}_2(Q)$.*

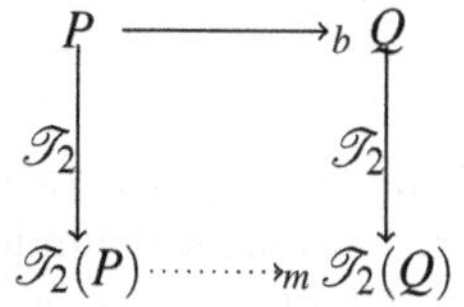

Proof. We proceed by structural induction.

If $P = pino(\rho).\sigma \mid \sigma_0(P_1)$, where P_1 is a brane system and $pino(\rho).\sigma \mid \sigma_0$ is a combination of brane actions, then from the definition of $\to_b$ we have that $Q = \sigma \mid \sigma_0(\rho(\ ) \circ P_1)$. Using the definition of $\mathcal{T}_2$ and $P = pino(\rho).\sigma \mid \sigma_0(P_1)$ we have that $\mathcal{T}_2(P) = [\mathcal{T}_2(P_1)]_{\mathcal{S}(pino(\rho).\sigma|\sigma_0)}$. From $Q = \sigma \mid \sigma_0(\rho(\ ) \circ P_1)$ and the definition of $\mathcal{T}_2$ we have that $\mathcal{T}_2(Q) = [[\mathcal{T}_2(P_1)]_{\mathcal{S}(\rho)}]_{\mathcal{S}(\sigma|\sigma_0)}$. From the definition of $\to_m$ we get $\mathcal{T}_2(P) \to_m \mathcal{T}_2(Q)$.

If $P = \overline{exo}_n.\tau \mid \tau_0(exo_n.\sigma \mid \sigma_0(P_1) \circ P_2)$, where P_1, P_2 are two brane systems and $\overline{exo}_n.\tau \mid \tau_0$, $exo_n.\sigma \mid \sigma_0$ are combinations of brane actions, then from the definition of $\to_b$ we have that $Q = P_1 \circ \sigma \mid \sigma_0 \mid \tau \mid \tau_0(P_2)$. Using the definition of $\mathcal{T}_2$ and $P = \overline{exo}_n.\tau \mid \tau_0(exo_n.\sigma \mid \sigma_0(P_1) \circ P_2)$ we have that $\mathcal{T}_2(P) = [[\mathcal{T}_2(P_1)]_{\mathcal{S}(exo_n.\sigma|\sigma_0)}$ $\mathcal{T}_2(P_2)]_{\mathcal{S}(\overline{exo}_n.\tau|\tau_0)}$. From $Q = P_1 \circ \sigma \mid \sigma_0 \mid \tau \mid \tau_0(P_2)$ and the definition of $\mathcal{T}_2$ we have that $\mathcal{T}_2(Q) = \mathcal{T}_2(P_1)[\mathcal{T}_2(P_2)]_{\mathcal{S}(\sigma|\sigma_0|\tau|_0)}$. From the definition of $\to_m$ we get $\mathcal{T}_2(P) \to_m \mathcal{T}_2(Q)$.

If $P = phago_n.\sigma \mid \sigma_0(P_1) \circ \overline{phago}_n(\rho).\tau \mid \tau_0(P_2)$, where P_1, P_2 are two brane systems and $phago_n.\sigma \mid \sigma_0$, $\overline{phago}_n(\rho).\tau \mid \tau_0$ are combinations of brane actions, then from the definition of $\to_b$ we have that $Q = \tau \mid \tau_0(\rho(\sigma \mid \sigma_0(P_1)) \circ P_2)$. Using the definition of $\mathcal{T}_2$ and $P = phago_n.\sigma \mid \sigma_0(P_1) \circ \overline{phago}_n(\rho).\tau \mid \tau_0(Q_2)$ we have that $\mathcal{T}_2(P) = [\mathcal{T}_2(P_1)]_{\mathcal{S}(phago_n.\sigma|\sigma_0)}[\mathcal{T}_2(P_2)]_{\mathcal{S}(\overline{phago}_n(\rho).\tau|\tau_0)}$. From $Q = \tau \mid \tau_0(\rho(\sigma \mid \sigma_0(P_1)) \circ P_2)$ and the definition of $\mathcal{T}_2$ we have that $\mathcal{T}_2(Q) = [[[\mathcal{T}_2(P_1)]_{\mathcal{S}(\sigma|\sigma_0)}]_{\mathcal{S}(\rho)}\ \mathcal{T}_2(P_2)]_{\mathcal{S}(\tau|\tau_0)}$. From the definition of $\rightarrow_m$ we get $\mathcal{T}_2(P) \rightarrow_m \mathcal{T}_2(Q)$.

Let $P = P_1 \circ P_3$ and $Q = P_2 \circ P_3$ such that $P_1 \rightarrow_b P_2$. From the definition of $\rightarrow_b$ it results that $P \rightarrow_b Q$. Using the definition of $\mathcal{T}_2$ we have that $\mathcal{T}_2(P) = \mathcal{T}_2(P_1)\mathcal{T}_2(P_3)$ and $\mathcal{T}_2(Q) = \mathcal{T}_2(P_2)\mathcal{T}_2(P_3)$. Using the structural induction, from $P_1 \rightarrow_b P_2$ it results that $\mathcal{T}_2(P_1) \rightarrow_m \mathcal{T}_2(P_2)$. From the definition of $\rightarrow_m$ we get $\mathcal{T}_2(P) \rightarrow_m \mathcal{T}_2(Q)$.

Let $P = \sigma(P_1)$ and $Q = \sigma(P_2)$ such that $P_1 \rightarrow_b P_2$. From the definition of $\rightarrow_b$ it results that $P \rightarrow_b Q$. Using the definition of $\mathcal{T}_2$ we have that $\mathcal{T}_2(P) = [\mathcal{T}_2(P_1)]_{\mathcal{S}(\sigma)}$ and $\mathcal{T}_2(Q) = [\mathcal{T}_2(P_2)]_{\mathcal{S}(\sigma)}$. Using the structural induction, from $P_1 \rightarrow_b P_2$ it results that $\mathcal{T}_2(P_1) \rightarrow_m \mathcal{T}_2(P_2)$. From the definition of $\rightarrow_m$ we get $\mathcal{T}_2(P) \rightarrow_m \mathcal{T}_2(Q)$.

Let $P \rightarrow_b Q$ such that $P \equiv_b P'$, $P' \rightarrow Q'$ and $Q \equiv_b Q'$. Using the definition of $\mathcal{T}_2$ we have that $\mathcal{T}_2(P) = \mathcal{T}_2(P')$ and $\mathcal{T}_2(Q) = \mathcal{T}_2(Q')$. Using the structural induction, from $P_1 \rightarrow_b P_2$ it results that $\mathcal{T}_2(P_1) \rightarrow_m \mathcal{T}_2(P_2)$. From the definition of $\rightarrow_m$ we get $\mathcal{T}_2(P) \rightarrow_m \mathcal{T}_2(Q)$.

Proposition 3.20. *If $\mathcal{T}_2(P) \rightarrow_m M$ then there exists Q such that $M = \mathcal{T}_2(Q)$.*

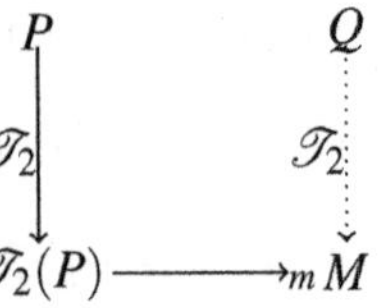

Proof. If $P = pino(\rho).\sigma \mid \sigma_0(P_1)$, where P_1 is a brane system and $pino(\rho).\sigma \mid \sigma_0$ is a combination of brane actions, then from the definition of $\mathcal{T}_2$ we have that $\mathcal{T}_2(P) = [\mathcal{T}_2(P_1)]_{\mathcal{S}(pino(\rho).\sigma|\sigma_0)}$. From the definition of $\rightarrow_m$ we get $\mathcal{T}_2(P) \rightarrow_m M$, where $M = [[\mathcal{T}_2(P_1)]_{\mathcal{S}(\rho)}]_{\mathcal{S}(\sigma|\sigma_0)}$. For this M there exists Q with $Q = \sigma_1(\rho_1(P_1))$ with $S(\sigma_1) = S(\sigma \mid \sigma_0)$ and $S(\rho_1) = S(\rho)$ such that $M = \mathcal{T}_2(Q)$.

If $P = \overline{exo}_n.\tau \mid \tau_0(exo_n.\sigma \mid \sigma_0(P_1) \circ P_2)$, where P_1, P_2 are two brane systems and $\overline{exo}_n.\tau \mid \tau_0$, $exo_n.\sigma \mid \sigma_0$ are combinations of brane actions, then from the definition of $\mathcal{T}_2$ we have that $\mathcal{T}_2(P) = [[\mathcal{T}_2(P_1)]_{\mathcal{S}(exo_n.\sigma|\sigma_0)}\mathcal{T}_2(P_2)]_{\mathcal{S}(\overline{exo}_n.\tau|\tau_0)}$. From the definition of $\rightarrow_m$ we get $\mathcal{T}_2(P) \rightarrow_m M$, where $M = \mathcal{T}_2(P_1)[\mathcal{T}_2(P_2)]_{\mathcal{S}(\sigma|\sigma_0|\tau|_0)}$. For this M there exists Q with $Q = P_1 \circ \sigma_1(P_2)$ and $S(\sigma_1) = S(\sigma \mid \sigma_0 \mid \tau \mid \tau_0)$ such that $M = \mathcal{T}_2(Q)$.

If $P = phago_n.\sigma \mid \sigma_0(P_1) \circ \overline{phago}_n(\rho).\tau \mid \tau_0(Q_2)$, where P_1, P_2 are two brane systems and $phago_n.\sigma \mid \sigma_0$, $\overline{phago}_n(\rho).\tau \mid \tau_0$ are combinations of brane actions, then from the definition of $\mathcal{T}_2$ we have that $\mathcal{T}_2(P) = [\mathcal{T}_2(P_1)]_{\mathcal{S}(phago_n.\sigma|\sigma_0)}$ $[\mathcal{T}_2(P_2)]_{\mathcal{S}(\overline{phago}_n(\rho).\tau|\tau_0)}$. From the definition of $\rightarrow_m$ we get $\mathcal{T}_2(P) \rightarrow_m M$, where $M = [[[\mathcal{T}_2(P_1)]_{\mathcal{S}(\sigma|\sigma_0)}]_{\mathcal{S}(\rho)}\ \mathcal{T}_2(P_2)]_{\mathcal{S}(\tau|\tau_0)}$. For this M there exists Q with $Q = \tau_1(\rho_1(\sigma_1(P_1)) \circ P_2)$ and $S(\tau_1) = S(\tau \mid \tau_0)$, $S(\rho_1) = S(\rho)$, $S(\sigma_1) = S(\sigma \mid \sigma_0)$ such that $M = \mathcal{T}_2(Q)$.

If $P = P_1 \circ P_3$ where P_1, P_3 are two brane systems, then from the definition of $\mathcal{T}_2$ we have $\mathcal{T}_2(P) = \mathcal{T}_2(P_1)\mathcal{T}_2(P_3)$. From the definition of $\rightarrow_m$ we get $\mathcal{T}_2(P) \rightarrow_m M$, where $M = \mathcal{T}_2(P_2)\mathcal{T}_2(P_3)$ for $\mathcal{T}_2(P_1) \to_m \mathcal{T}_2(P_2)$. For this M there exists Q with $Q = P_2 \circ P_3$ such that $M = \mathcal{T}_2(Q)$.

If $P = \sigma(P_1)$ where P_1 is a brane system, then from the definition of $\mathcal{T}_2$ we have $\mathcal{T}_2(P) = [\mathcal{T}_2(P_1)]_{\mathcal{S}(\sigma)}$. From the definition of $\to_m$ we get $\mathcal{T}_2(P) \to_m M$, where $M = [\mathcal{T}_2(P_2)]_{\mathcal{S}(\sigma)}$ for $\mathcal{T}_2(P_1) \to_m \mathcal{T}_2(P_2)$. For this M there exists Q with $Q = \sigma_1(P_2)$ and $S(\sigma_1) = S(\sigma)$ such that $M = \mathcal{T}_2(Q)$.

If $P \equiv_b P'$ where P' is a brane system, then from the definition of $\mathcal{T}_2$ we have that $\mathcal{T}_2(P) = \mathcal{T}_2(P')$. From the definition of $\to_m$ we get $\mathcal{T}_2(P) \to_m M$, where $M = \mathcal{T}_2(Q')$ for $\mathcal{T}_2(P') \to_m \mathcal{T}_2(Q')$. For this M there exists Q with $Q \equiv_b Q'$ such that $M = \mathcal{T}_2(Q)$.

The next remark is a consequence of the fact that we translate a formalism with an interleaving semantics into a formalism with a parallel semantics.

Remark 3.7. In Proposition 3.20 it is possible that $P \not\to_b Q$. Consider the process $P = \overline{exo}_n.\overline{exo}_n(exo_n.phago_n(\,))$. By translation $M = [[\,]_{exo_n\, phago_n}]_{\overline{exo}_n\, \overline{exo}_n}$, such that $M \to_m [\,]_{phago_n \overline{exo}_n}$. We observe that there exists $Q = phago_n.\overline{exo}_n(\,)$ such that $N = \mathcal{T}_2(Q)$, but $P \not\to_b Q$.

The PEP calculus may be extended as in [40] to contain also molecules inside the membranes. A new reduction simulates the exchanging of molecules simultaneously between the interior and exterior of a membrane. In this case the translation can be easily extended by introducing objects in membranes as in [31] and an antiport evolution rule in the definition of $\to_m$.

3.5 Mobile Membranes with Objects on Surface into Petri Nets

Systems biology is an emerging field which arises from the interaction of biology, mathematics and computer science. In order to cope with ensembles and quantities in biology, new formal approaches are required. Biologists increasingly recognize that mathematics and computational methods have become powerful enough to model the complexity of biological entities and systems. In this sense, "mathematics is biology's next microscope, only better"[75]. In biology, new ensemble behaviours emerge from the interactions of biological elements, and new formalisms are required to cope with these properties. In this sense, "biology is mathematics' next physics, only better"[75]. Formal models are used for many purposes, and each purpose influences the degree of detail. If we provide greater detail, the number of systems to which our model applies will decrease. Moreover, a formal model should have three properties, and each of these properties trades off against the other two [113]: *generality* (the number of systems and situations to which the model correctly applies), *realism* (the degree to which the model mimics the real world), *power and precision* (collection of revealed properties, and the accuracy of the models' predictions).

In what follows we use two formalisms: mobile membranes (more realism, being inspired by cell biology) and coloured Petri nets (power and precision provided also by complex software tools). A relation can be established between these two formalisms by providing an encoding of mobile membranes into coloured Petri nets. By considering the endocytic pathway for low-density lipoprotein degradation, we show how mobile membranes can be used to model such a biological phenomenon, while coloured Petri nets can be used to analyze and automatically verify some behavioural properties of the pathway.

There exist already some formal approaches to describe the endocytic pathway for low-density lipoprotein (LDL) degradation. In [35] is presented a model of the LDL degradation pathway in brane calculi and in symport-antiport membrane systems. Another paper investigating such a pathway is [136], where the LDL degradation pathway is modelled using bioambients, and then static analysis techniques are applied. However, none of the previous descriptions of the pathway is translated into a formalism having a software tool able to automatically check complex behavioural properties.

Some connections between membrane systems and Petri nets are presented for the first time in [78] and [137]. In [101, 102], a direct structural relationship between these two formalisms is established by defining a new class of Petri nets called Petri nets with localities. This new class of Petri nets has been used to show how maximal evolutions from membrane systems are faithfully reflected in the maximally concurrent step sequence semantics of their corresponding Petri nets with localities.

3.5.1 Coloured Petri Nets

Coloured Petri nets (CPN) represent a graphical language used to describe systems in which communication, synchronization and resource sharing play an important role [99]. The CPN model contains places (drawn as ellipses or circles), transitions (drawn as rectangular boxes), a number of directed arcs connecting places and transitions, and finally some textual inscriptions located near the places, transitions and arcs.

The places are used to represent the state of the modelled system, and this state is given by the number of tokens of all the places. Such a state is called a marking of the CPN model. By convention, we write the names of the places inside the ellipses. The names have no formal meaning, but they have a practical importance for the readability of a CPN model, just like the use of mnemonic names in traditional programming.

The arc expressions on the input arcs of a transition determine when the transition is enabled, i.e., activated by a certain marking. A transition is enabled whenever it is possible to find a binding of the variables that appear in the surrounding arc expressions of the transition such that the arc expression of each input arc evaluates to a multiset of tokens that is present in the corresponding input place. When a transition occurs with a given binding, it removes from each input place the multiset of tokens to which the corresponding input arc expression evaluates. Analogously, it adds to each output place the multiset of tokens to which the corresponding output arc expression evaluates.

Coloured Petri nets also have a mathematical representation with a well defined syntax and semantics. This formal representation is the framework for the study of different behavioural properties. We denote by $EXPR$ the set of expressions provided by the inscription language (which is ML in the case of CPN Tools), and by $Type[e]$ we denote the type of an expression $e \in EXPR$, i.e., the type of the values obtained when evaluating e. The set of free variables in an expression e is denoted $Var[e]$, and the type of a variable x is denoted $Type[x]$. We denote the set of variables by X; the set of expressions $e \in EXPR$ such that $Var[e] \subseteq X$ is denoted by $EXPR_X$. The set of all multisets over S, i.e., the multiset type over S is denoted S_{MS}. The following definition differs from that presented in [99] in that simultaneous parallel arcs from the same place to the same transition are not allowed (it is enough to have only one arc).

Definition 3.14. A **non-hierarchical Coloured Petri Net** is a nine tuple

$$CPN = (P,T,A,\Sigma,X,C,G,E,I), \text{ where}$$

1. P is a finite set of **places**;
2. T is a finite set of **transitions** T such that $P \cap T = \emptyset$;
3. $A \subseteq (P \times T) \cup (T \times P)$ is a set of directed **arcs**;
4. Σ is a finite set of non-empty **colour sets**;
5. X is a finite set of **typed variables** such that $Type[x] \in \Sigma$ for all $x \in X$;
6. $C : P \to \Sigma$ is a **colour set function** that assigns a colour set to each place;
7. $G : T \to EXPR_X$ is a **guard function** that assigns a guard to each transition t such that $Type[G(t)] = Bool$;
8. $E : A \to EXPR_X$ is an **arc expression function** that assigns a guard to each arc a such that $Type[E(a)] = C(p)_{MS}$, where p is the place connected to the arc a;
9. $I : P \to EXPR_\emptyset$ is an **initialization function** that assigns an initialization expression to each place p such that $Type[I(p)] = C(p)_{MS}$.

A distribution of tokens over the places of a net is called a marking. A set U of transitions is called enabled at a marking m if all its transitions are enabled at m (each transition requires a number of tokens that cannot be shared with any other transition). We use the notation $m[U\rangle m'$ to express that m enables the set U of transitions, and the marking m' results from m after applying all the transitions of U.

3.5.2 Mobile Membranes as Coloured Petri Nets

We denote by $\Pi = (M_0, R)$ a system of mobile membranes with a set R of rules having an initial membrane configuration $M_0 = (w_1^0, \ldots, w_n^0, \mu)$, where w_i^0 denotes the initial multiset of objects placed on membrane i, and μ the initial membrane structure. We consider that the system has at any point of evolution at most $k > 0$ membranes. Given such a system of mobile membranes, the corresponding coloured

Petri net is denoted by $CPN_\Pi = (P,T,A,\Sigma,X,C,G,E,I)$, where the components are defined as follows:

- $P = \{1,\ldots,k\} \cup \{structure\}$, where *structure* is a place that contains the structure of the corresponding membrane system, namely the pairs (i,j);
- $T = \bigcup_{1 \le k \le s} t_k$, where each t_k represents a distinct transition for a rule of R; since the rules over mobile membranes contains no explicit label for membranes, it means that:
  - a *pino* rule can be instantiated at most k times in each step;
  - a *phago* rule can be instantiated at most $\dfrac{k!}{2!(k-2)!}$ times in each step;
  - an *exo* rule can be instantiated at most $\dfrac{k!}{2!(k-2)!}$ times in each step; 2 represents the number of membranes from the left hand side of an *exo* rule, and $\dfrac{k!}{2!(k-2)!}$ represents all the possible combinations of membranes;

  Thus $s = s_1 * k + s_2 * \dfrac{k!}{2!(k-2)!} + s_3 * \dfrac{k!}{2!(k-2)!}$, where s_1, s_2 and s_3 represent the numbers of pino, phago and exo rules from R.
- A contains input arcs $(P \times T)$ and output arcs $(T \times P)$; for a rule r and its associated transition t, we build the arcs as follows:
  - the input arcs are from the places that represent the membranes appearing in the left hand side of the evolution rule r, and from the place *structure*, to the transition t;
  - the output arcs are from the transition t to the places that represent the membranes appearing in the right hand side of the evolution rule r and to the place *structure*;
- $\Sigma = U \cup L$, where U is the colour set containing all the objects from O, and $L = \{1,\ldots,k\} \times \{1,\ldots,k\}$ is a colour set containing the membrane structure;
- $X = \{x,y,z,\ldots\}$ is a set of variables used when modifying the content of place *structure*;
- $C(p) = \begin{cases} U, \text{ if } p \in \{1,\ldots,k\} \\ L, \text{ if } p = structure \end{cases}$
- $G(t) = \begin{cases} [x = y], & \text{if } t \text{ is a transition simulating a phago rule; it checks if} \\ & \text{both membranes from the left hand side of a phago rule} \\ & \text{have the same parent;} \\ true, & \text{otherwise.} \end{cases}$
- For a rule r and its associated transition t, we build E as follows:

- we place the multiset of objects u on an input arc from a place that represents a membrane appearing in the left hand side of the evolution rule r (being marked with a multiset of objects u) to the transition t;
- we place all the pairs (i, j) describing the membrane structure appearing in the left hand side of the evolution rule r on the input arc from the place *structure* to the transition t;
- we place the multiset of objects v on an output arc from a transition t to a place that represents a membrane appearing in the right hand side of the evolution rule r (being marked with a multiset of objects v);
- we place all the pairs (i, j) describing the membrane structure appearing in the right hand side of the evolution rule r on the output arc from the transition t to the place *structure*;

∘ $I(p) = \begin{cases} w_p^0, & \text{if } p \in \{1,\dots,k\} \\ \{(i,j) \mid i,j \in \{1,\dots,k\}, (i,j) \in \mu\}, & \text{if } p = structure. \end{cases}$

We formally prove the relationship between the dynamics of the mobile membrane Π and that of the corresponding coloured Petri net CPN_Π.

Theorem 3.5. *If M and M' are two membrane configurations of Π, then*

$$M \overset{R'}{\Rightarrow} M' \text{ if and only if } \phi(M)\,[\psi(R')\rangle\,\phi(M'),$$

where

$$\phi(M)(i) = \begin{cases} w_i, \text{ for all places } i \in P; \\ \mu, \quad i{=}structure. \end{cases} \quad , \text{ and}$$

$$\psi(R) = \bigcup_{r_i \in R} \psi(r_i) \text{ with } \psi(r_i) = t_i.$$

Proof. The function ϕ represents a bijection between the multisets of objects of Π and the markings of CPN_Π based on the corresponding links between objects and tokens, and between membranes and places, respectively. Let $(w_1,\dots,w_k,\mu)$ be the multisets of objects from the membrane configuration M, together with its structure μ. Similarly, for a set of rules $R' = \{r_1,\dots,r_i\}$ of Π, the function ψ is a bijection constructing the set $\psi(R') = \{t_1,\dots,t_i\}$ of transitions of CPN_Π from the set R of rules.

A membrane configuration M_1 can evolve to a membrane configuration M_2 by applying an evolution rule r from R' if and only if, given the marking $\phi(M_1)$, we can obtain the marking $\phi(M_2)$ by firing a transition t in CPN_Π, where $\psi(R')(t) = r$. Overall, this is a direct consequence of the fact that ψ and ϕ are bijections. □

From the construction above it results that the initial configuration of Π corresponds through ϕ to the initial marking of CPN_Π. Moreover, according to Theorem 3.5 it results that the computation of Π coincides with the evolution of the CPN_Π.

3.5.3 LDL Degradation Pathway in Mobile Membranes

LDL is one of several complexes carrying cholesterol through the bloodstream. An LDL particle is a lipoprotein complex that contains one thousand or more cholesterol molecules in the form of cholesteryl esters. A monolayer of phospholipid surrounds the cholesterol and contains a single molecule of a large protein apolipoprotein B (known as apoB). In a receptor-mediated endocytosis, a cell engulfs a particle of low-density lipoprotein from the outside. To do this, the cell uses receptors that specifically recognize and bind to the LDL particle. The receptors are clustered together. By this mechanism cells acquire from the bloodstream the cholesterol required for the membrane synthesis that occurs during cell growth.

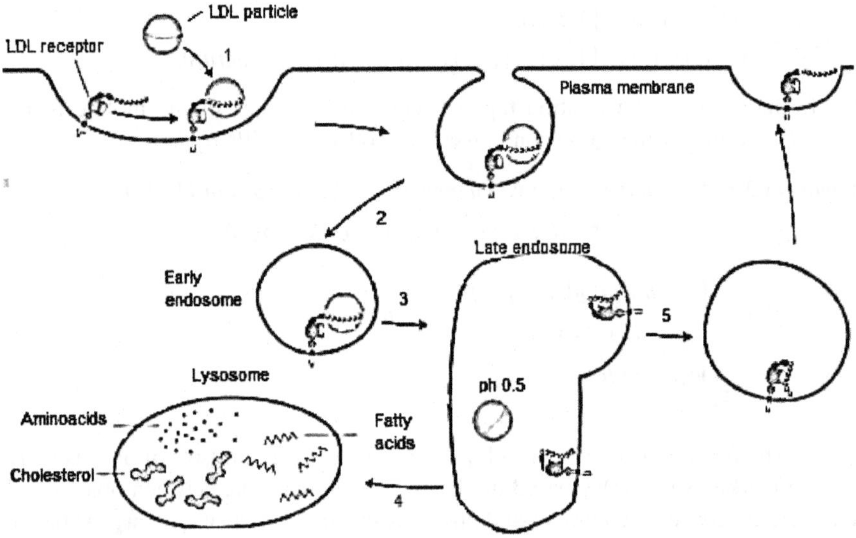

Fig. 3.3 Endocytic Pathway for Low-Density Lipoprotein

The degradation of LDL particles is realized in five steps (see Figure 3.3):

1. Cell-surface LDL receptors bind to an apoB protein of an LDL particle forming a receptor-ligand complex.
2. Clathrin-coated pits containing receptor-LDL complexes are pinched off.
3. After the vesicle coat is shed, the uncoated endocytic vesicle (early endosome) fuses with the late endosome. The acidic pH in this compartment causes a conformational change in the LDL receptor that leads to freeing the bound LDL particle.
4. The late endosome fuses with the lysosome, and the proteins and lipids of the free LDL particle are broken down into their constituent parts by enzymes in the lysosome.

5. The LDL receptor recycles to the cell surface where at the neutral pH of the exterior medium the receptor undergoes a conformational change so that it can bind another LDL particle.

In what follows we show how to model the LDL degradation pathway in terms of mobile membranes with objects on surface, by simulating the five steps presented in Figure 3.3. We describe an LDL particle in membrane systems as $[\,]_{apoB\ cho}$ representing the monolayer of phospholipid that contains a single *apoB* protein and cholesterol *cho*. The cell engulfing the LDL particle is described as

$[[\,]_{lyso} \parallel [\,]_{late\ aux}]_{recep\ recep}$, where

:

- $[\,]_{recep\ recep}$ represents the cell containing on its surface two receptors *recep* able to recognize an *apoB* protein; we do not use clathrin and other receptors of the cell since we are not interested in their evolution;
- $[\,]_{lyso}$ represents the lysosome;
- $[\,]_{late\ aux}$ represents the late endosome, and *aux* is an auxiliary object in creating new membranes by pino and phago rules.

This means that the initial configuration of the system is

$$M_1 = [\,]_{apoB\ cho} \parallel [\,[\,]_{lyso\ aux} \parallel [\,]_{late\ aux}]_{recep\ recep}$$

The steps depicted in Figure 3.3 are simulated using the following rules:

1. $[\,]_{apoB} \parallel [\,]_{recep\ recep} \rightarrow [[[\,]_{apoB}]_{recep}]_{recep}$ (phago) $(recep = \overline{apoB})$
2. $[\,]_{recep} \parallel [\,]_{late\ aux} \rightarrow [\,[\,[\,]_{recep}]_{aux}]_{late}$ (phago) $(aux = \overline{recep})$
3. $[\,[\,]_{recep}]_{aux} \rightarrow [\,]_{recep1\ aux}$ (exo) $(aux = \overline{recep})$
4. $[[\,]_{aux}]_{late} \rightarrow [\,]_{aux4\ late}$ (exo) $(late = \overline{aux})$
5. $[\,]_{lyso\ aux} \parallel [\,]_{late} \rightarrow [\,[\,[\,]_{late}]_{aux1}]_{lyso}$ (phago)$(aux = \overline{late})$
6. $[\,[\,]_{recep1}]_{aux1} \rightarrow [\,]_{recep2\ aux2}$ (exo) $(aux1 = \overline{recep1})$
7. $[\,]_{late\ recep2\ aux2\ aux4} \rightarrow [[\,]_{late\ recep3\ aux4}]_{aux3}$ (pino) $(aux2 = \overline{recep2})$
8. $[\,[\,]_{aux3}]_{lyso} \rightarrow [\,]_{lyso\ aux}$ (exo) $(lyso = \overline{aux3})$
9. $[\,[\,]_{apoB}]_{lyso} \rightarrow [\,]_{lyso\ apoB}$ (exo) $(lyso = \overline{apoB})$
10. $[\,]_{late\ recep3\ aux4} \rightarrow [\,[\,]_{recep4\ aux4}]_{late}$ (pino) $(late = \overline{recep3})$
11. $[\,]_{aux4\ recep4} \rightarrow [\,[\,]_{recep5}]_{aux5}$ (pino) $(aux4 = \overline{recep4})$
12. $[\,[\,]_{aux5}]_{late} \rightarrow [\,]_{late\ aux}$ (exo) $(late = \overline{aux5})$
13. $[\,[\,]_{recep5}]_{recep} \rightarrow [\,]_{recep\ recep}$ (exo) $(recep = \overline{recep5})$

where by writing $recep = \overline{apoB}$ we mean that an object *recep* is complementary to an object *apoB*.

The evolution of the system can be represented graphically as in Figure 3.4. By $M_1, \ldots, M_{24}$ we denote the possible configurations of the system, and on each arrow from an M_i to an M_j is placed the number of the rule which is applied in order to evolve from M_i to M_j. To denote that an object *recep* changes its position and interacts with different objects, we use different notations to denote it (namely, $recep, recep_1, \ldots, recep_5$) in the evolution of the system.

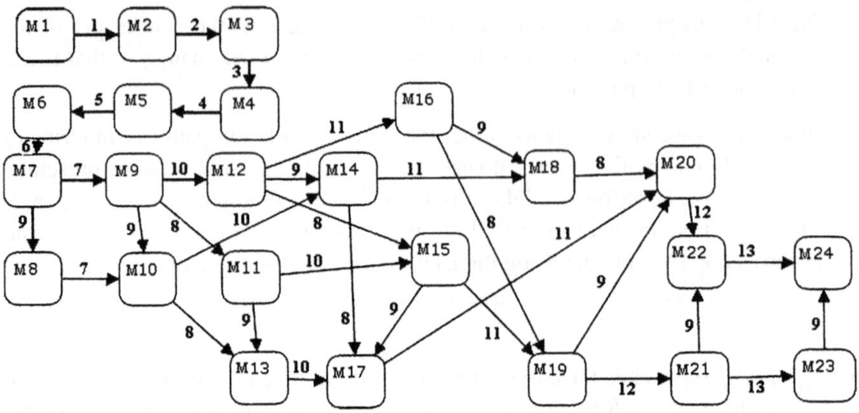

Fig. 3.4 Evolution of the Membrane System

Remark 3.8. The number of rules applied to reach the configuration M_{24} starting from the configuration M_1 is always 13.

3.5.4 Simulating LDL Degradation by CPN Tools

Now the LDL degradation pathway description by using mobile membranes is encoded in coloured Petri nets. The translation is provided in order that one may use a complex software tool able to automatically verify some important behavioural properties of the biological systems. For coloured Petri nets a complex software called *CPN Tools* is available in which simulations can be performed, and certain decidability results can be checked automatically: reachability, boundedness, deadlock, liveness, fairness, etc. CPN Tools (cpntools.org) is a tool for editing, simulating, state space analysis, and performance analysis of systems described as coloured Petri nets. In what follows we show how the rules of mobile membranes used to model the LDL degradation pathway can be simulated using CPN Tools. To make it easier to observe how the evolution takes place using CPN Tools, we simplify the system and use only the transitions that eventually occur.

A CPN model is always created in CPN Tools as a graphical drawing. Figure 3.5 describes the LDL degradation pathway model, namely the membrane configuration M_1 from Subsection 3.5.3. The diagram contains eight places (drawn as ellipses or circles), four substitution transitions (drawn as double-rectangular boxes), a number of directed arcs connecting places and transitions, and finally some textual inscriptions next to the places, transitions and arcs. The inscriptions are written in ML, the programming language of CPN Tools. Places and transitions are called nodes. Together with the directed arcs, they constitute the net structure. An arc always connects a place to a transition or a transition to a place; it is illegal to have an arc between two nodes of the same kind, i.e., between two places or two transitions. The arc expressions are built from variables, constants, operators, and functions. When all variables in an expression are bound to values of the correct type, the expression can be evaluated. In general, arc expressions may evaluate to a multiset of token colours. Next to each place is an inscription which determines the set of token colours (data values) that the tokens on that place are allowed to have. The set of possible token colours is specified by means of a type (familiar from programming languages) which is called the colour set of the place. By convention, the colour set is written below the place. The place $structure1$ has the colour set P, while all the others have the colour set U. The colour set P is used to model the structure of a membrane configuration (pairs of integer numbers of the form (i, j)), while the colour set U is used to model the set of objects from a mobile membrane.

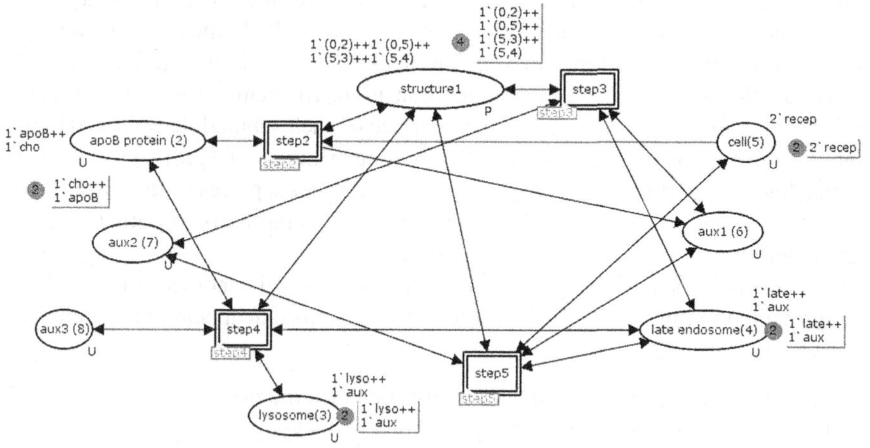

Fig. 3.5 LDL Degradation Pathway in CPN Tools

Colour sets are defined using the CPN ML keyword $colset$:

$colset\ I = int;$

$colset\ P = product\ I * I;$

$colset\ U = with\ cho \mid apoB \mid lyso \mid late \mid aux \mid aux1 \mid aux2 \mid aux3 \mid aux4 \mid$
$aux5 \mid recep \mid recep1 \mid recep2 \mid recep3 \mid recep4 \mid recep5;$

The inscription on the upper side of a place specifies the *initial marking* of that place. The inscription of the place $late\ endosome(4)$ is $1`late + +1`aux$ specifying that the initial marking of this place consists of two tokens with the values $late$ and aux. The symbols $++$ and ` are operators used to construct a multiset of tokens. The infix operator ` takes a non-negative integer as its left argument, specifying the number of appearances of the element provided as the right argument. The operator $++$ takes two multisets as arguments and returns their union (sum of their multiplicities). The absence of an inscription specifying the initial marking means that the place initially contains no tokens. The marking of each place is indicated next to the place. The number of tokens on the place is shown in a small circle, and the detailed token colours are indicated in a box positioned next to the small circle.

The four transitions drawn as rectangles represent the events that can take place in the system. The names of the transitions are written inside the rectangles. The transition names have no formal meaning, but they are very important for the readability of the model. In Figure 3.5 the transition names are $step2$, $step3$, $step4$ and $step5$ symbolizing that each of these transitions simulate the corresponding steps of the LDL degradation pathway described in Figure 3.3.

A transition with a double-line border is a substitution transition. Each of them has a *substitution tag* positioned next to it. The substitution tag contains the name of a submodule which is related to the substitution transition. Intuitively, this means that the submodule presents a more detailed view of the behaviour represented by the substitution transition, and this is particularly useful when modelling large systems. The input places of substitution transitions are called input sockets, and the output places are called output sockets. The socket places of a substitution transition constitute the interface of the substitution transition. To obtain a complete hierarchical model, it must be specified how the interface of each submodule is related to the interface of its substitution transition. This is done by means of a port-socket relation which links the port places of the submodule to the socket places of the substitution transition. Input ports are related to input sockets, output ports to output sockets, and input/output ports to input/output sockets.

For instance, behind the substitution transition $step4$ is another coloured Petri net presented in Figure 3.6. The substitution transitions that appear in this coloured Petri net are:

- the substitution transition *phago*1-*step*4 simulates the mobile membrane rule 5 from the description of the LDL degradation pathway;
- the substitution transition *exo*1-*step*4 simulates the mobile membrane rule 6 from the description of the LDL degradation pathway;
- the substitution transition *pino-step*4 simulates the mobile membrane rule 7 from the description of the LDL degradation pathway;
- the substitution transition *exo*2-*step*4 simulates the mobile membrane rule 8 from the description of the LDL degradation pathway;
- the substitution transition *exo*3-*step*4 simulates the mobile membrane rule 9 from the description of the LDL degradation pathway;

We may observe that the marking of places appearing in Figure 3.6 is similar to the one of the corresponding places in Figure 3.5. The substitution transition $exo1 - step4$ is replaced by the Petri net presented in Figure 3.7.

In Figure 3.8, transition $phagostep2$ is surrounded by a green shadow indicating that it is enabled. When the transition occurs, it removes a token *apoB* from place $apoBprotein(2)$, a token *recep* from place $cell(5)$, and two tokens $(0,2)$ and $(0,5)$ from place $structure1$. The arc expressions of the input arc from the place $structure1$ are $(x,2)$ and $(y,5)$; they are tested using the test expression $[x=0, y=0]$. The test

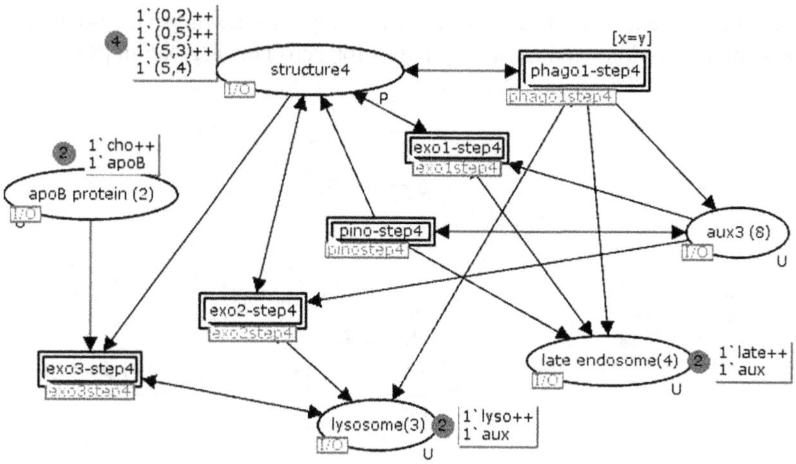

Fig. 3.6 Step 4 Transition

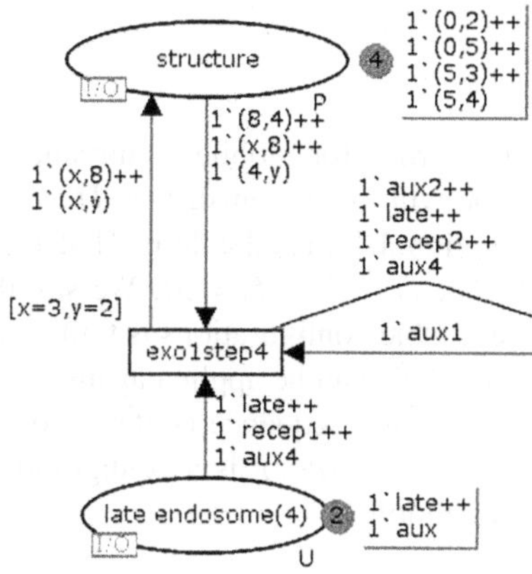

Fig. 3.7 Exo1 Step 4 Transition

is performed in order to see that the simulated membranes 2 and 5 have the same parent 0. After firing the transition, a token *recep* is added to the place $aux1(6)$, a token *apoB* is added to the place $apoBprotein(2)$, and three tokens $(0,5)$, $(6,2)$ and $(5,6)$ are added to the place $structure1$.

A state space is a directed graph where we have a node for each reachable marking and an arc for each occurring transition. The state space of a CPN model can be computed fully automatically which makes it possible to automatically analyze and verify an abundance of properties concerning the behaviour of the model: the minimum and maximum numbers of tokens in a place, reachability, boundedness, etc. When working with Petri nets, some behavioural properties (e.g., reachability, boundedness, liveness, fairness) are easier to study once a state space is calculated. A good survey for known decidability issues for Petri nets is given in [83].

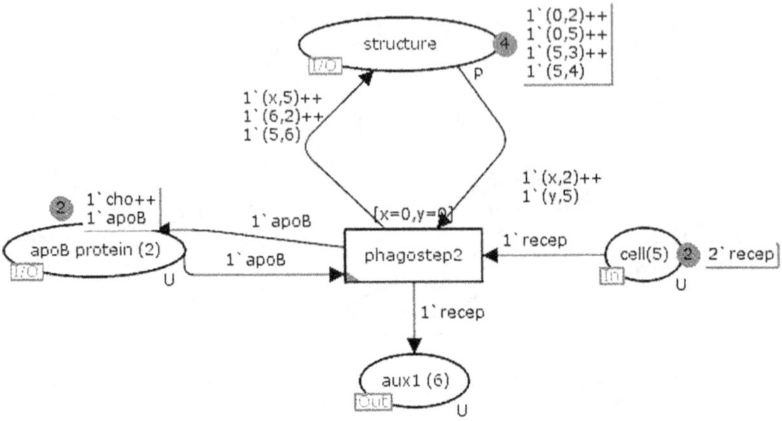

Fig. 3.8 Phago Step 2 Transition

We can now define similar properties for mobile membranes with objects on surface. Given a mobile membrane with object on surface Π with initial configuration M_0, we say that a configuration M is reachable in Π if there exists sets of transitions $U_1, \ldots, U_n$ such that $M_0[U_1\rangle \ldots [U_n\rangle M_n = M$. We say that a membrane system is *bounded* if the set of reachable configurations is finite. A membrane system has the *liveness* property if each rule can be applied again in another evolution step, and it is *fair* if no infinite execution sequence contains some configurations which occur only finitely. A *home configuration* is a configuration which can be reached from any reachable configuration.

3.5.5 Preservation of Properties Through Translation

By considering a coloured Petri net CPN_Π obtained from a mobile membrane Π, we have the following results:

Proposition 3.21. *If the reachability problem is decidable for* CPN_Π*, then the reachability problem is also decidable for* Π*.*

Proof (Sketch). The initial marking of CPN_Π is the same as the initial configuration of Π according to the construction presented in Subsection 3.5.1, and each step of the Petri net corresponds to an evolution of the mobile membranes with objects on surface (according to Theorem 3.5). Thus the reachability problem becomes decidable for mobile membranes with objects on surface as soon as it is decidable for coloured Petri nets. □

In a similar way, we can prove several properties for mobile membranes with objects on surface as soon as they hold for their corresponding coloured Petri nets.

Proposition 3.22.

- *If CPN_Π is bounded, then Π is bounded.*
- *If CPN_Π has the liveness property, then Π has the liveness property.*
- *If CPN_Π is fair, then Π is fair.*

Since the properties of reachability, boundedness, liveness and fairness can be derived automatically by using CPN Tools, these results are of great help when studying similar properties for mobile membranes with objects on surface. For instance, using CPN Tools and the model for the LDL degradation pathway, we can check whether we can reach the configuration in which the membrane marked by *apoB* is inside the membrane marked by *lyso*, for instance.

Applying CPN Tools on this system we obtain the following output file:

Home Markings: [24] Dead Markings: [24];
Dead Transition Instances: None Live Transition Instances: None
Fairness Properties: No infinite occurrence sequences

meaning that we always reach configuration M_{24} (home marking), the computation stops here (dead marking), and that there are no infinite occurrence sequences.

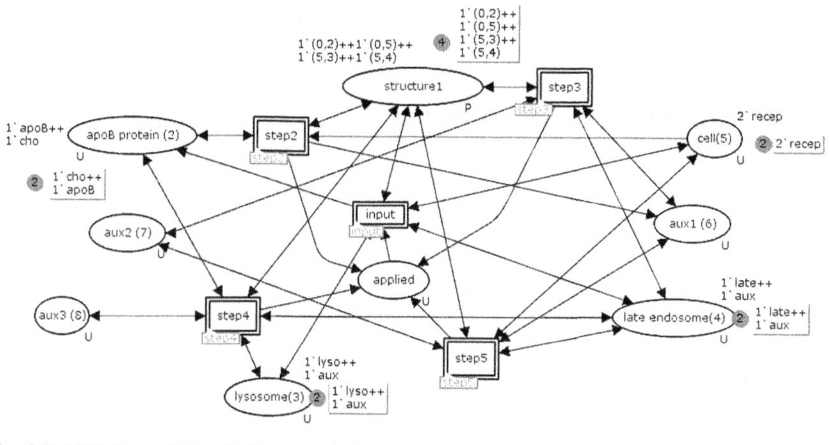

Fig. 3.9 LDL Degradation Pathway with Input

This simulation is not entirely correct from a biological point of view since a cell is able to process more than one LDL molecule. An arbitrary number of LDL molecules cannot be simulated in mobile membranes, but it can be simulated in coloured Petri nets by adding a new transition *input* and a new place *applied* as in Figure 3.9.

In Figure 3.10 we show how the transition *input* is built, namely what are the input arcs and output arcs together with their inscriptions. This transitions works as follows: if the cell has the initial structure less the initial LDL molecule, than a new LDL molecule is added to the system in order to reiterate the entire process. Applying the CPN Tool on this extended system we obtain the following output file:

Home Markings: All   Dead Markings: None;
Dead Transition Instances: None   Live Transition Instances: All
Fairness Properties: All

meaning that from any reachable configuration M_i we can always reach any configuration M_j (home marking), the computation never stops (dead marking), and that there are infinite occurrence sequences.

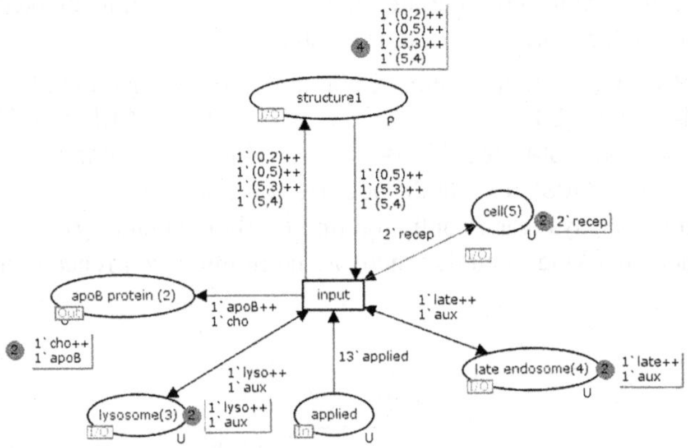

Fig. 3.10 LDL Degradation Pathway with Input

Summary

This book contains the results obtained and published recent last years on mobility aspects in process algebra and membrane computing. It presents new ideas concerning several formalisms (π-calculus, mobile ambients, brane calculi, membrane systems), new properties and relationships. The emphasis is mainly on the computational properties of the models. Moreover, the formalisms are used to model and analyze biological systems.

What is mobility in process algebra? The first formalism in computer science able to describe mobility was the π-calculus. It was followed by ambient calculus. A biologically-inspired version of ambient calculus is given by bioambients and several brane calculi. When expressing mobility, it should be mentioned what entities move and in what space they move. There are several possibilities: processes moving in a physical space of computing locations, processes moving in a virtual space of linked processes, links moving in a virtual space of linked processes, etc.

What is mobility in membrane systems? Mobile membranes represent a formalism able to describe the movement of membranes inside a spatial structure. When they are considered as computing devices, two main research direction are considered in order to prove that they are both powerful, mostly equivalent to Turing machines, and efficient, membrane system algorithms have been developed which provide efficient solutions to NP-complete problems through the generation of an exponential space in polynomial time.

What is the relationship between these two research areas? The difference between the two research areas (process algebra and membrane computing) is the fact that process algebras provide a tool for the high-level description of interactions, communications, and synchronizations between a collection of independent agents or processes, providing also algebraic laws that allow process descriptions to be manipulated and analyzed, and permit formal reasoning about equivalences between processes (e.g., using bisimulation), while membrane computing uses techniques from languages, automata, complexity, and dynamical systems. Several links between these two fields are established in order to be able to use techniques from one area in the other one. We consider our encodings as the first efforts towards bridging the gap between process calculi and systems of mobile membranes.

B. Aman, G. Ciobanu, *Mobility in Process Calculi and Natural Computing*,
Natural Computing Series, DOI 10.1007/978-3-642-24867-2,

What are other research directions into these fields? Several papers have been devoted to process algebra, by defining faithful π-nets [70], abstract structures for distributed systems [62], a π-calculus machine [71] and Markov abstractions for probabilistic π-calculus [21]. A simplified timed distributed π-calculus called TiMo is defined in [58, 59]. PerTiMo extends TiMo by working with processes having appropriate access rights to communicate [60]. On the subject of complex systems with interaction, there are a few papers using process algebra to model biological phenomena [49, 72]. A concrete description of a biological system by using P systems is provided in [26]. Starting in 2002, several papers have been devoted to membrane computing and distributed computing [50, 54], to P systems implementations on a single computer [27, 64] and on clusters of computers [74]. A software platform for timed mobility and timed interaction is presented in [56], and a high-level language for mobile agents with timers in [57]. The formal semantics for P systems was introduced and studied in [22, 24, 51]. Some papers have been devoted to various aspects relating to membrane systems: P transducers [66], Mealy multiset automata [55], control mechanisms of membranes [23] and cellular modelling [140]. A distributed evolutionary algorithms using membrane systems [154] and a P system with minimal parallelism [63] represent two important contributions to membrane computing. Some complexity aspects in membrane systems are presented in [6, 68, 69]. The problem of causality is of great importance for any computational model that aims to be supported by good analysis methods and tools, and even more significant for membrane computing where such approaches are in many cases based on methods developed for other computational models. Research work on causality in membrane systems is presented in [3, 4, 5].

References

1. M. Abadi, A. Gordon. The Spi-calculus. *Computer and Communication Security*, 36–47, 1997.
2. M. Abadi, C. Fournet. Mobile Values, New Names, and Secure Communication. In *Symposium on Principles of Programming Languages*, ACM, 104–115, 2001.
3. O. Agrigoroaiei, G. Ciobanu. Reversing Computation in Membrane Systems. *Journal of Logic and Algebraic Programming*, vol.79(3-5), 278–288, 2010.
4. O. Agrigoroaiei, G. Ciobanu. Rule-Based and Object-Based Event Structures for Membrane Systems. *Journal of Logic and Algebraic Programming*, vol.79(6), 295–303, 2010.
5. O. Agrigoroaiei, G. Ciobanu. Quantitative Causality in Membrane Systems. *Lecture Notes in Computer Science*, accepted, to appear in 2011.
6. O. Agrigoroaiei, G. Ciobanu, A. Resios. Evolving by Maximizing the Number of Rules: Complexity Study. *Lecture Notes in Computer Science*, vol.5957, 2010, 149–157.
7. B. Alberts, A. Johnson, J. Lewis, M. Raff, K. Roberts, P. Walter. *Molecular Biology of the Cell* - Fifth Edition. Garland, 2008.
8. R. Alur, D.L. Dill. A Theory of Timed Automata. *Theoretical Computer Science*, vol.126, 183–235, 1994.
9. B. Aman, G. Ciobanu. Timers and Proximities for Mobile Ambients. *Lecture Notes in Computer Science*, vol.4649, 33–43, 2007.
10. B. Aman, G. Ciobanu. Mobile Ambients with Timers and Types. *Lecture Notes in Computer Science*, vol.4711, 50–63, 2007.
11. B. Aman, G. Ciobanu. On the Reachability Problem in P Systems with Mobile Membranes. *Lecture Notes in Computer Science*, vol.4860, 113–123, 2007.
12. B. Aman, G. Ciobanu. Translating Mobile Ambients into P Systems. *Electronic Notes in Theoretical Computer Science*, vol.171(2), 11–23, 2007.
13. B. Aman, G. Ciobanu. Structural Properties and Observability in Membrane Systems. *Proceedings of SYNASC07: 9th International Symposium on Symbolic and Numeric Algorithms for Scientific Computing*, IEEE Computer Society, 74–81, 2007.
14. B. Aman, G. Ciobanu. On the Relationship Between Membranes and Ambients. *Biosystems*, vol.91(3), 515–530, 2008.
15. B. Aman, G. Ciobanu. Timed Mobile Ambients for Network Protocols. *Lecture Notes in Computer Science*, vol.5048, 234–250, 2008.
16. B. Aman, G. Ciobanu. Describing the Immune System Using Enhanced Mobile Membranes. *Electronic Notes in Theoretical Computer Science*, vol.194(3), 5–18, 2008.
17. B. Aman, G. Ciobanu. Simple, Enhanced and Mutual Mobile Membranes. *Transactions on Computational Systems Biology XI*, LNBI vol.5750, 26–44, 2009.
18. B. Aman, G. Ciobanu. Turing Completeness Using Three Mobile Membranes. *Lecture Notes in Computer Science*, vol.5715, 42–55, 2009.
19. B. Aman, G. Ciobanu. Mobile Membranes and Mobile Ambients. In [130], 645–653.

B. Aman, G. Ciobanu, *Mobility in Process Calculi and Natural Computing*,
Natural Computing Series, DOI 10.1007/978-3-642-24867-2,

20. B. Aman, G. Ciobanu. Membrane Systems with Surface Objects. *Natural Computing*, vol.10(2), 777–793, 2011.
21. H. Anderson, G. Ciobanu. Markov Abstractions for Probabilistic π-calculus. *Electronic Communication of the European Association of Software Science and Technology*, vol.22, 2009. http://journal.ub.tu-berlin.de/eceasst/article/view/317
22. O. Andrei, G. Ciobanu, D. Lucanu. Structural Operational Semantics of P Systems. *Lecture Notes in Computer Science*, vol.3850, 32–49, 2006.
23. O. Andrei, G. Ciobanu, D. Lucanu. Expressing Control Mechanisms of Membranes by Rewriting Strategies. *Lecture Notes in Computer Science*, vol.4361, 154–169, 2006.
24. O. Andrei, G. Ciobanu, D. Lucanu. A Rewriting Logic Framework for Operational Semantics of Membrane Systems. *Theorerical Computer Science*, vol.373(3), 163–181, 2007.
25. J. Bengtsson, W. Yi. Timed Automata: Semantics, Algorithms and Tools. *Lecture Notes in Computer Science*, vol.3098, 87–124, 2004.
26. D. Besozzi, G. Ciobanu. A P System Description of the Sodium-Potassium Pump. *Lecture Notes in Computer Science*, vol.3365, 210–223, 2005.
27. C. Bonchiş, G. Ciobanu, C. Izbaşa, Dana Petcu. A Web-Based P Systems Simulator and Its Parallelization. *Lecture Notes in Computer Science*, vol.3699, 58–69, 2005.
28. I. Boneva, J.-M. Talbot. When Ambients Cannot be Opened! *Lecture Notes in Computer Science*, vol.2620, 169–184, 2003.
29. M. M. Bonsangue, J. N. Kok, G. Zavattaro. Comparing Coordination Models and Architectures Using Embeddings. *Science of Computer Programming*, vol.46, 31–69, 2003.
30. A. Borodin. On Relating Time and Space to Size and Depth. *SIAM Journal of Computing*, vol.6(4), 733–744, 1977.
31. R. Brijder, M. Cavaliere, A. Riscos-Núñez, G. Rozenberg, D. Sburlan. Membrane Systems with Proteins Embedded in Membranes. *Theoretical Computer Science*, vol.404, 26–39, 2008.
32. L. Brodo, P. Degano, C. Priami. Reflecting Mobile Ambients into the π-calculus. *Lecture Notes in Computer Science*, vol.2874, 25–56, 2003.
33. A. Brogi, J. M. Jacquet. On the Expressiveness of Coordination Via Shared Dataspaces. *Science of Computer Programming*, vol.46, 71–98, 2003.
34. M. Bugliesi, G. Castagna, S. Crafa. Boxed Ambients. *ACM Transactions on Programming Languages and Systems*, vol.26(1), 57–124, 2004.
35. N. Busi. On the Computational Power of the Mate/Bud/Drip Brane Calculus: Interleaving vs. Maximal Parallelism. *Lecture Notes in Computer Science*, vol.3850, 144–158, 2006.
36. N. Busi. Expressiveness Issues in Brane Calculi: A Survey. *Electronic Notes in Theoretical Computer Science*, vol.209, 107–124, 2008.
37. N. Busi, R. Gorrieri. On the Computational Power of Brane Calculi. *Third Workshop on Computational Methods in Systems Biology*, 106–117, 2005.
38. N. Busi, G. Zavattaro. On the Expressive Power of Movement and Restriction in Pure Mobile Ambients. *Theoretical Computer Science*, vol.322, 477–515, 2004.
39. N. Busi, G. Zavattaro. Deciding Reachability Problems in Turing-Complete Fragments of Mobile Ambients. *Mathematical Structures in Computer Science*, vol.19, 1223–1263, 2009.
40. L. Cardelli. Brane Calculi. Interactions of Biological Membranes. *Lecture Notes in Bioinformatics*, vol.3082, 257–278, 2004.
41. L. Cardelli, G. Ghelli, A. Gordon. Types for the Ambient Calculus. *Information and Computation*, vol.177(2), 160–194, 2002.
42. L. Cardelli, A. Gordon. Mobile Ambients. *Theoretical Computer Science*, vol.240(1), 177–213, 2000.
43. L. Cardelli, A. Gordon. Anytime, Anywhere: Modal Logics for Mobile Ambients. *Symposium on Principles of Programming Languages*, 365—377, 2000.
44. L. Cardelli, S. Pradalier. Where Membranes Meet Complexes. *BioConcur*, 2005.
45. L.Cardelli, Gh. Păun. An Universality Result for a (Mem)Brane Calculus Based on Mate/Drip Operations. *International Journal of Foundations of Computer Science*, vol.17(1), 49–68, 2006.

46. M. Cavaliere, P. Leupold. Evolution and Observation - A New Way to Look at Membrane Systems. *Lecture Notes in Computer Science*, vol.2933, 70–87, 2004.
47. M. Cavaliere, S. Sedwards. Membrane Systems with Peripherial Proteins: Transport and Evolution. *Electronic Notes in Theoretical Computer Science*, vol.171(2), 37–53, 2007.
48. W. Charatonik, A. Gordon, J.-M. Talbot. Finite-Control Mobile Ambients. *Lecture Notes in Computer Science*, vol.2305, 295–313, 2002.
49. G. Ciobanu. On a Formal Description of the Molecular Processes. *Recent Topics in Mathematical and Computational Linguistics*, 82–96, Romanian Academy Publishing House, 2000.
50. G. Ciobanu. Distributed Algorithms over Communicating Membrane Systems. *Biosystems*, vol.70(2), 123–133, 2003.
51. G. Ciobanu. Semantics of P Systems. In [130], 413–436.
52. G. Ciobanu. Mobility in Computer Science and in Membrane Systems. *Lecture Notes in Computer Science*, vol.6501, 7–17, 2010.
53. G. Ciobanu, V. Ciubotariu, B. Tanasă. A π-Calculus Model of the Na Pump. *Genome Informatics*, 469–472, 2002.
54. G. Ciobanu, R. Desai, A. Kumar. Membrane Systems and Distributed Computing. *Lecture Notes in Computer Science*, vol.2597, 187–202, 2002.
55. G. Ciobanu, V.M. Gontineac. Mealy Multiset Automata. *International Journal of Foundations of Computer Science*, vol.17(1), 111–126, 2006.
56. G. Ciobanu, C. Juravle. A Software Platform for Timed Mobility and Timed Interaction. *Lecture Notes in Computer Science*, vol.5522, 106–121, 2009.
57. G. Ciobanu, C. Juravle. Mobile Agents with Timers, and Their Implementation. *Intelligent Distributed Computing IV, Studies in Computational Intelligence*, vol.315, 229–239, 2010.
58. G. Ciobanu, M. Koutny. Modelling and Verification of Timed Interaction and Migration. *Lecture Notes in Computer Science*, vol.4961, 215–229, 2008.
59. G. Ciobanu, M. Koutny. Timed Mobility in Process Algebra and Petri Nets. *Journal of Logic and Algebraic Programming*, vol.80(7), 377–391, 2011.
60. G. Ciobanu, M. Koutny. Timed Migration and Interaction with Access Permissions. *Lecture Notes in Computer Science*, vol.6664, 293–307, 2011.
61. G. Ciobanu, S.N. Krishna. Enhanced Mobile Membranes: Computability Results. *Theory of Computing Systems*, vol.48(3), 715–729, 2011.
62. G. Ciobanu, E.F. Olariu. Abstract Structures for Communication between Processes. *Lecture Notes in Computer Science*, vol.1755, 221–227, 2000.
63. G. Ciobanu, L. Pan, Gh. Păun, M. J. Pérez-Jiménez. P Systems with Minimal Parallelism. *Theoretical Computer Science*, vol.378(1), 117–130, 2007.
64. G. Ciobanu, D. Paraschiv. P System Software Simulator. *Fundamenta Informaticae*, vol.49(1-3), 61–66, 2002.
65. G. Ciobanu, Gh. Păun, M. J. Pérez-Jiménez (Eds.). *Applications of Membrane Computing*. Springer, 2006.
66. G. Ciobanu, Gh. Păun, Gh. Ştefănescu. P Transducers. *New Generation Computing*, vol.24(1), 1–28, 2006.
67. G. Ciobanu, C. Prisacariu. Timers for Distributed Systems. *Electronic Notes in Theoretical Computer Science*, vol.164(3), 81–99, 2006.
68. G. Ciobanu, A. Resios. Computational Complexity of Simple P Systems. *Fundamenta Informaticae*, vol.87(1), 49–59, 2008.
69. G. Ciobanu, A. Resios. Complexity of Evolution in Maximum Cooperative P Systems. *Natural Computing*, vol.8(4), 807–816, 2009.
70. G. Ciobanu, M. Rotaru. Faithful π-nets - A Graphical Representation of the Asynchronous π-calculus. *Electronic Notes in Theoretical Computer Science*, vol.18, 24–45, 1998.
71. G. Ciobanu, M. Rotaru. A π-calculus Machine. *Journal of Universal Computer Science*, vol.6(1), 39–59, 2000.
72. G. Ciobanu, M. Rotaru. Molecular Interaction. *Theoretical Computer Science*, vol.289, 801–827, 2002.

73. G. Ciobanu, V. Zakharov. Encoding Mobile Ambients into π-calculus. *Lecture Notes in Computer Science*, vol.4378, 148–161, 2006.
74. G. Ciobanu, G. Wenyuan. P Systems Running on a Cluster of Computers. *Lecture Notes in Computer Science*, vol.2933, 2003, 123–139.
75. J.E. Cohen. Mathematics is Biology's Next Microscope, Only Better; Biology is Mathematics' Next Physics, Only Better. *PLoS Biology*, vol.12, e439, 2004.
76. J. Dacks, M. Field. Evolution of the Eukaryotic Membrane-Trafficking System: Origin, Tempo and Mode. *Journal of Cell Science*, vol.120, 2977–2985, 2007.
77. S. Dal Zilio. Mobile Processes: A Commented Bibliography. *Lecture Notes in Computer Science*, vol.2067, 206–222, 2000.
78. S. Dal Zilio, E. Formenti. On the Dynamics of PB Systems: a Petri Net View. *Lecture Notes in Computer Science*, vol.2933, 153–167, 2004.
79. V. Danos, C. Laneve. Core Formal Molecular Biology. *Lecture Notes in Computer Science*, vol.2618, 302–318, 2003.
80. J. Dassow, Gh. Păun. *Regulated Rewriting in Formal Language Theory*. Springer-Verlag, 1990.
81. F. S. de Boer, C. Palamidessi. Embedding as a Tool for Language Comparison. *Information and Computation*, vol.115, 128–157, 1994.
82. G. Delzanno, L. Van Begin. On the Dynamics of PB Systems with Volatile Membranes. *Lecture Notes in Computer Science*, vol.4860, 240–256, 2007.
83. J. Esparza, M. Nielson. Decidability Issues for Petri Nets - A Survey. *Journal of Information Processing and Cybernetics*, vol.30(3), 143–160, 1994.
84. G. Ferrari, S. Gnesi, U. Montanari, N. Pistore, R. Gioia. Verifying Mobile Processes in the HAL Environment. *Lecture Notes in Computer Science*, vol.1427, 511–515, 1998.
85. A. Finkel, Ph. Schnoebelen. Well-Structured Transition Systems Everywhere! *Theoretical Computer Science*, vol.256, 63–92, 2001.
86. C. Fournet, G. Gonthier. The Reflexive Chemical Abstract Machine and the Join-Calculus. *Symposium on Principles of Programming Languages*, 372—385, 1996.
87. C. Fournet, G. Gonthier, J.-J. Levy, L. Maranget, D. Remy. A Calculus of Mobile Agents. *Lecture Notes in Computer Science*, vol.1119, 406–421, 1996.
88. C. Fournet, J.-J. Levy, A. Schmitt. An Asynchronous Distributed Implementation of Mobile Ambients. *Lecture Notes in Computer Science*, vol.1872, 348–364, 2000.
89. R. Freund. Asynchronous P Systems and P Systems Working in the Sequential Mode. *Lecture Notes in Computer Science*, vol.3365, 36–62, 2005.
90. R. Freund, Gh. Păun. On the Number of Non-Terminals in Graph-Controlled, Programmed, and Matrix Grammars. *Lecture Notes in Computer Science*, vol.2055, 214–225, 2001.
91. M. Hennessy. *A Distributed Pi-calculus*. Cambridge University Press, 2007.
92. M. Hennessy, T. Regan, A Process Algebra for Timed Systems. *Information and Computation*, vol.117, 221–239, 1995.
93. M. Hennessy, J. Riely. Resource Access Control in Systems of Mobile Agents. *Information and Computation*, vol.173(1), 82–120, 2002.
94. D. Hirschkoff, D. Teller, P. Zimmer. Using Ambients to Control Resources. *Lecture Notes in Computer Science*, vol.2421, 288–303, 2002.
95. K. Honda, M. Tokoro. An Object Calculus for Asynchronous Communication. *Lecture Notes in Computer Science*, vol.512, 133–147, 1991.
96. O. Ibarra, A. Păun, Gh. Păun, A. Rodríguez-Patón, P. Sosík, S. Woodworth. Normal Forms for Spiking Neural P Systems. *Theoretical Computer Science*, vol.372, 196–217, 2007.
97. J. M. Jacquet, K. De Bosschere, A. Brogi. On Timed Coordination Languages. *Lecture Notes in Computer Science*, vol.1906, 81–98, 2000.
98. C. A. Janeway, P. Travers, M. Walport, M. J. Shlomchik. *Immunobiology - The Immune System in Health and Disease*, 5th Edition. Garland, 2001.
99. K. Jensen. *Coloured Petri Nets; Basic Concepts, Analysis Methods and Practical Use. Vol. 1,2,3*. Monographs in Theoretical Computer Science, Springer, 1992, 1994, 1997.
100. K. Kabnick, D. Peattie. Giardia: A Missing Link Between Prokaryotes and Eukaryotes. *American Scientist*, vol.79, 34–43, 1991.

101. J. Kleijn, M. Koutny. Petri Nets and Membrane Computing. In [130], 389–412.
102. J. Kleijn, M. Koutny, G. Rozenberg. Towards a Petri Net Semantics for Membrane Systems. *Lecture Notes in Computer Science*, vol.3850, 292–309, 2006.
103. S. N. Krishna. The Power of Mobility: Four Membranes Suffice. *Lecture Notes in Computer Science*, vol.3526, 242–251, 2005.
104. S. N. Krishna. On the Efficiency of a Variant of P Systems with Mobile Membranes. *Cellular Computing: Complexity Aspects*, ESF PESC Exploratory Workshop, Fenix Editora, Sevilla, 237–246, 2005.
105. S. N. Krishna. Universality Results for P Systems Based on Brane Calculi Operations. *Theoretical Computer Science*, vol.371, 83–105, 2007.
106. S. N. Krishna. Membrane Computing with Transport and Embedded Proteins. *Theoretical Computer Science*, vol.410, 355–375, 2009.
107. S. N. Krishna, G. Ciobanu. On the Computational Power of Enhanced Mobile Membranes. *Lecture Notes in Computer Science*, vol.5028, 326–335, 2008.
108. S. N. Krishna, Gh. Păun. P Systems with Mobile Membranes. *Natural Computing*, vol.4(3), 255–274, 2005.
109. C. Laneve, F. Tarissan. A Simple Calculus for Proteins and Cells. *Electronic Notes in Theoretical Computer Science*, vol.171(2), 139–154, 2007.
110. F. Levi, S. Maffeis. An abstract interpretation framework for mobile ambients. *Lecture Notes in Computer Science*, vol.2126, 395–411, 2001.
111. F. Levi, D. Sangiorgi. Controlling Interference in Ambients. *Symposium on Principles of Programming Languages*, 352–364, 2000.
112. F. Levi, D. Sangiorgi. Mobile Safe Ambients. *ACM TOPLAS*, vol.25, 1–69, 2003.
113. R. Levins. The Strategy of Model Building in Population Biology. *American Scientist*, vol.54, 421–431, 1966.
114. S. Maffeis, I. Phillips. On the Computational Strength of Pure Mobile Ambients. *Theoretical Computer Science*, vol.330, 501–551, 2004.
115. E. W. Mayr. An Algorithm for the General Petri Net Reachability Problem. *SIAM Journal of Computing*, vol.13(3), 441–460, 1984.
116. M. Merro, M. Hennessy. Bisimulation Congruences in Safe Ambients. *Symposium on Principles of Programming Languages*, 71—80, 2002.
117. M. Merro, F. Zappa Nardelli. Behavioural Theory for Mobile Ambients. *Journal of ACM*, vol.52(6), 961–1023, 2005.
118. D. Mills. A Brief History of NTP Time: Memoirs of an Internet Timekeeper. *Computer Communication Review*, vol.33, 9–21, 2003.
119. R. Milner. Functions as Processes. *Mathematical Structures in Computer Science*, vol.2(2), 119–141, 1992.
120. R. Milner. Elements of Interaction. Turing Award Lecture. *Communications of the ACM*, vol.36(1), 78–89, 1993.
121. R. Milner. *Communicating and Mobile Systems: The π-calculus*. Cambridge University Press, 1999.
122. R. Milner. *The Space and Motion of Communicating Agents*. Cambridge University Press, 2009.
123. R. Milner, D. Sangiorgi. Barbed Bisimulation. *Lecture Notes in Computer Science*, vol.623, 685–695, 1992.
124. R. Milner, M. Tofte. Co-induction in Relational Semantics. *Theoretical Computer Science*, vol.87, 209–220, 1991.
125. M. Minsky. *Finite and Infinite Machines*. Prentice Hall, Englewood Cliffs, 1967.
126. Ch. H. Papadimitriou. *Computational Complexity*. Addison-Wesley, Reading, 1994.
127. Gh. Păun. Computing with Membranes. *Journal of Computer and System Sciences*, vol.61(1), 108–143, 2000.
128. Gh. Păun. *Membrane Computing. An Introduction*. Springer-Verlag, Berlin, 2002.
129. Gh. Păun. Membrane Computing and Brane Calculi (Some Personal Notes). *Electronic Notes in Theoretical Computer Science*, vol.171, 3–10, 2007.

130. Gh. Păun, G. Rozenberg, A. Salomaa. *Handbook of Membrane Computing*. Oxford University Press, 2010.
131. M. J. Pérez-Jiménez, A. Riscos-Núñez, A. Romero-Jiménez, D. Woods. Complexity - Membrane Division, Membrane Creation. In [130], 302–336.
132. M. J. Pérez-Jiménez, A. Romero-Jiménez, and F. Sancho-Caparrini. Complexity Classes in Models of Cellular Computing with Membranes. *Natural Computing*, vol.2(3), 265–285, 2003.
133. I. Petre, L. Petre. Mobile Ambients and P Systems. *Journal of Universal Computer Science*, vol.5(9), 588–598, 1999.
134. I. Phillips, M. Vigliotti. On Reduction Semantics for the Push and Pull Ambient Calculus. *Proceedings of IFIP International Conference on Theoretical Computer Science*, 550–562, 2002.
135. B. C. Pierce, D. N. Turner. Pict: A Programming Language Based on the π-calculus. *Proof, Language and Interaction: Essays in Honour of Robin Milner*, MIT Press, 1997.
136. H. Pilegaard, F. Nielson, H. Riis Nielson. Static Analysis of a Model of the LDL Degradation Pathway. *Proceedings of Computational Methods in System Biology*, 14–26, 2005.
137. Z. Qi, J. You, H. Mao. P Systems and Petri Nets. *Lecture Notes in Computer Science*, vol.2933, 286–303, 2004.
138. A. Regev, E. M. Panina, W. Silverman, L. Cardelli, E. Shapiro. BioAmbients: An Abstraction for Biological Compartments. *Theoretical Computer Science*, vol.325, 141–167, 2004.
139. A. Regev, E. Shapiro. The π-calculus as an Abstraction for Biomolecular Systems. In G. Ciobanu, G. Rozenberg (Eds.): *Modelling in Molecular Biology*, Natural Computing Series, Springer, 219–266, 2004.
140. F. Romero-Campero, M. Gheorghe, G. Ciobanu, J. Auld, M. Pérez-Jiménez. Cellular Modelling Using P Systems and Process Algebra. *Progress in Natural Science*, vol.17(4), 375–383, 2007.
141. G. Rozenberg, A. Salomaa. *The Mathematical Theory of L Systems*. Academic, 1980.
142. A. Salomaa. *Formal Languages*. Academic, 1973.
143. D. Sangiorgi. *Expressing Mobility in Process Algebras: First-Order and Higher-Order Paradigms*. Ph.D. Thesis, University of Edinburgh, 1992.
144. D. Sangiorgi, D. Walker. On Barbed Equivalence in π-calculus. *Lecture Notes in Computer Science*, vol.2154, 292–304, 2001.
145. D. Sangiorgi, D. Walker. *The π-calculus - A Theory of Mobile Processes*. Cambridge University Press, 2003.
146. R. Schroeppel. A Two Counter Machine Cannot Calculate 2^N. Massachusetts Institute of Technology, A. I. Laboratory, Artificial Intelligence Memo #257, 1972.
147. E. Y. Shapiro. Separating Concurrent Languages With Categories of Language Embeddings. *STOC'91*, 198–208, 1991.
148. E. Y. Shapiro. Embeddings Among Concurrent Programming Languages. *Lecture Notes in Computer Science*, vol.630, 486–503, 1992.
149. J. C. Shepherdson, J. E. Sturgis. Computability of Recursive Functions. *Journal of the ACM*, vol.10(2), 217–255, 1963.
150. A. Sorkin, M. von Zastrow. Endocytosis and Signalling: Intertwining Molecular Networks. *Nature Reviews Molecular Cell Biology*, vol.10, 609–622, 2009.
151. R.J. Van Glabbeek. Linear Time - Branching Time Spectrum - The Semantics of Concrete, Sequential Processes. *Handbook of Process Algebra*, Elsevier, 3–99, 2001.
152. B. Victor, F. Moller. The Mobility Workbench – A Tool for the π-calculus. *Lecture Notes in Computer Science*, vol.818, 428–440, 1994.
153. A. Wright, M. Felleisen. A Syntactic Approach to Type Soundness. *Information and Computation*, vol.115, 38–94, 1994.
154. D. Zaharie, G. Ciobanu. Distributed Evolutionary Algorithms Inspired by Membranes in Solving Continuous Optimization Problems. *Lecture Notes in Computer Science*, vol.4361, 536–553, 2006.
155. P. Zimmer. On the Expressiveness of Pure Safe Ambients. *Mathematical Structures in Computer Science*, vol.13(5), 721–770, 2003.

List of Figures

B. Aman, G. Ciobanu, *Mobility in Process Calculi and Natural Computing*, Natural Computing Series, DOI 10.1007/978-3-642-24867-2,

List of Tables

B. Aman, G. Ciobanu, *Mobility in Process Calculi and Natural Computing*,
Natural Computing Series, DOI 10.1007/978-3-642-24867-2,

Index

B. Aman, G. Ciobanu, *Mobility in Process Calculi and Natural Computing*,
Natural Computing Series, DOI 10.1007/978-3-642-24867-2,